Welcome to *Twenty First Century Science*

GCSE Physics

This is a course in three parts.

Modules P1–P3 explore questions that matter to everyone and that physics can help to answer. They will help you to understand more about the Earth and its place in the Universe, about different forms of radiation, and about global warming. In these modules you will also learn about how scientific explanations develop, and how to assess risks and make decisions about science and technology.

Modules P4–P6 introduce more physics explanations and take a more quantitative approach. Here you will encounter ways of describing and explaining the motions of many things, from traffic to space rockets, from atoms to stars. You will study simple electric circuits used in electrical devices of many kinds, and how electricity is generated. Appreciating the behaviour common to all waves will enable you to understand how modern communications systems use radio and microwaves to carry information.

The final module (P7) has four topics. The first topic introduces observatories and telescopes. The second topic describes the mapping of the heavens – what astronomers have been able to discover using both naked-eye observation and telescopes. The third topic looks inside stars to understand what they are made of and how they work. The last topic looks at the life cycle of stars, describing how they change and why. You will see that dying stars can be extremely violent and that some produce black holes. Stories of physics in action in this module take you to the frontiers of astronomy today.

How to use this book

If you want to find a particular topic, use the **Contents** and **Index** pages. You can also use the **Glossary**. This explains all the key words used in the book.

Each module has two introduction pages, which tell you the main ideas you will study. They look like this:

Why study the wave model of radiation?
Why it is useful to know about this topic.

The science
The scientific information you will learn about in this module.

Physics in action
What you will learn from this module about how science works.

Find out about
The main ideas explored in this module.

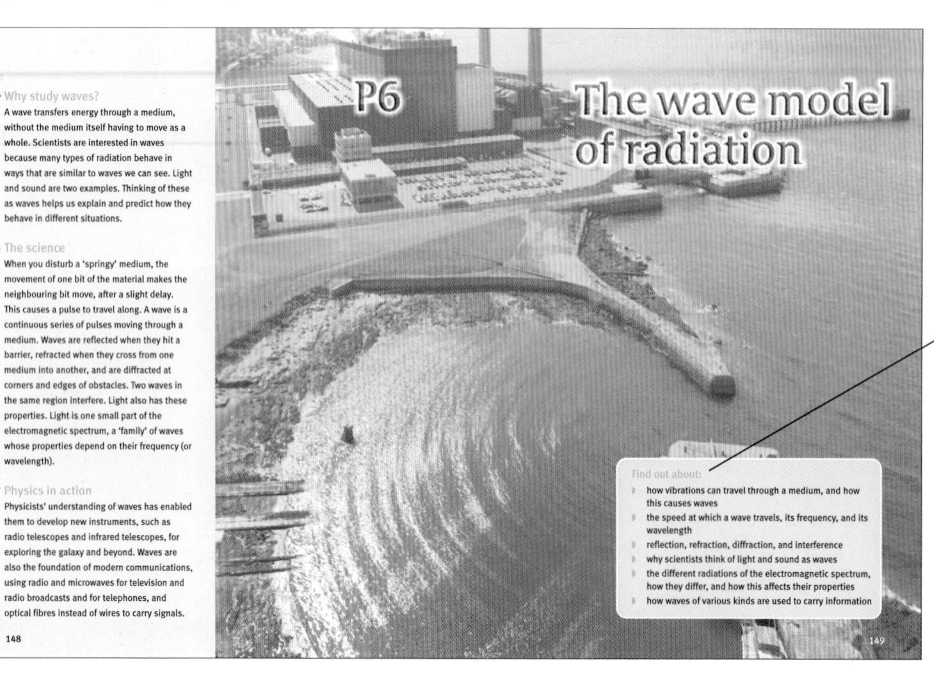

Why study waves?
A wave transfers energy through a medium, without the medium itself having to move as a whole. Scientists are interested in waves because many types of radiation behave in ways that are similar to waves we can see. Light and sound are two examples. Thinking of these as waves helps us explain and predict how they behave in different situations.

The science
When you disturb a 'springy' medium, the movement of one bit of the material makes the neighbouring bit move, after a slight delay. This causes a pulse to travel along. A wave is a continuous series of pulses moving through a medium. Waves are reflected when they hit a barrier, refracted when they cross from one medium into another, and are diffracted at corners and edges of obstacles. Two waves in the same region interfere. Light also has these properties. Light is one small part of the electromagnetic spectrum, a 'family' of waves whose properties depend on their frequency (or wavelength).

Physics in action
Physicists' understanding of waves has enabled them to develop new instruments, such as radio telescopes and infrared telescopes, for exploring the galaxy and beyond. Waves are also the foundation of modern communications, using radio and microwaves for television and radio broadcasts and for telephones, and optical fibres instead of wires to carry signals.

148

P6 The wave model of radiation

Find out about:
- how vibrations can travel through a medium, and how this causes waves
- the speed at which a wave travels, its frequency, and its wavelength
- reflection, refraction, diffraction, and interference
- why scientists think of light and sound as waves
- the different radiations of the electromagnetic spectrum, how they differ, and how this affects their properties
- how waves of various kinds are used to carry information

149

Each module is split into sections. Pages in a section look like this:

Heading
Each section looks at a different part of the module.

Find out about
The key points explored in the section.

Higher Tier
The 'H' flag next to something on the page means that it refers to Higher Tier material in the specification.

Example
Examples boxes show you how to do calculations.

Questions
Each section has questions for you to try. You can answer most of the questions using the book.

Circled questions
For a few questions, your teacher may give you some help. These questions have a circle around the question number.

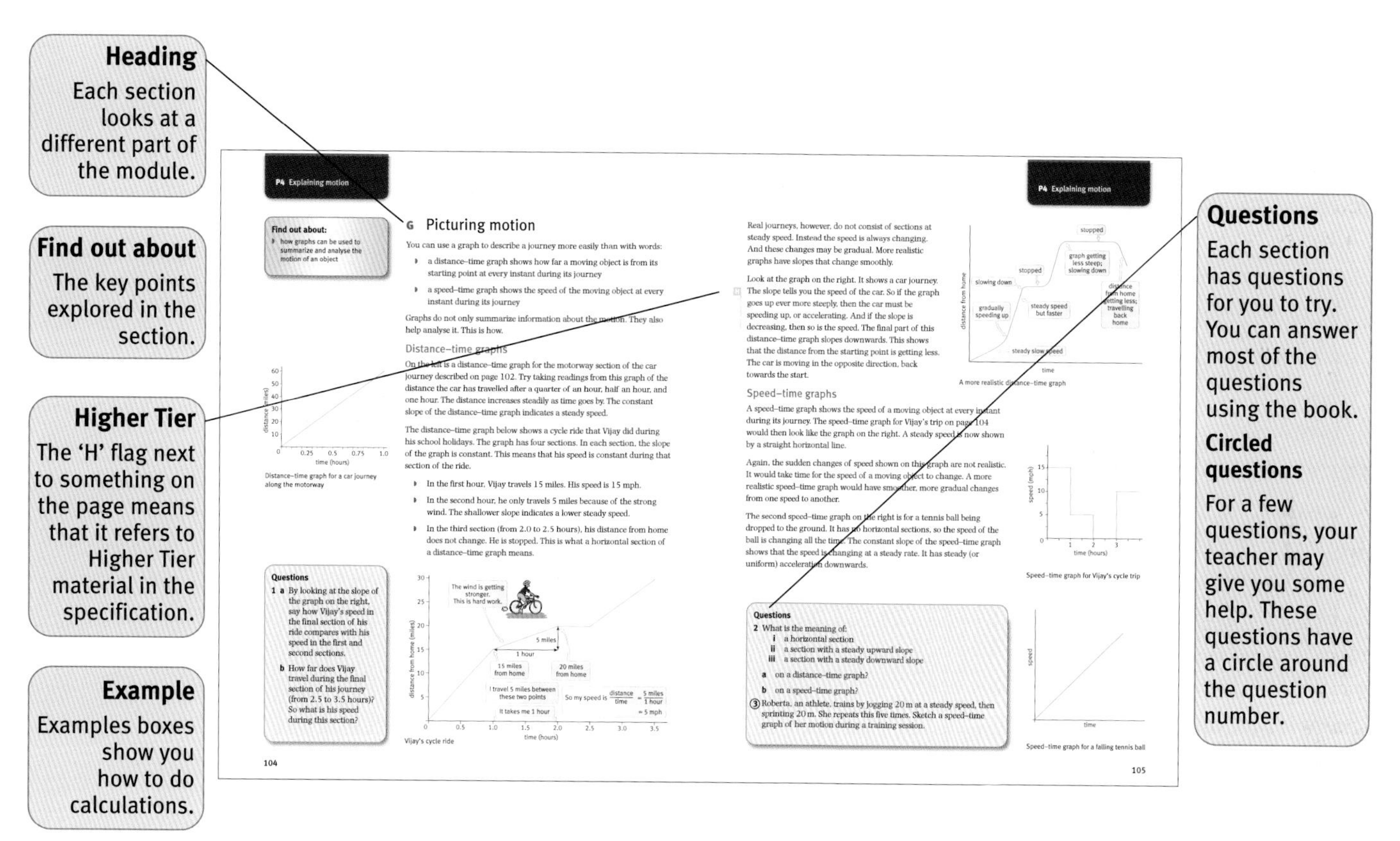

P4 Explaining motion

Find out about:

- how graphs can be used to summarize and analyse the motion of an object

G Picturing motion

You can use a graph to describe a journey more easily than with words:

- a distance–time graph shows how far a moving object is from its starting point at every instant during its journey
- a speed–time graph shows the speed of the moving object at every instant during its journey

Graphs do not only summarize information about the motion. They also help analyse it. This is how.

Distance–time graphs

On the left is a distance–time graph for the motorway section of the car journey described on page 102. Try taking readings from this graph of the distance the car has travelled after a quarter of an hour, half an hour, and one hour. The distance increases steadily as time goes by. The constant slope of the distance–time graph indicates a steady speed.

Distance–time graph for a car journey along the motorway

The distance–time graph below shows a cycle ride that Vijay did during his school holidays. The graph has four sections. In each section, the slope of the graph is constant. This means that his speed is constant during that section of the ride.

- In the first hour, Vijay travels 15 miles. His speed is 15 mph.
- In the second hour, he only travels 5 miles because of the strong wind. The shallower slope indicates a lower steady speed.
- In the third section (from 2.0 to 2.5 hours), his distance from home does not change. He is stopped. This is what a horizontal section of a distance–time graph means.

Vijay's cycle ride

Questions

1 a By looking at the slope of the graph on the right, say how Vijay's speed in the final section of his ride compares with his speed in the first and second sections.

b How far does Vijay travel during the final section of his journey (from 2.5 to 3.5 hours)? So what is his speed during this section?

104

P4 Explaining motion

Real journeys, however, do not consist of sections at steady speed. Instead the speed is always changing. And these changes may be gradual. More realistic graphs have slopes that change smoothly.

Look at the graph on the right. It shows a car journey. The slope tells you the speed of the car. So if the graph goes up ever more steeply, then the car must be speeding up, or accelerating. And if the slope is decreasing, then so is the speed. The final part of this distance–time graph slopes downwards. This shows that the distance from the starting point is getting less. The car is moving in the opposite direction, back towards the start.

A more realistic distance–time graph

Speed–time graphs

A speed–time graph shows the speed of a moving object at every instant during its journey. The speed–time graph for Vijay's trip on page 104 would then look like the graph on the right. A steady speed is now shown by a straight horizontal line.

Again, the sudden changes of speed shown on this graph are not realistic. It would take time for the speed of a moving object to change. A more realistic speed–time graph would have smoother, more gradual changes from one speed to another.

The second speed–time graph on the right is for a tennis ball being dropped to the ground. It has no horizontal sections, so the speed of the ball is changing all the time. The constant slope of the speed–time graph shows that the speed is changing at a steady rate. It has steady (or uniform) acceleration downwards.

Speed–time graph for Vijay's cycle trip

Speed–time graph for a falling tennis ball

Questions

2 What is the meaning of:
 i a horizontal section
 ii a section with a steady upward slope
 iii a section with a steady downward slope

 a on a distance–time graph?
 b on a speed–time graph?

③ Roberta, an athlete, trains by jogging 20 m at a steady speed, then sprinting 20 m. She repeats this five times. Sketch a speed–time graph of her motion during a training session.

105

Each module ends with a summary, and some also have questions. Here is an example:

Summary
A checklist of key points explained in this module.

Questions
Some further questions to help bring together what you have learnt in this module.

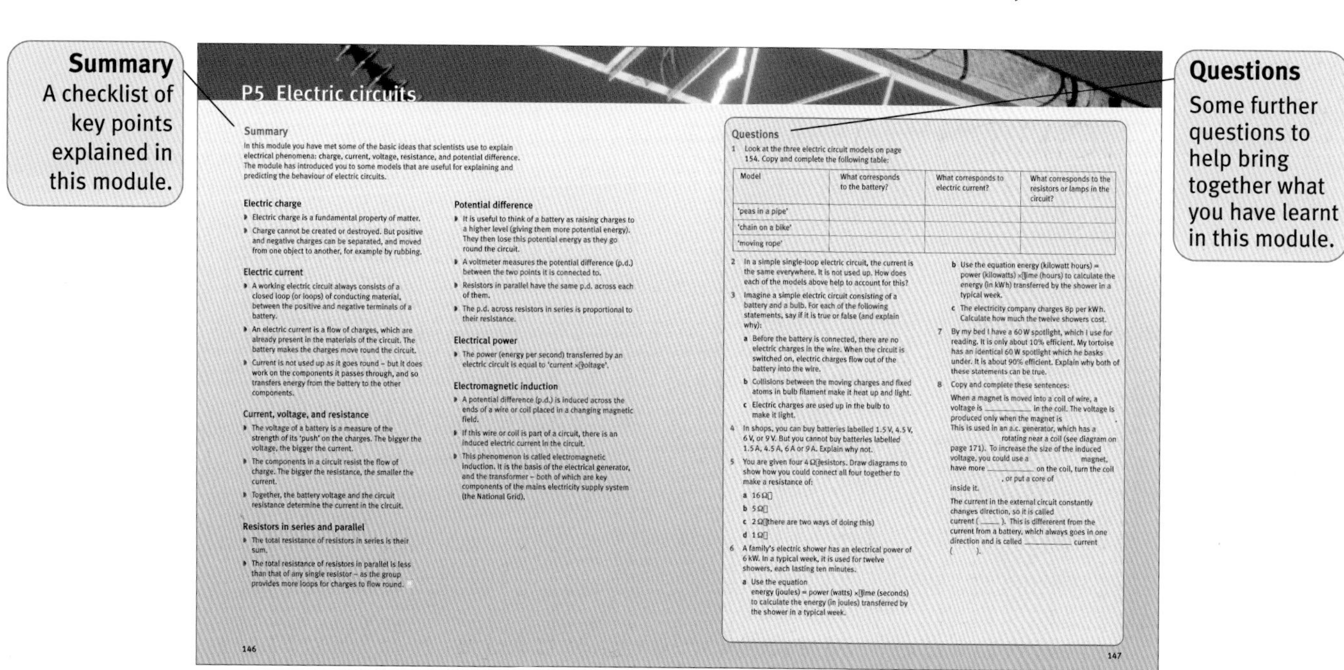

P5 Electric circuits

Summary

In this module you have met some of the basic ideas that scientists use to explain electrical phenomena: charge, current, voltage, resistance, and potential difference. The module has introduced you to some models that are useful for explaining and predicting the behaviour of electric circuits.

Electric charge

- Electric charge is a fundamental property of matter.
- Charge cannot be created or destroyed. But positive and negative charges can be separated, and moved from one object to another, for example by rubbing.

Electric current

- A working electric circuit always consists of a closed loop (or loops) of conducting material, between the positive and negative terminals of a battery.
- An electric current is a flow of charges, which are already present in the materials of the circuit. The battery makes the charges move round the circuit.
- Current is not used up as it goes round – but it does work on the components it passes through, and so transfers energy from the battery to the other components.

Current, voltage, and resistance

- The voltage of a battery is a measure of the strength of its 'push' on the charges. The bigger the voltage, the bigger the current.
- The components in a circuit resist the flow of charge. The bigger the resistance, the smaller the current.
- Together, the battery voltage and the circuit resistance determine the current in the circuit.

Resistors in series and parallel

- The total resistance of resistors in series is their sum.
- The total resistance of resistors in parallel is less than that of any single resistor – as the group provides more loops for charges to flow round.

Potential difference

- It is useful to think of a battery as raising charges to a higher level (giving them more potential energy). They then lose this potential energy as they go round the circuit.
- A voltmeter measures the potential difference (p.d.) between the two points it is connected to.
- Resistors in parallel have the same p.d. across each of them.
- The p.d. across resistors in series is proportional to their resistance.

Electrical power

- The power (energy per second) transferred by an electric circuit is equal to 'current × voltage'.

Electromagnetic induction

- A potential difference (p.d.) is induced across the ends of a wire or coil placed in a changing magnetic field.
- If this wire or coil is part of a circuit, there is an induced electric current in the circuit.
- This phenomenon is called electromagnetic induction. It is the basis of the electrical generator, and the transformer – both of which are key components of the mains electricity supply system (the National Grid).

146

Questions

1 Look at the three electric circuit models on page 154. Copy and complete the following table:

Model	What corresponds to the battery?	What corresponds to electric current?	What corresponds to the resistors or lamps in the circuit?
'peas in a pipe'			
'chain on a bike'			
'moving rope'			

2 In a simple single-loop electric circuit, the current is the same everywhere. It is not used up. How does each of the models above help to account for this?

3 Imagine a simple electric circuit consisting of a battery and a bulb. For each of the following statements, say if it is true or false (and explain why):

 a Before the battery is connected, there are no electric charges in the wire. When the circuit is switched on, electric charges flow out of the battery into the wire.
 b Collisions between the moving charges and fixed atoms in bulb filament make it heat up and light.
 c Electric charges are used up in the bulb to make it light.

4 In shops, you can buy batteries labelled 1.5 V, 4.5 V, 6 V, or 9 V. But you cannot buy batteries labelled 1.5 A, 4.5 A, 6 A or 9 A. Explain why not.

5 You are given four 4 Ω resistors. Draw diagrams to show how you could connect all four together to make a resistance of:

 a 16 Ω
 b 5 Ω
 c 2 Ω (there are two ways of doing this)
 d 1 Ω

6 A family's electric shower has an electrical power of 6 kW. In a typical week, it is used for twelve showers, each lasting ten minutes.

 a Use the equation energy (joules) = power (watts) × time (seconds) to calculate the energy (in joules) transferred by the shower in a typical week.
 b Use the equation energy (kilowatt hours) = power (kilowatts) × time (hours) to calculate the energy (in kWh) transferred by the shower in a typical week.
 c The electricity company charges 8p per kWh. Calculate how much the twelve showers cost.

7 By my bed I have a 60 W spotlight, which I use for reading. It is only about 10% efficient. My tortoise has an identical 60 W spotlight which he basks under. It is about 90% efficient. Explain why both of these statements can be true.

8 Copy and complete these sentences:

When a magnet is moved into a coil of wire, a voltage is ________ in the coil. The voltage is produced only when the magnet is ________. This is used in an a.c. generator, which has a ________ rotating near a coil (see diagram on page 171). To increase the size of the induced voltage, you could use a ________ magnet, have more ________ on the coil, turn the coil ________, or put a core of ________ inside it.

The current in the external circuit constantly changes direction, so it is called ________ current (____). This is differerent from the current from a battery, which always goes in one direction and is called ________ current (____).

147

Internal assessment

In *GCSE Physics* your internal assessment counts for 33.3% of your total grade. Marks are given for:

- *either* a practical investigation
- *or* a case study and a data analysis

Your school or college will decide on the type of internal assessment. You may be given the marking schemes to help you understand how to get the most credit for your work.

Internal assessment (33.3% of total marks)

EITHER: Investigation (33.3%)

Investigations are carried out by scientists to try and find the answers to scientific questions. The skills you learn from this work will help prepare you to study any science course after GCSE.

To succeed with any investigation you will need to:

- choose a question to explore
- select equipment and use it appropriately and safely
- design ways of making accurate and reliable observations

Your investigation report will be based on the data you collect from your own experiments. You may also use information from other people's work. This is called secondary data.

Marks will be awarded under five different headings.

Strategy

- Choose the task for your investigation.
- Decide how much data you need to collect.
- Choose a procedure to give you reliable data.

Collecting data

- Take careful, accurate measurements safely.
- Collect enough data and check its reliability.
- Collect data across a wide enough range.
- Control factors that might affect the results.

Interpreting data

- Present your data to make clear patterns in the results.
- State your conclusions from the results.
- Use chemical knowledge to explain your conclusion.

Evaluation

- Say how you could improve your method.
- Explain how reliable your evidence is.
- Suggest ways to increase the confidence in your conclusions.

Presentation

- Write a full report of your investigation.
- Lay out your report clearly and logically.
- Describe you apparatus and procedure.
- Show all units correctly.
- Take care with spelling, grammar, and scientific terms.

OR: Case study and data analysis (33.3%)

A **case study** is a report which weights up evidence about a scientific question. You find out what different people have said about the issue. Then you evaluate the information and make your own conclusions.

You choose a topic from one of these categories:

- A question where the scientific knowledge is not certain.
- A question about decision-making using scientific information.
- A question about a personal issue involving science.

Selecting information

- Collect information from a range of sources.
- Decide how reliable each source is.
- Choose relevant information.
- Say where your information came from.

Understanding the question

- Use science knowledge to explain your topic.
- Report on the scientific evidence used by people with views on the issue.

Making your own conclusion

- Compare different evidence and points of view.
- Weigh the risks and benefits of different courses of action.
- Say what you think should be done based on the evidence.

Presenting your study

- Set out your report clearly and logically.
- Use an appropriate style of presentation.
- Illustrate your report.
- Take care with spelling, grammar, and scientific terms.

A **data analysis** task is based on a practical experiment which you carry out. You may do this alone or work in groups and pool all your data. Then you interpret and evaluate the data.

Interpreting data

- Present your data in tables, charts, or graphs.
- State your conclusions from the data.
- Use chemical knowledge to explain your conclusions.

Evaluation

- Say how you could improve your method.
- Explain how reliable your evidence is.
- Suggest ways to increase the confidence in your conclusions.

Why study the Earth in the Universe?

Many people want to understand more about the Earth and its place in the Universe. Natural disasters, such as volcanoes and earthquakes, can be life-threatening. Can anything be done to predict them? The Earth is very fragile. It is a very, very small place in a huge and almost empty Universe. Some scientists think that an asteroid collision made the dinosaurs extinct. Could another big asteroid hit the Earth?

The science

Science can explain changes to the Earth. Some changes happen very quickly, and some happen very slowly. For example, over millions of years, whole mountain ranges grow, and then disappear. Astronomers study changes in stars and galaxies. These changes can take thousands of millions of years. Stars made the atoms found in everything: including everything on Earth and everything in your body.

Ideas about science

How can scientists be sure? Partly they depend on data and careful observations of the Earth and Universe. But scientists need to interpret the evidence they collect. So, imagination is also important.

How are scientific ideas tested? Often there are many arguments put forward before scientists accept new data or agree with new explanations.

The Earth in the Universe

Find out about:

- evidence of the Earth's history found in rocks
- the movement of the Earth's continents
- how scientists develop explanations of the Earth and space
- the history of the Universe

Find out about:

- what is known about the Earth and the Universe

A Time and space

Our rocky planet was made from the scattered dust of ancient stars. It may or may not be the only place in the whole Universe with life.

As the graphics on these two pages show, scientists know a lot about:

- the history of the Earth
- where and how the Earth moves through space

But there are many things that we still do not know. And there are some we may never know.

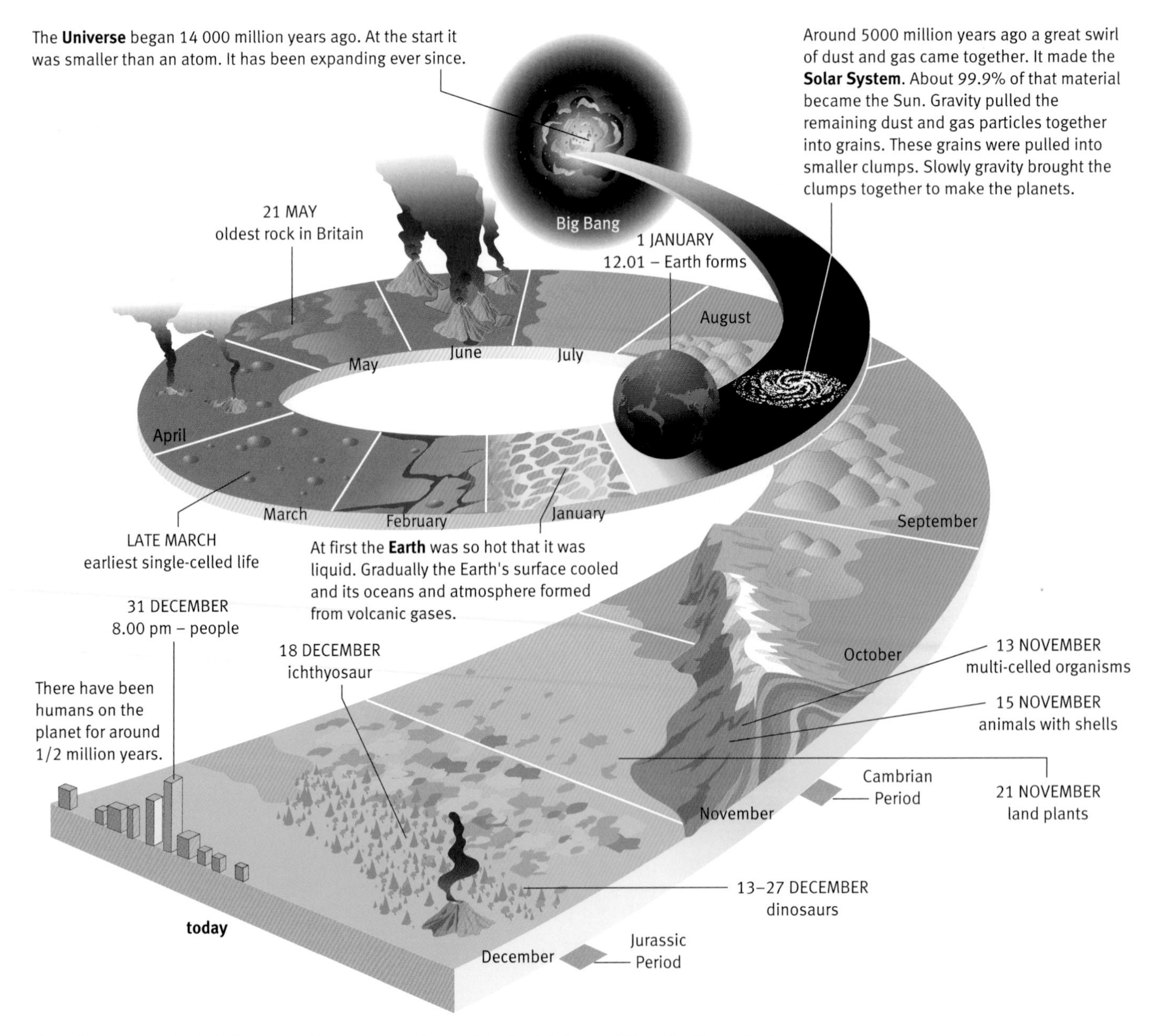

Timeline: from the big bang to the present day, with the history of the Earth scaled as if it took place in a single year

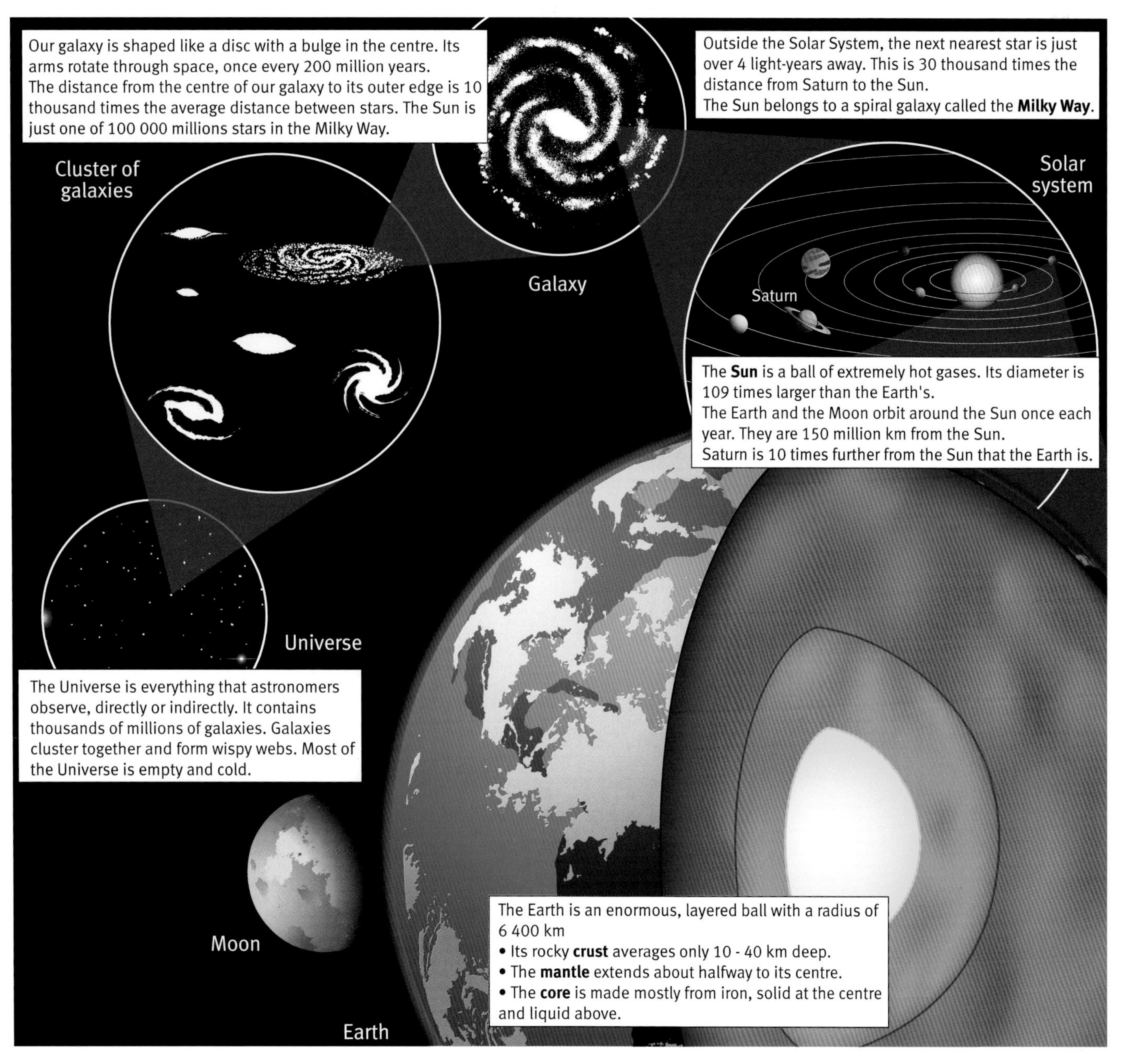

Nested structures in the Universe

Questions

1 The timeline on page 10 shows the age of the Earth.

a Redraw it as if it happened over a period of 15 years (roughly your lifetime).

b On this scale, how long ago did the dinosaurs die out?

2 Make a list of the ways that scientists explore earlier times, or places, that they cannot visit and observe directly.

Key words

Universe
Solar System
Milky Way
Sun
crust
mantle
core

Find out about:

- James Hutton's explanation for the variety of rocks he found
- how old rocks are and how scientists date them

B Deep time

James Hutton and the stories that rocks tell

Without some way of building new mountains, erosion would wear the continents flat.

Rivers carry sediment to the oceans, where it settles at the bottom as sand and silt.

Sediments are compressed and cemented to form sedimentary rocks. In some places, layers of sedimentary rocks are tilted or folded.

Around 300 years ago, people started asking new questions about the Earth. They found fossils of seashells and other marine organisms in rocks at the tops of mountains. 'Why here?' They wondered.

James Hutton was an unusually well-educated and observant farmer. He watched heavy rains wash valuable soil off farmers' fields. He also noticed that many rocks are made up of eroded and deposited material (now called sedimentary rocks). Travelling around England and Scotland, he studied rock formations and collected rock specimens (samples). Slowly an idea formed in his mind, as he learned to interpret rocks.

Using the present to interpret the past

In 1785 Hutton explained his startling new theory of Earth cycles at a meeting of the Royal Society of Edinburgh, which at the time was like a scientific club. The Society published his theory in its *Transactions*, a kind of newsletter. In this way, his ideas reached an audience right across Europe.

What Hutton described is today called the rock cycle. **Erosion** and deposition of sediment take place, very slowly. Over enormous periods of time, these processes add up to huge changes in the Earth's surface. They also make new soil and are therefore essential to human survival. The Earth has a history – it was not created all at once.

The millions of years over which the Earth has changed are now called 'deep time'.

Most Europeans in Hutton's time believed that the Earth had been created exactly as they saw it, just 6000 years earlier. This figure for the Earth's age came from an interpretation of the Christian Bible. They rejected Hutton's theory. It took another century and the support of a leading British geologist, Charles Lyell, before Hutton's ideas became accepted.

Dating rocks

Gradually, geologists learned to work out the history recorded in rocks. They used clues like these:

- deeper is older – in layered rocks, the youngest rocks are usually on top of older ones.
- fossils are time markers – many species lived at particular times and later became extinct.
- cross-cutting features – if one type of rock cuts across another rock type, it is younger. For example, hot magma can fill cracks and solidify as rock.

But these clues only tell you which rocks are older than others. They don't tell you how old the rocks are.

Some rocks are radioactive. Scientists today estimate their age by measuring the radiation that these rocks emit (give off). This is called **radioactive dating**. The Earth's oldest rocks were made 3900 million years ago.

The development of scientific ideas

This first case study about James Hutton, contains examples of:

- data
- expanations
- the role of imagination

Data

Fossils, rocks of different types, the way that rock types are layered, folded, or joined.

Explanations

Hutton's idea of a rock cycle, different ways of dating rocks.

Imagination

Hutton could imagine the millions of years needed for familiar processes to slowly change the landscape.

Which layer has the fallen rock come from?

Key words

erosion radioactive dating

Questions

1 In what time order did the creatures shown in the cliff above live?

② Hutton called his rock specimens 'God's books'. To his mind, why was this an appropriate name for them?

Find out about:
- a scientific debate started by Alfred Wegener
- evidence that the continents are very slowly moving

C Continental drift

How are mountains formed?

A hundred years after Hutton, scientists wanted to know how mountains form. Most geologists believed that the Earth began hot. They compared the Earth with a drying apple, which wrinkles as it shrinks. If the Earth had cooled and shrunk, its surface would have wrinkled too. They claimed that chains of mountains are those wrinkles.

Moving continents?

Scientists discovered radioactivity around 1900. The heating effect of radioactive materials inside the Earth prevents the Earth from cooling. So a new theory of mountain building was needed.

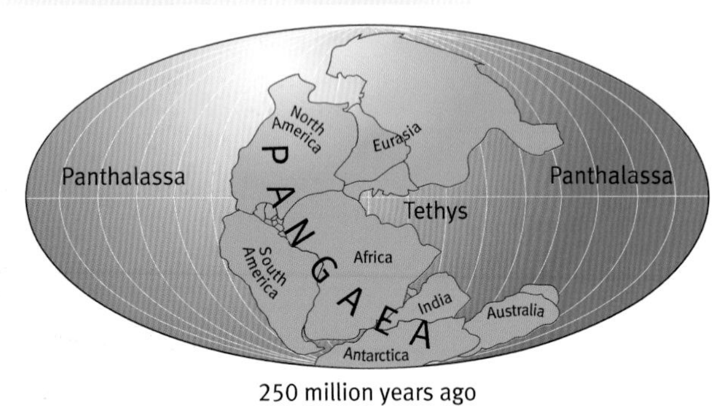

250 million years ago

Wegener showed how all the continents could once have formed a single continent, called Pangaea.

Many people can spot the match between the shapes of South America and Africa. The two continents look like pieces of a jigsaw. Alfred Wegener thought this meant that the continents were moving. They had once been joined together. He looked for evidence, recorded in their rocks.

In 1912 Wegener presented his idea of **continental drift**, and his supporting evidence, to a meeting of the Geological Society of Frankfurt. Geologists around the world read the English translation of his book, *The Origin of Continents and Oceans*, published in 1922.

Key words
continental drift

POLAR EXPLORER DIES

The frozen body of the German meteorologist and polar explorer Alfred Wegener was found on 12 May 1931. Wegener had been leading an expedition in Greenland and went missing just a day after his 50th birthday on 1 November 1930. Unfortunately he is likely to remembered for being too bold in his science.

Wegener claimed that continents move, by ploughing across the ocean floor. That, he said, explains why there are mountain chains at the edges of continents.

As evidence of continental drift, he found some interesting matches between mountain chains, rocks and fossils on different continents. But most geologists reject such a grand and unlikely explanation for these observations.

Questions

1 In this case study, identify examples of
   a data **b** explanations
2 Which continents have mountains at their edge?
3 'Peer review' involves scientists commenting on the work of other scientists. How did other scientists learn about Wegener's ideas?

Good morning. Let me tell you about continental drift.
Wegener, we have all heard of your crazy idea.

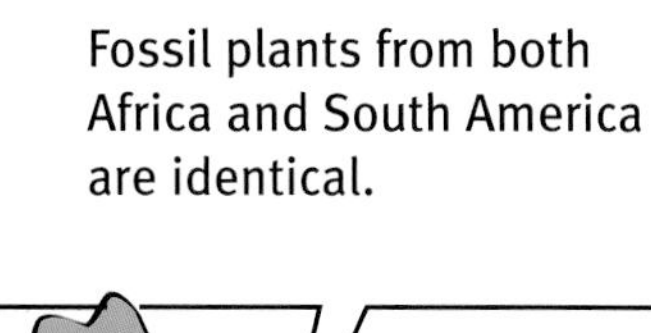
Fossil plants from both Africa and South America are identical.

Reptile fossils match too.

You can't draw wild conclusions from just a few fossils.

There could once have been a land bridge joining Africa and South America.

Where is the evidence for a land bridge?
Er, Er.

The rock types on each continent fit like pictures on a jigsaw. The continents were once joined together.
North America
Eurasia
South America
Africa

A key principle of geology is 'use the present to interpret the past.' You don't see continents moving today, do you?
No, but only because continents move so slowly.

I'm not prepared to listen to all this nonsense. What force could move a continent?

Two new scientific tools

- New instruments, called magnetometers, could measure tiny variations in the Earth's magnetic field.
- Detecting seismic waves. Earthquakes, small or large, produce vibrations that travel through the Earth. From the 1930s onwards, scientists used seismic waves to map the Earth's internal structure.

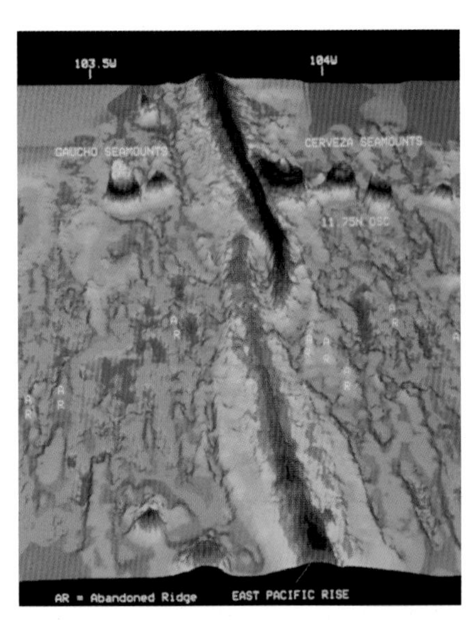

This computer-generated model shows part of the Pacific Ocean floor. (Water is not shown.)

Mapping the seafloor

During the 1950s the US Navy paid for research at three ocean science research centres. The Navy wanted to know how to:

- use magnetism to detect enemy submarines, and
- move its own submarines near the ocean floor, where they could avoid detection

A few dozen scientists at these three centres, plus two universities, organized many expeditions. They gathered huge amounts of data, and published thousands of scientific papers. Their thinking completely changed our understanding of Earth processes.

From zebra stripes to seafloor spreading

Scientists started to make maps of the ocean floor. To their great surprise, they found a chain of mountains under most oceans. This is now called an **oceanic ridge**. In 1960 a scientist called Harry Hess suggested that the seafloor moves away from either side of an oceanic ridge. This process, called **seafloor spreading**, could move continents.

Beneath a ridge, material from the Earth's solid mantle rises slowly, like warm toffee. As it approaches the ridge, pressure falls. So some of the material melts to form magma. Movements in the mantle pull the ridge apart, like two conveyor belts. Hot magma erupts and cools to make new rock.

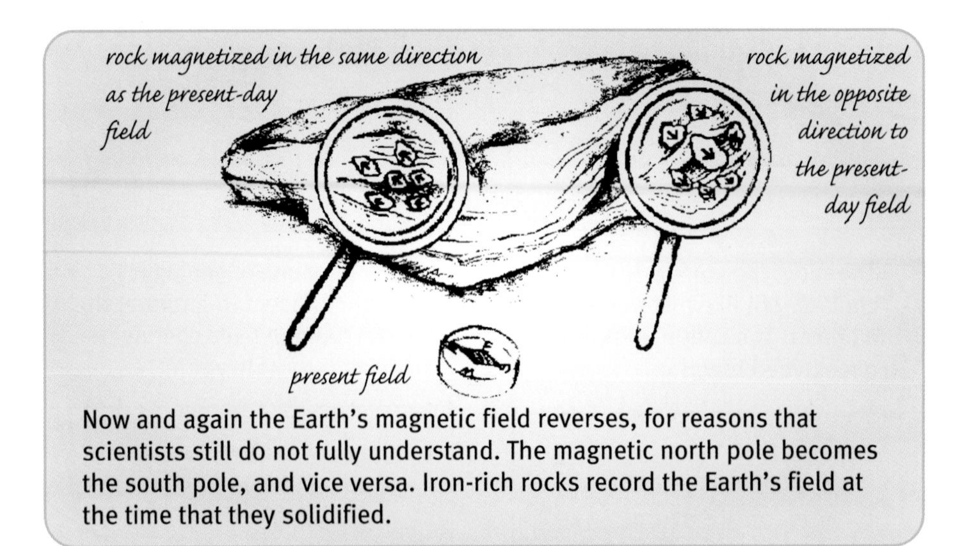

Now and again the Earth's magnetic field reverses, for reasons that scientists still do not fully understand. The magnetic north pole becomes the south pole, and vice versa. Iron-rich rocks record the Earth's field at the time that they solidified.

A young British research student, Fred Vine, explained the symmetrical 'stripe' pattern found in rock magnetism across an oceanic ridge. If hot magma rises at a ridge and cools to make new rock, said Vine, then the rock will be magnetized in the direction of the Earth's field at the time. The science journal *Nature* published his explanation in 1963.

By 1966 an independent group of scientists had found a clearer pattern of symmetrical stripes in magnetic data either side of another ridge. This forced other scientists to accept the idea of seafloor spreading.

Tanya Atwater was at university studying geology at that time. She describes a meeting of scientists late in 1966. Fred Vine had shown them an especially clear pattern of magnetic stripes.

'[The pattern] made the case for seafloor spreading. It was as if a bolt of lightning had struck me. My hair stood on end. ... Most of the scientists [went into that meeting] believing that continents were fixed, but all came out believing that they move.'

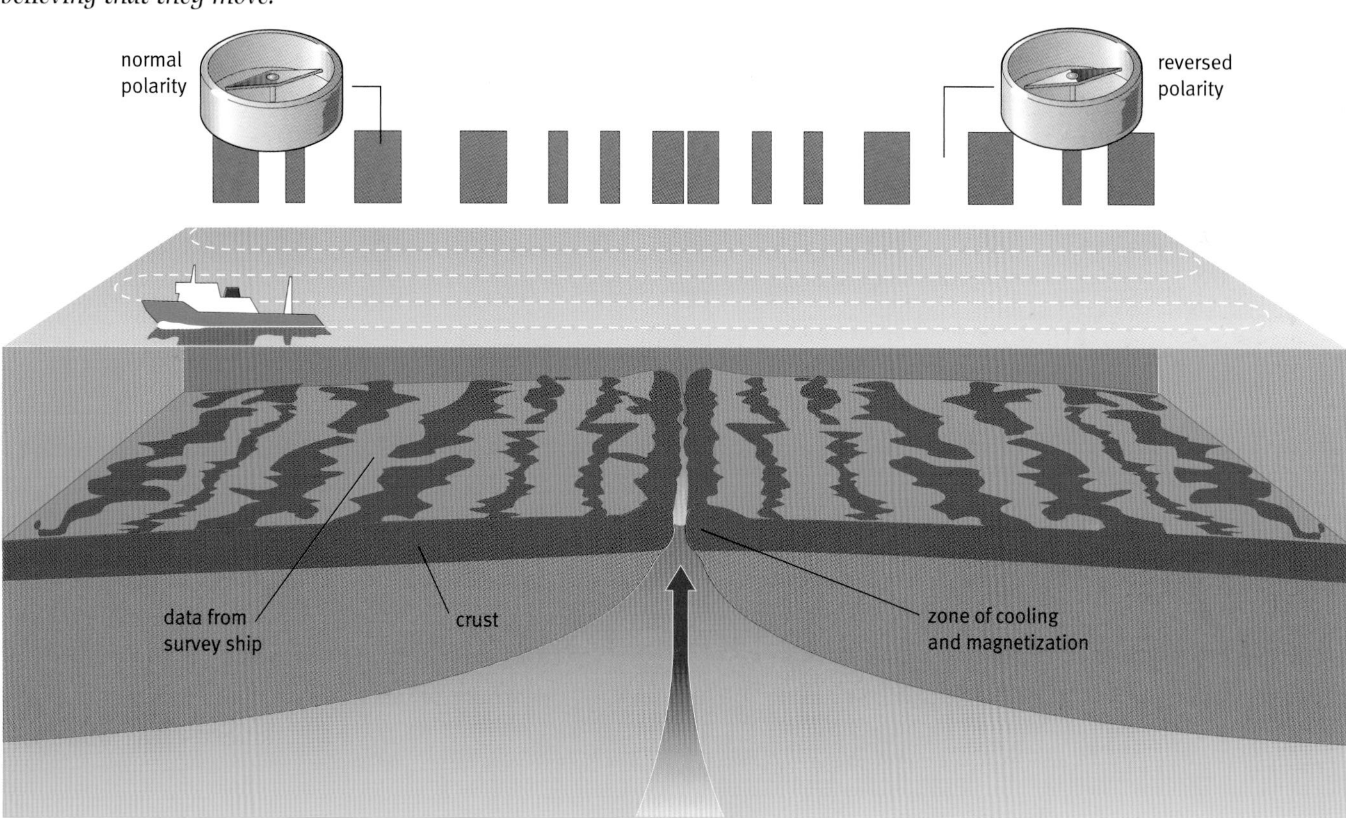

New ocean floor is being made all the time at oceanic ridges. Rock magnetism either side of an oceanic ridge shows the same zebra stripe pattern.

Ocean sediments confirm seafloor spreading

Seafloor drilling in 1969 provided further evidence of seafloor spreading. Sediments further away from oceanic ridges are thicker. This shows that the ocean floor is youngest near oceanic ridges, and oldest far away from ridges.

Key words
oceanic ridge
seafloor spreading

Questions

4 In this case study, identify examples of:

a data

b explanations

c prediction

5 Describe carefully how a zebra stripe pattern provides evidence for the seafloor spreading idea.

Find out about:

- a big explanation for many Earth processes
- ways to limit the damage caused by volcanoes and earthquakes

In 1973, the geologist Tanya Atwater wrote:

'I think I spend half of my time just talking and listening to people from many fields, searching together for how it might all fit together. And when something does fall into place, there is that mental explosion and the wondrous excitement. I think the human brain must love order.'

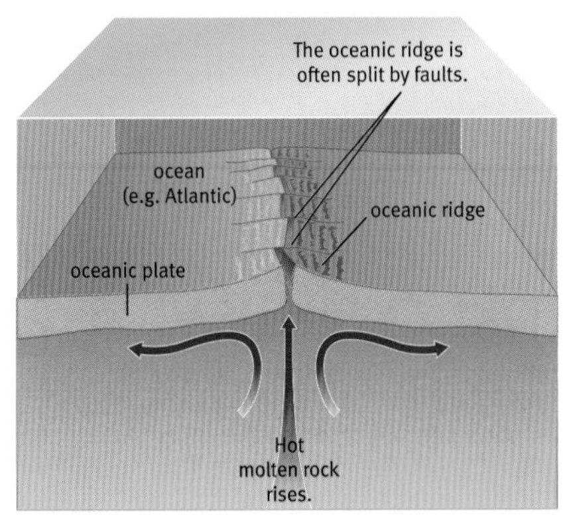

Constructive margin

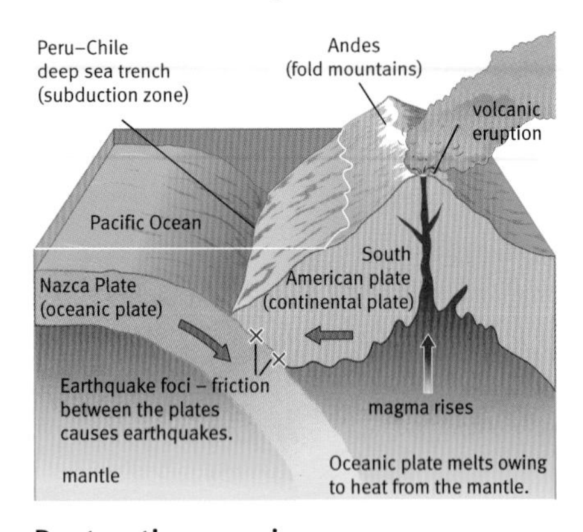

Destructive margin

D The theory of plate tectonics

By 1967, seafloor spreading and several other Earth processes were linked together in one big explanation. It was called plate tectonics.

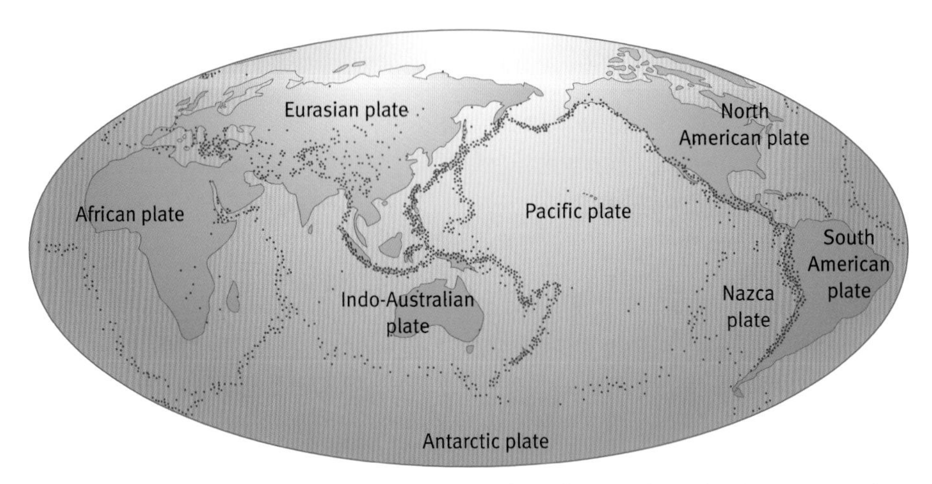

Each red dot on this map represents an earthquake. Earthquakes happen at the boundaries between tectonic plates.

This is the plate tectonics explanation of the Earth's outer layer:

- The Earth's outermost zone, or lithosphere, consists of crust plus the rigid upper mantle.
- It is made up of about a dozen giant slabs of rock, and many smaller ones. These are called **tectonic plates**.
- Currents in the Earth's solid mantle carry the plates along.
- The ocean floor gradually grows wider at oceanic ridges by seafloor spreading. These are called constructive margins.
- Ocean floor is destroyed where plates dip down beneath an **oceanic trench**. These are called a subduction zones, or destructive margins.
- The rigid plates are moved slowly, sometimes moving apart, sometimes pushing together, and sometimes sliding past each other.

Global Positioning Satellites (GPS) detect the movement of continents. The Atlantic is growing wider by 2.5 cm every year, on average. This is roughly how fast your fingernails grow. In some places, seafloors spread as fast as 20 cm each year.

Questions

1. **a** How far does the Atlantic spread in 100 years (a lifetime)?
   b How far has it spread in 10 000 years (all of human history)?
   c How far has it moved in 100 million years?
2. How does the answer to **c** compare with the present width of the Atlantic Ocean?

Plate tectonics explanations

The movement of tectonic plates causes continents to drift. It also explains:

- parts of the **rock cycle**
- mountain-building
- most earthquakes
- most volcanoes

Making mountains

Collisions between tectonic plates cause mountains to be formed. There are three ways that this can happen.

1 Where an ocean plate dives back down into the Earth, volcanic peaks may form at the surface.

2 The pushing movement at destructive margins can also cause rocks to buckle and fold, forming a **mountain chain.**

3 Sometimes an ocean closes completely, and two continents collide in slow motion. The edges of the continents crumple together and pile up, making mountain chains. This is happening today in the Himalayas and Tibet.

The Grampian mountains in Scotland are the eroded roots of mountains that were created some 400 million years ago. Scotland and Northern Ireland slowly crashed into England, Wales, and Southern Ireland.

The rock cycle

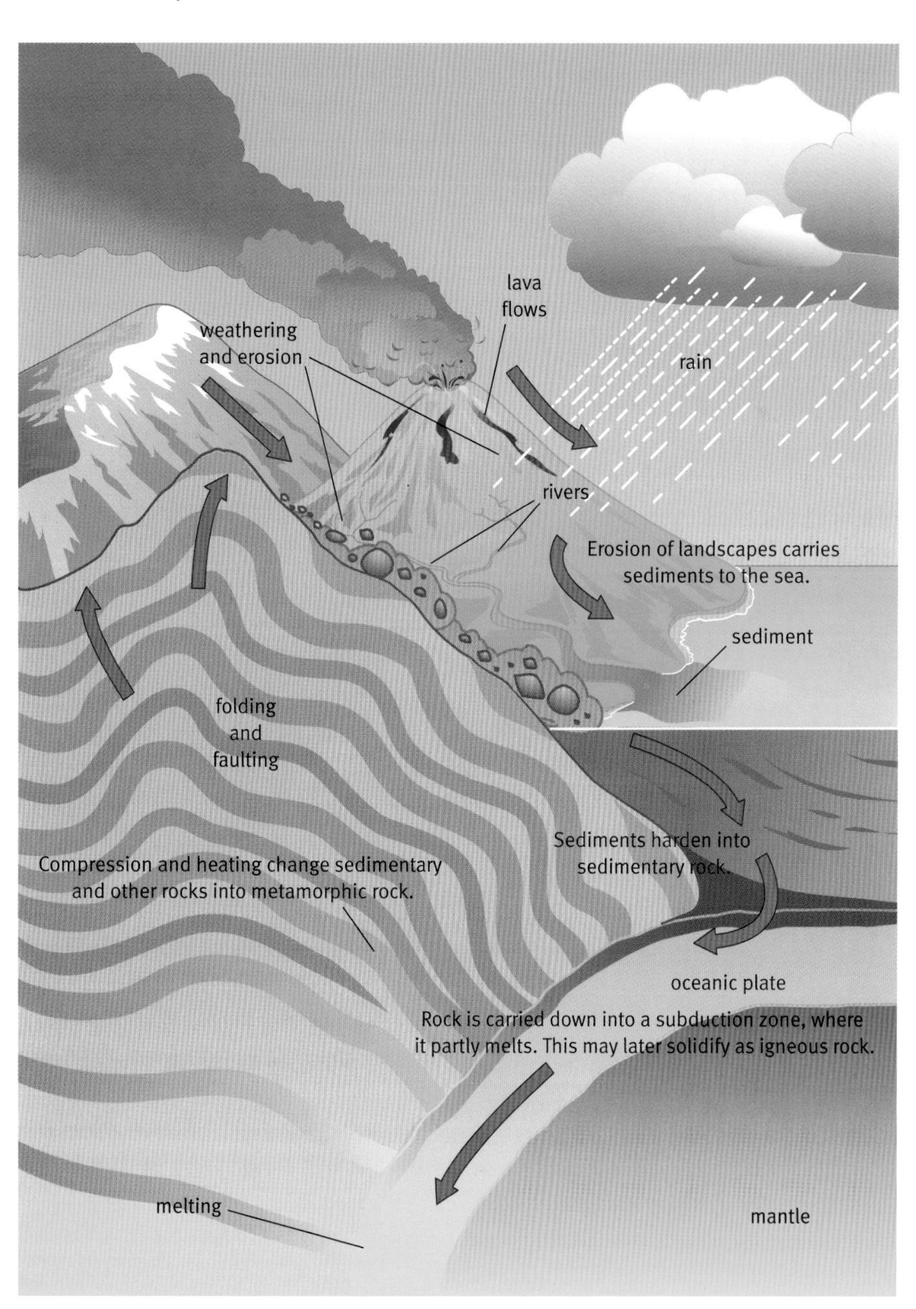

The movement of tectonic plates also plays a part in the rock cycle.

Key words
tectonic plates
rock cycle
mountain chain

Earthquakes

Earth scientists record more than 30 000 earthquakes a year. On average, one of these is hugely destructive.

There are three ways that plates can move against each other:

- move apart in a stretching movement, as at oceanic ridges
- push together, in a squashing movement, as at the Himalayas
- slide past each other, as at the San Andreas Fault in California

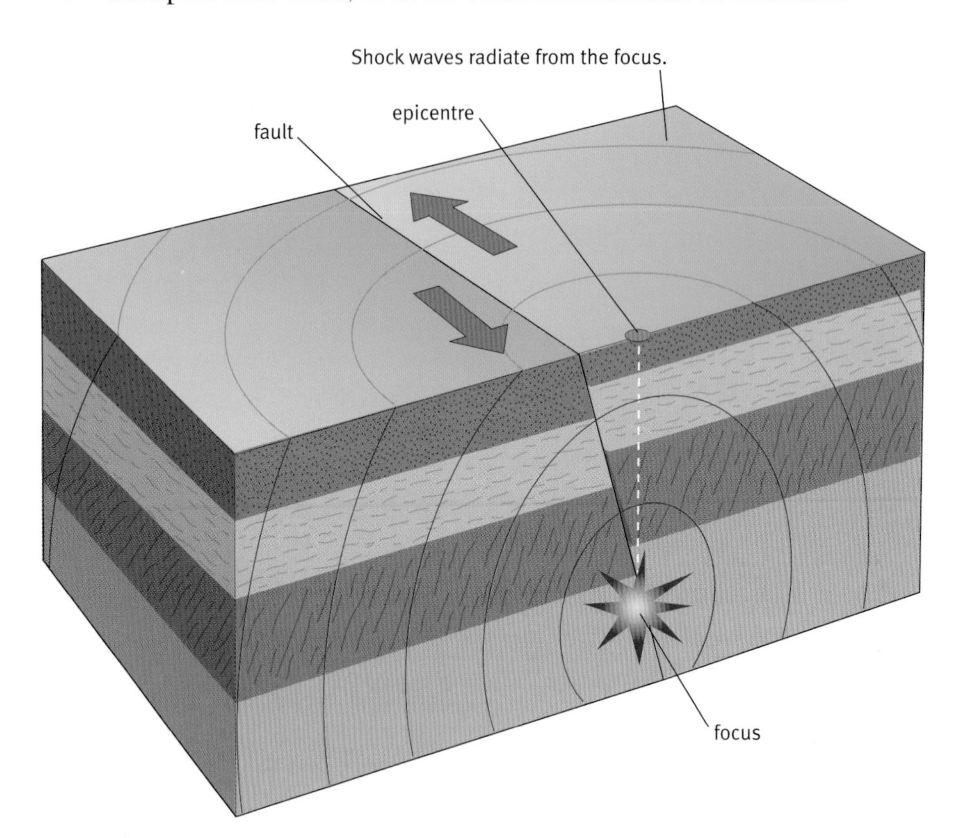

Most earthquakes happen along previous breaks, called faults. The shunting of the Earth's plates causes forces to build up along fault lines. Eventually the forces are so great that rocks locked together break, and allow plate movement. The ground shakes, making an **earthquake**.

Earthquakes are common at all moving plate boundaries. The most destructive happen at sliding boundaries on land, or undersea, causing tsunamis.

Volcanoes

A **volcano** is simply a vent in the Earth's surface that erupts magma (molten rock). The magma then forms lava or ash as it releases its gases. Each year there are about 50 eruptions from the world's 500 active volcanoes. They are common at plate boundaries, where the Earth's crust is being stretched, compressed, or uplifted.

The island of Montserrat was successfully evacuated before its volcano erupted in 1996. This volcano is located at a destructive margin. Volcanoes like this are especially explosive and dangerous.

Predicting disasters

Scientists know where earthquakes might happen. But they still cannot predict when.

Some volcanoes erupt regularly. Others store up pressure for thousands of years and then go off with a huge bang. Scientists monitor volcanoes and watch for warning signs. They know a volcano may erupt if:

- there is a change in the amount and type of gases it gives off
- there is local earthquake activity
- the sides of the volcano swell, as the inside fills with molten magma

Reducing the damage

If a volcanic eruption is predicted, people need to be evacuated from the affected area.

To be ready for an earthquake or tsunami, governments can:

- educate people, so that they know what to do
- organize public drills
- enforce building regulations that reduce the chance of buildings collapsing
- prepare emergency plans and ensure that trained staff can respond quickly

Questions

3 a Describe four major effects of tectonic plate movement.

b Where are these effects most common?

4 Where are the biggest earthquakes in the world expected?

5 What can public authorities do to reduce the damage caused by volcanoes, earthquakes, and tsunamis?

Key words

earthquake
volcano

Find out about:

- reasons for studying craters
- possible explanations for the extinction of dinosaurs

E The Solar System - danger!

Attack from space

Look up into a starry night. You might see a streak of light dash across the sky. That's a meteor. Most meteors are just tiny grains of dust. They shower down from space all the time. There will be quite a few micrometeorites on your school roof. They have diameters of less than 1 mm. Occasionally bigger ones, called meteorites, hit the ground. And several times during the Earth's history, a massive **asteroid** or **comet** has struck.

Impact craters

There is a huge crater in Arizona, USA. It is named after a mining engineer called Daniel Barringer. The first scientists to see it thought it was made by a volcano.

But in 1902 Barringer suggested another explanation. The crater rim contains many fragments of iron. He knew meteorites contain iron. He concluded that a violent impact had caused the crater.

But, other scientists wanted further evidence to support the theory. They found quartz dust particles that are only produced by huge pressures. They also looked at the layers of rock surrounding the crater. They found they were in the reverse order to the layers of rock in the surrounding desert. These two observations supported the impact explanation.

The Moon What could have made all these craters: volcanoes? violent impacts? The Moon is covered in craters. Even with the naked eye you can see large dark areas.

The Barringer crater, Arizona, USA. The crater floor is 200 m below the rim. It is the the size of 20 football pitches.

The Aorounga crater is in the Sahara Desert, Chad (Africa). It is much more eroded and weathered than the Barringer crater.

Crater	Diameter (km)	Age measured using radioactive dating (millions of years - MOY)
Barringer, USA	1	0.05
Silverpit, North Sea	3	60
Chicxulub, Mexico	170	65
Manicouagan, Chad	100	210
Aorounga, Chad	17	360
Sudbury, Canada	250	1850

This table gives the ages of some of the world's craters.

An iron-rich meteorite hit this car in Peekskill, New York. Fortunately no one was injured. Meteorites hit the Earth's surface with speeds of 12 to 70 km/s.

The age of the Solar System

Scientists have examined rock samples from the Earth, the Moon, Mars, and meteorites. Radioactive dating shows that none are older than 5000 million years. So, the Solar System is probably about 5000 million years old.

Crater work

Michelle Boast studied geology at university in the UK. Now she's doing scientific research at the Sudbury crater in Canada. Plate tectonics has changed the Earth a lot since the crater was created. Michelle and her colleagues are trying to trace the crater's geological history. They are trying to find copper and nickel deposits formed by the ancient collision.

Michelle works with a team based in Canada's University of New Brunswick. With its support she writes scientific papers about her research. She sends these to journals for publication. They are read by other geologists all over the world.

Michelle has no ordinary job. She has to travel widely to conferences. At the crater site she often has to camp. And she has an assistant with her for safety as one of the main hazards there is wild bears!

Michelle (right), and her assistant Tamara after a day's work at the Sudbury crater.

Questions

1 Look at the photograph of the Moon. What evidence is there that the craters were not made all at once?

2 Daniel Barringer thought an impact made his crater. What evidence was found later to support this idea?

3 Use the table above. Is there a correlation between the diameters of the craters and their ages? (HINT: Draw a graph to check this.)

(4) How do scientific papers and conferences help in making scientific knowledge more reliable?

Key words

asteroid
comet

What killed off the dinosaurs?

A **mass extinction** is dramatic. A lot of the world's plants and animals die out. Fossils show that there have been several mass extinction events over the last 600 million years. The most famous was 65 million years ago, when the dinosaurs disappeared.

What caused these extinctions is something that scientists cannot yet agree about. It is still an area of scientific uncertainty.

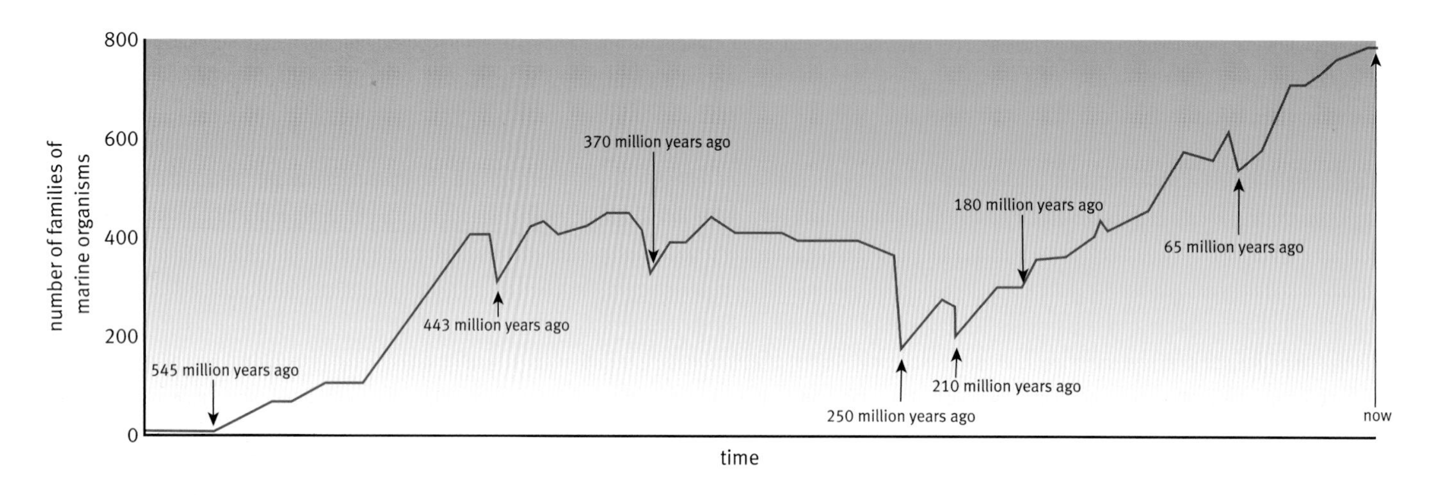

Asteroid collision – evidence and explanation

Luis and Walter Alvarez, father and son, found a thin layer of clay in rocks in Italy. The clay contained a metal called iridium. The iridium was in amounts that are common for meteorites or asteroids.

The rocks above and below the layer told them that the clay arrived there 65 million years ago.

In 1980 they published a scientific paper. They suggested that their layer of clay was the dust from an asteroid collision. They also said that that this could be the explanation for the extinction of the dinosaurs.

Publishing the paper could have ruined their scientific reputations. At the time, there was no evidence of a gigantic crater of the right age. Nobody had found an iridium-rich layer like this anywhere else.

But in 1991, some other scientists dated a huge impact crater at Chicxulub, Mexico. It was 65 million years old. Others found iridium-rich deposits at different places, all around the world. The impact must have been so violent that it partly vaporized the ground and the asteroid. Wind would have carried the material all around the planet. Over the following months and years it would have settled into a layer of dust on the ground.

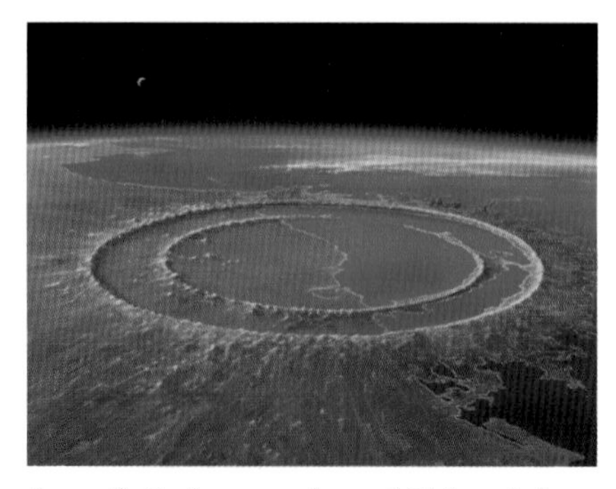

An artist's impression of Chicxulub crater in mexico.

The big BUT

So, an asteroid hit the Earth around the time that the last dinosaurs died out. But that is not enough evidence to be confident that one event directly caused the other.

There are two main problems with the asteroid explanation.

- Many dinosaurs (and animals and plants) had started to die out before the asteroid struck.
- There have been other major impacts which did not cause mass extinctions.

Another explanation – enormous eruptions

A third of the land surface of India has layers of black rock called basalt. It must have arrived there in floods of molten rock. There were hundreds of lava flows from a super-volcano. And eruptions release a lot of poisonous gases.

The eruptions that made India's basalt were at their most intense 65 million years ago. But they started before then. These eruptions could explain why extinctions began before the 65 million year mark.

There are flood basalts in Siberia, too. Those are much older – 250 million years. In the worst mass extinction ever, 95% of all the world's species died out at that time.

Basalt in India now. It took many eruptions to produce this rock, all between 63 and 68 million years ago.

Another big BUT

But there were also flood basalt events that did not cause a mass extinction. Some of the rocks of Scotland and Northern Ireland are flood basalts. They became solid 58 million years ago. There was no mass extinction then.

Questions

5 These pages present two possible explanations for dinosaur extinction.

a Make a table or chart to list the points for and against each explanation.

b Which do you think is more likely? Give your reasons.

⑥ How could scientists tell that the layer of iridium-rich clay around the world was all deposited at the same time?

⑦ What might have happened to the asteroid explanation of dinosaur distinction if

a the Chicxulub crater had not been found?

b iridium-rich layers had turned out to have different ages at different places?

Key words

mass extinction

Find out about:

- how scientists know what stars are made from
- the process that releases energy in stars and produces new elements

F What are we made of?

Everything on Earth is made from just 92 kinds of atom, or elements. Salt, soil, ants, trees, and humans are all made from the same stuff. Atoms simply get recycled as things grow and die.

Stars and Earth stuff

Scientists can spread light into spectra and study the colours present. If they shine light through different chemical elements, then each element produces a unique pattern. Fine, dark lines in the spectrum show where that element absorbs light. Analytical scientists use this technique to identify what chemicals are present in a sample.

The Sun is a light source. Its light tells us how hot it is, and what it is made of. When astronomers first looked at the spectrum of sunlight, they were amazed to find similar patterns to those seen in the laboratory. They looked at other stars. Exactly the same 92 elements, everywhere.

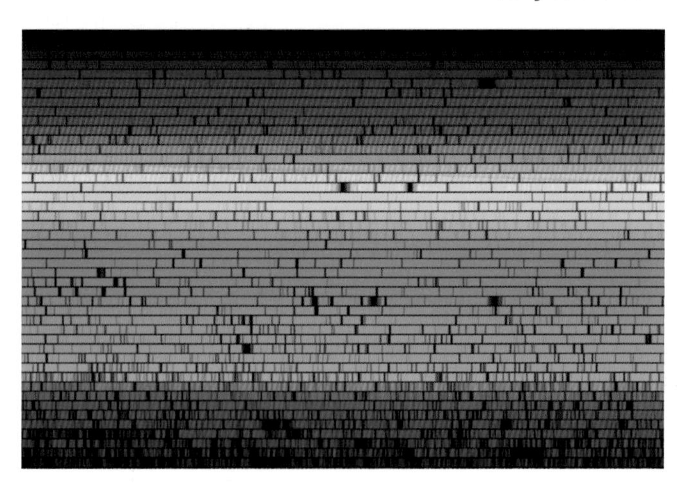

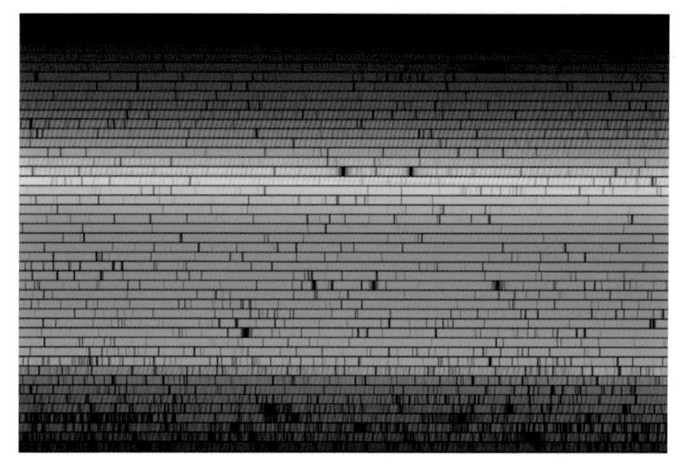

The spectra of the bright star Arcturus (left) and the Sun (right) – similar but not identical. The colours in its light show that the Sun is hotter.

Nuclear fusion

Scientists once struggled to understand the Sun. It could not be a great ball of fire. Fire is a chemical process, requiring fuel and oxygen. Any oxygen would have run out long ago.

Then they found that atoms have a central core, called a nucleus. Joining small nuclei together releases energy. New elements are created. This is called **nuclear fusion**.

Nuclear fusion does not happen easily. Nuclei all exert repulsive electric force on each other. Before small nuclei will join together they have to be colliding together so fast that they overcome this repulsion. Only at extremely high temperatures do nuclei have enough energy. This happens in stars.

The Sun fuses hydrogen to make helium.

Star birth and star death

How a star is born:

- gravity pulls a star together in the first place, from hydrogen gas spread out in space
- the hydrogen gas collapses, faster and faster
- some of the nuclei in the gas collide hard enough to overcome their repulsion
- fusion starts to happen
- fusion releases energy to keep the temperature high and the nuclei moving fast

Sooner or later a star runs out of small nuclei. None of them last forever. Stars, like people, frogs, and trees, have **life cycles**. The mass of a star determines how long it lives, and how it dies.

This photograph, taken by the Hubble Space Telescope, shows new solar systems forming in a dense gas region called the Eagle nebula. Dusty discs surround baby stars.

Heavy elements are made in stars

The most common element in the Universe is hydrogen. In stars, fusion continues to make bigger and bigger elements. When fusion stops, big stars explode as supernovae. Their debris, containing all 92 elements, is scattered through space.

When our Solar System formed, it gathered debris from dead stars. Except for hydrogen and helium, the chemical elements that make up everything on Earth come from stars. We are made of stardust.

In 1987, a star exploded. The second image shows its gassy remains. The supernova behaved as scientists had predicted. But the type of star that exploded was unexpected. Ideas about star processes needed some fine tuning.

Questions

1. How can scientists find out about **a** the temperature and **b** the chemical elements in the outer layers of a star?
2. How do you know that the Sun is not a ball of gas on fire?
3. Scientists believe that hydrogen, helium, and a little lithium were the only elements in the Universe before there were stars. There are 92 natural elements on Earth. Where did the other elements come from?

Key words

nuclear fusion
life cycles (star)

Find out about:

- ways of measuring the distance to stars
- planets around stars other than the Sun

G Are we alone?

In good conditions, you can see more than 2000 stars at a time with your unaided eyes. With so many stars in the sky, people have talked for centuries about other possible worlds. Now a scientific search for aliens has begun.

The Sun, Moon, and planets appear to move against a fixed background of stars. This means that stars are not part of the Solar System.

Looking back in time

Light moves fast. It could travel the length of Britain in just 6 millionths of a second. At 300 000 km/s, light from the Sun takes just over 8 minutes to reach Earth. This means that you see the Sun as it was 8 minutes ago. You see other stars as they were many years ago.

And it works the other way too. Any star beyond the Solar System is much further away. If aliens are spying on Earth, what they see is history. They see the world as it once was: perhaps Roman times.

Star distances

Here are two ways of working out the distances to stars.

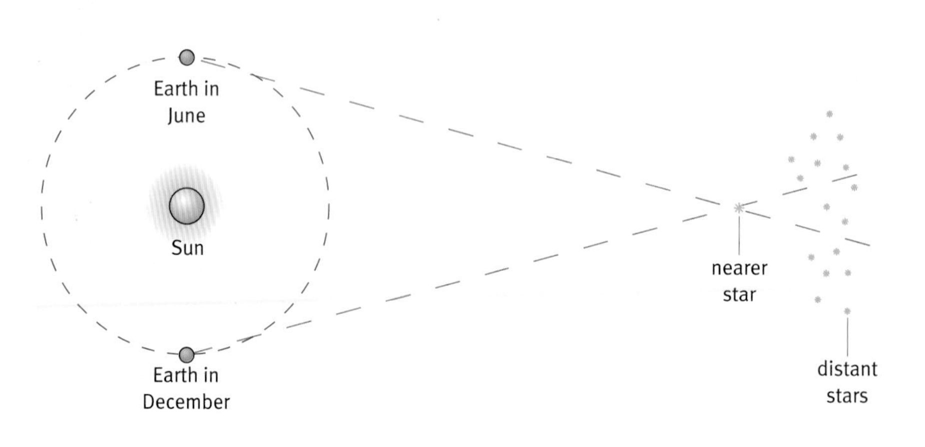

1 Parallax: The Earth moves from one side of the Sun to the other, every six months. Seen through an Earth-based telescope, a nearby star will shift its position against the background of more distant stars. The nearer a star is, the more it shifts.

This effect is called **parallax**. It provides a way of measuring distance.

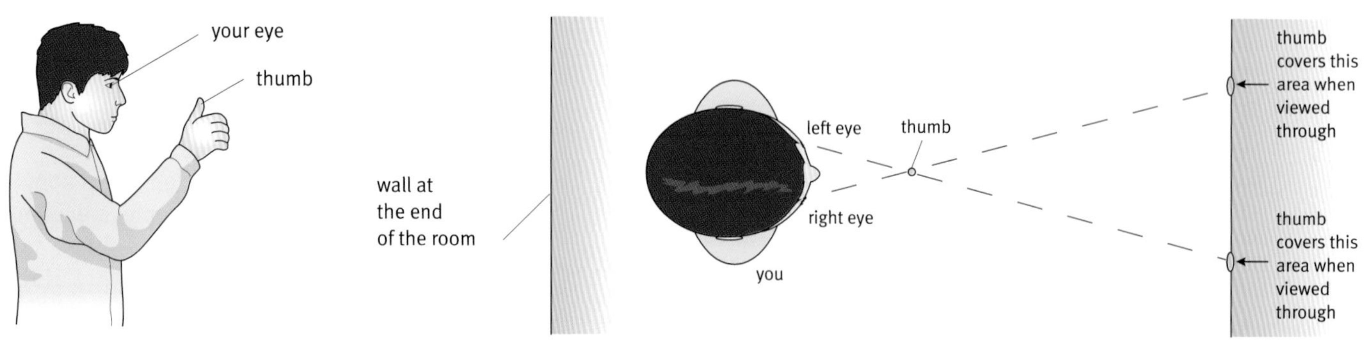

To see the parallax effect, hold up your thumb and look at it with each eye in turn.

2 Brightness: Imagine a large number of lights of different brightness. Some are much further away than others. It could be hard to tell the difference between a nearby torch and a distant searchlight. But if you know whether a light is a torch or a searchlight, then you can judge its distance.

That's how it is with stars. If you know what kind of star it is, then you can use its apparent brightness to estimate its distance. The nearer a star is, the brighter it seems.

These streetlights all shine with the same brightness. But the further away a streetlight is, the fainter it appears.

Light-years away

Proxima Centauri is not bright enough to see without a telescope. But it is the closest star outside the Solar System. Parallax measurement shows that it is 4.22 light-years away.

A **light-year** is a unit of distance used by astronomers. It is the distance travelled by light in one year.

Arcturus is another of the nearer stars. Arcturus is 36.7 light-years away.

The SETI project

Between 1990 and 2005 astronomers found over 130 stars that have planets, and they are still finding more. They detect these planets by clever techniques, like small dips in the brightness of a star as its planet passes in front of it. Or by the wobbling motion of the star caused by the gravity of a planet. Planets around other stars are called **exoplanets**.

Exoplanet 2M1207

In 2004, astronomers made the first ever image of an exoplanet. They called it 2M1207.

In 1992, NASA began a Search for Extra-Terrestrial Intelligence. It looks for radio signals that might be produced by aliens, checking one star at a time. Using SETI@home, some 50 000 people around the world use their home computers to help process the data that SETI collects.

So far there is no evidence of life elsewhere.

Questions

1 Some light from Alpha Centuari is reaching Earth as you read this. How old were you when that light left Alpha Centauri?

② Suppose you are an alien on a planet 10 light-years from Earth. Describe any possible evidence you could have to suggest that the Earth exists, and that there is life on it.

3 There may be intelligent life forms on exoplanets. What risks and benefits could there be in communicating with them?

Key words

parallax
light-year
exoplanets

New telescopes

By the early 20th century, astronomers were starting to use some really big telescopes, especially in America. Such telescopes are usually on mountaintops, where the atmosphere has less effect on the light. And there is less **light pollution** from sources such as streetlights.

Harlow Shapley worked in California, investigating faint patches of light called nebulae. He could see that some nebulae are dense clusters of stars. Thanks to Henrietta Leavitt, at the Harvard Observatory, he had a new way to measure their distances. The results were shocking. Some of them seemed to be more than 100 000 light-years away. Shapley suggested that they were part of a gigantic star system: the Milky Way. The Solar System too, is part of the Milky Way.

The great debate of 1920

Some nebulae have a spiral shape. One of these is called Andromeda. Some astronomers thought that these were 'island universes', outside the Milky Way. There were heated arguments, but not enough evidence to reach an agreed conclusion.

Harlow Shapley suggested that they were part of the Milky Way. Perhaps they were gas clouds. In 1920, he took part in a public debate in Washington. It was a head-to-head discussion with Heber D Curtis, another astronomer. Curtis claimed that spiral nebulae are star systems outside the Milky Way. They called it 'The Great Debate'.

On the night, Shapley came off better. His evidence seemed stronger. A few years later, the argument was finally settled by new evidence. Curtis had been right.

Satellites produced this night image of the Earth. Many parts of the Earth are affected by light pollution.

More than one galaxy

Edwin Hubble used a new telescope to try to find out how far away Andromeda is. He used the same method as Harlow Shapley had done to study nebulae. The result was surprising. It seemed that Andromeda was a *million* light-years away. It is another huge collection of stars held together by gravity, another **galaxy**.

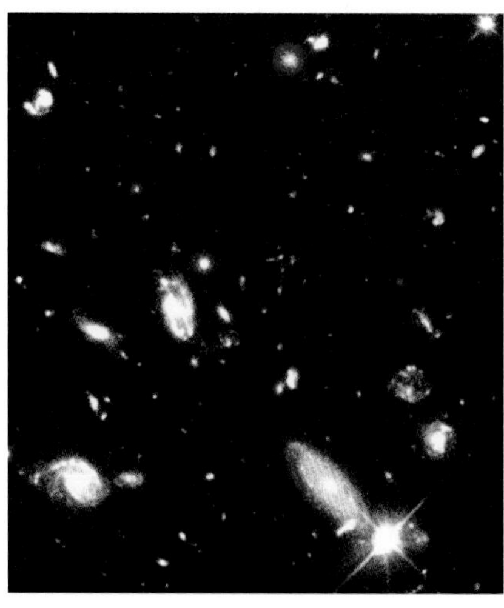

An image made by light at the end of a long, long journey.

Generations of stars

The Hubble Space Telescope image (right) shows extremely distant galaxies. The light that made this image left its stars over 10 000 million years ago. That was long before the Sun and the rest of the Solar System existed.

Scientists can only suppose that, in the time it took for their light to reach the telescope, the stars in these galaxies will have changed. Some may have exploded as supernovae. Debris from these first generation stars would include atoms of oxygen, carbon, and iron. Those atoms are now likely to be inside second and third generation stars and planets.

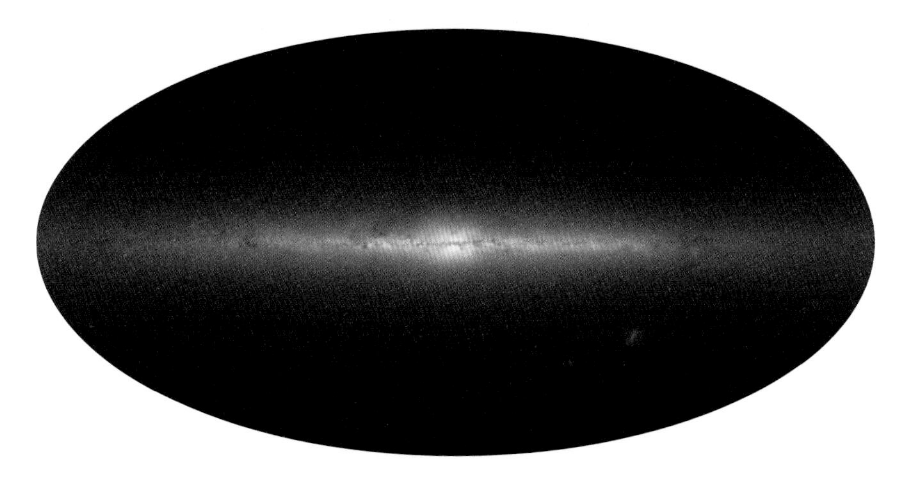

This is the view towards the centre of our own Milky Way galaxy, made with infra-red radiation. Each speck of light is a star. The diameter of the Milky Way is 100 000 light-years.

Key words
light pollution
galaxy

Questions

4 What was Harlow Shapley's new observation about the nebulae?

5 Read about the Great Debate of 1920.

a Copy the table below and write the main ideas of the two astronomers in two columns.

Harlow Shapley	Heber D Curtis

b Who seemed to have stronger evidence?

c Who was proved right by later evidence?

3 What new observation showed that there were objects outside the Milky Way?

Find out about:
- the age of the universe
- an explanation called 'big bang'

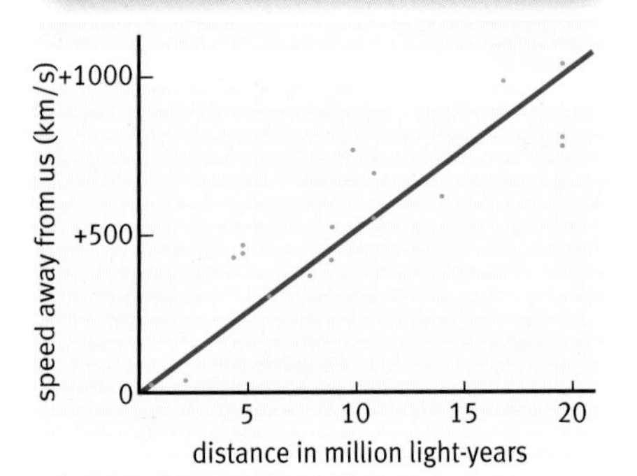

Edwin Hubble published a famous paper in 1929. It showed that more distant galaxies are moving away from us faster.

Imagine yourself on the surface of a very big balloon, looking along a line with galaxies at one-metre intervals. If the balloon is expanding, every metre is growing larger. Let's say the nearest galaxy moves half a metre away from you. In the same time, the second galaxy seems to move away by a metre and the third galaxy by 1.5 metres. The more distant the galaxy, the faster it moves away from you.

H How did the Universe begin?

The Universe is everything. It's stars and galaxies. It's clouds and oceans. It's bacteria and birds. You are part of the Universe.

A big bang

Until the 20th century, most people thought that the Universe was eternal – it never changed.

Big telescopes changed everything. Light from distant galaxies tells astronomers that clusters of galaxies are all moving away from each other. The Universe is big and getting bigger. Space itself is expanding.

Scientists now imagine that the Universe was once incredibly hot, tiny, and dense. This explanation is called **big bang** theory.

The theory passed a major test in 1965. A group of scientists had predicted, in 1948, that an afterglow of the big bang event should still fill the whole Universe. Years later, two radio engineers in New Jersey tried hard to get rid of an annoying background hiss in their antenna. Eventually they gave up, and reported the noise. Arno Penzias and Robert Wilson won Nobel Prizes for this work. They had discovered cosmic microwave background radiation.

Switch on your TV without tuning it in to any channel and you get a screen full of nonsense. Part of what your TV aerial picks up is microwave radiation from the big bang.

The age of the Universe

If you look at the light from distant galaxies the position of dark lines in their spectrum (see page 26) is shifted towards the red end. This is called redshift. The amount of their redshift shows how fast galaxies are moving away.

Fifty years ago scientists used the speed and distance of galaxies to estimate how long ago all galaxies were in the same place. They had to assume that the galaxies have always moved away at the same rate, which may or may not be true. So it was only a rough estimate. The age of the Universe came out at somewhere between 10 000 million years and 20 000 million years.

Then in 2003, new observations of the cosmic microwave background gave a much more precise answer. The Universe is 13 700 million years old, plus or minus 200 million years.

Other lines of evidence too support big bang theory. Among them:

- A hot big bang explains why the early Universe was about 24% helium by mass.
- The oldest stars (12 000 million years old) are younger than the Universe.

Models of the Universe

The scientific study of the Universe is called cosmology. Like other scientists, cosmologists often use mathematics to help their thinking. Computer models use maths, for example, to simulate the formation of galaxies. Cosmologists also use another kind of maths, involving visual modelling, called topology.

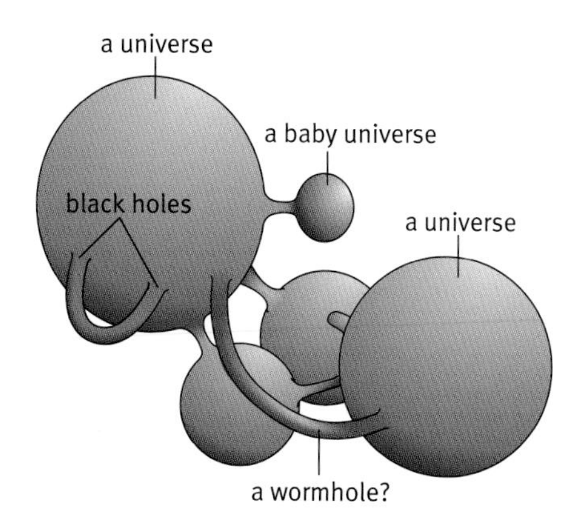

Topological models

Cosmologists at work

Around the world there are thousands of cosmologists. Most of them work in universities. They usually work in groups. When a group develops a new idea, they write a paper for a scientific journal.

Before a scientific paper is published, other experts must first review it. They try to make sure that what gets published has something useful and new to say. This process is called **peer review**.

Model-independent dark energy test with sigma8 using results from the Wilkinson Microwave Anisotrpohy Probe

M Kunz, P-S Corasaniti, D Parkinson, and E J Copeland,

Physical Review D **70** 041301 (R) (2004) *ICG 04/30*

Scientific papers are written for other scientists. You have to be an expert to understand them. Through papers like this cosmologists around the world share their ideas.

Will the Universe expand forever?

As recently as the 1990s it was thought that gravity would eventually cause the Universe to collapse again in a big 'crunch'. But, recent observations of supernovae seem to show that the rate of expansion of the Universe is increasing. Scientists do not yet agree on the evidence or how to explain it. The ultimate fate of the Universe is difficult to predict.

Subrahmanyan Chandrasekhar's studies of the structure and evolution of stars won him the Nobel Prize in 1983. NASA's Chandra X-ray telescope was named in his honour. It was launched in 1999.

Key words

big bang
peer review

Questions

1 List four observations that support the big bang theory.

② Some cosmological theories produce predictions that are impossible to test (for example, that there might be other Universes besides our own). In your opinion, should these theories be rejected as 'unscientific'?

Science explanations

In this Module you have seen how scientists gather evidence (data and observations) and try to persuade others of their explanations of it.

You should know:

- about how rocks provide evidence for changes in the Earth
- Alfred Wegener's explanation of mountain-building
- that the Earth must be older than its oldest rocks
- about evidence for continental drift and tectonic plates
- where earthquakes, volcanoes, and mountain building generally occur
- several things that public authorities can do to reduce the damage caused by geohazards
- what is in our Solar System
- that fusion of hydrogen nuclei is the source of the Sun's energy
- the possible consequences of an asteroid collision with the Earth
- that light travels at very high speed
- that distant objects in the night sky are observed as younger than they are now
- that the Sun is a star in the Milky Way galaxy.
- that all chemical elements with more mass than helium were made in earlier stars.
- that distant galaxies are moving away from us
- that the Universe began with a 'big bang' about 14 000 million years ago.
- the relative ages of the Earth, the Sun, and the Universe; and the relative diameters of the Earth, the Sun, and the Milky Way

Ideas about science

New scientific data and explanations become more reliable after other scientists have critically evaluated them. This process is called peer review. Scientists communicate with other scientists through conferences, books, and journals.

Scientists test new data and explanations by trying to repeat experiments and observations that others have reported.

The chapter includes many Case Studies. From these you should be able to identify:

- statements that are data
- statements that are all or part of an explanation
- data or observations that an explanation can account for
- data or observations that don't agree with an explanation

Scientific explanations should lead to predictions that can be tested. You should know:

- how observations that agree or disagree with a prediction can make scientists more or less confidence about an explanation

Scientists don't always come to the same conclusion about what some data means. The debate about Wegener's idea of continental drift provides an example of this. You should know:

- that working out an explanation takes creativity and imagination
- why Wegner's explanation was rejected at the time
- some scientific questions have not been answered yet
- distances to many stars and galaxies are not known exactly, because they are so difficult to measure
- the ultimate fate of the Universe is difficult to predict

These ideas are illustrated through Case Studies, including: James Hutton; Alfred Wegener; Fred Vines; Daniel Barringer; Michelle Boast; Luis and Walter Alvarez; Edwin Hubble; and Subrahmanyan Chandrasekhar.

Why study radiation?

Human eyes see one type of radiation - visible light. But there are many other types of 'invisible' radiation. Some radiations are harmful. You hear a lot about the health risks of different radiations: for example, from natural sources such as sunlight, and from devices like mobile phones. Radiation is involved in climate change, and this is the biggest risk of all.

The science

Science shows that microwaves, X-rays, visible light, and other kinds of radiation all belong to one family. This is called the electromagnetic spectrum.

The Earth's atmosphere may seem transparent to sunlight. But its ozone layer absorbs the UV radiation in sunlight, protecting life on Earth. Science can explain how radiation warms the atmosphere, and uses computer modeling to predict global warming.

Ideas about science

To make sense of media stories about radiation you need to understand a few things about correlation and cause. It will also help if you know how to evaluate reports from health studies, and how to interpret statements about risks.

Radiation and life

gamma rays

x-rays

ultraviolet

visible

infrared

microwave

radio

Find out about:

- how radiation affects living cells
- microwave radiation from mobile phones
- weighing up risks against benefits
- the evidence of global warming, and its possible effects

Find out about:

- benefits and risks of exposure to sunlight
- how the ozone layer protects life on Earth

A Sunlight, the atmosphere, and life

Skin colour

The **ultraviolet radiation** (UV), in sunlight can cause skin cancer. Skin cancer can kill.

Melanin, a brown pigment in skin, provides some protection from UV radiation. People whose ancestors lived in sunnier parts of the world are more likely to have protective brown skin.

Fair skin is good at making vitamin D. But it gives less protection against UV radiation. Melanoma is the worst kind of skin cancer. One severe sunburn in childhood doubles the risk of melanoma in later life.

Vitamins from sunlight

Human skin absorbs sunlight to make vitamin D. This nutrient strengthens bones and muscles. It also boosts the immune system, which protects you from infections. Recent research suggests that vitamin D can also prevent the growth and spread of cancers in the breast, colon, ovary, and other organs.

Darker skin makes it harder for the body to make vitamin D. So in regions of the world that are not so sunny there is an advantage in having fair skin.

Nowadays the links between UV radiation, skin cancer, and vitamin D are clear. People with fair skin can keep healthy in sunny countries by being careful not to expose their skin to too much UV radiation. People with dark skin can keep healthy in less sunny countries if they get enough vitamin D from their food.

Feeling good

People like sunshine. It can alter your mood chemically and reduce the risk of depression.

A suntan is the body's attempt to protect itself against UV radiation and skin cancer. A tanned skin has more UV-absorbing melanin. But the protection is only weak.

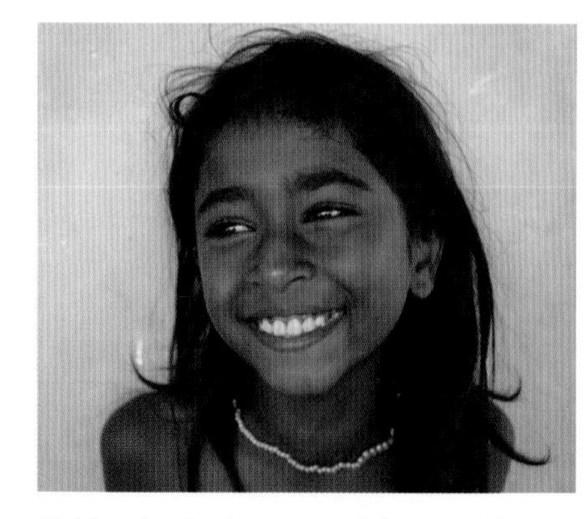

Sri Lanka is the one of the world's sunniest places. A high level of protection from UV radiation is important.

Balancing risks and benefits

Is sunlight good for you? There is no simple answer. Over a lifetime, the risk of developing one type of skin cancer, malignant melanoma, is 1 in 147 (UK males) or 1 in 117 (UK females). There are also risks from staying indoors all the time.

Protecting your health involves reducing risks, whenever possible. And balancing risks against benefits.

Skin cancer warnings ignored

Too much exposure to the sun is dangerous. A Cancer Research UK survey found a worrying gap between how much people know about skin cancer and how little they actually do to protect themselves in the sun.

Among 16–24-year-olds, 73% believed that exposure to the sun might cause skin cancer. But only a quarter of this age-group apply high-factor sunscreen as protection. And fewer than 20% cover up or seek shade from the sun.

Correlation or cause?

A study of 2600 people found that people who were exposed to high levels of sunlight were up to four times more likely to develop a cataract (clouding of the eye lens). Exposure to sunlight is a **factor**. Eye cataracts are an **outcome**. There is a **correlation** between exposure to sunlight and eye cataracts. But doctors do not say that exposure to sunlight will **cause** cataracts. There are other risk factors involved, such as age and diet.

Questions

1 Exposure to sunlight increases your risk of developing skin cancer. List some benefits of staying indoors and avoiding direct sunlight. List some risks as well.

② Describe at least three ways that a person could reduce the risk of skin cancer while enjoying a beach holiday.

③ In what way is sunbathing safer than crossing the road? In what way might crossing the road be safer than sunbathing?

Key words

ultraviolet radiation (UV)
factor
outcome
correlation
cause

Chlorophyll, a chemical in leaf cells, is essential for photosynthesis. It absorbs red and blue light but does not absorb green light.

Sunlight and life

When a material **absorbs** light, or any kind of electromagnetic radiation, it takes its energy from it. The radiation then ceases to exist.

Absorbing energy from sunlight

Sunlight falls on leaves, which absorb the red and blue light from it. This is selective absorption. Leaves reflect other colours. They **reflect** green light most strongly, so leaves look green.

When leaves absorb light from the Sun, they gain energy. Leaf cells use the energy to combine water and carbon dioxide. They make starch and release oxygen. This chemical process is called **photosynthesis**. Roots gather water, and leaves take carbon dioxide from the air.

Respiration

Plants store the starch they make. They can use it later to produce energy, through a process called **respiration**. Leaves take in oxygen from the air and release carbon dioxide. The process of respiration is the reverse of photosynthesis.

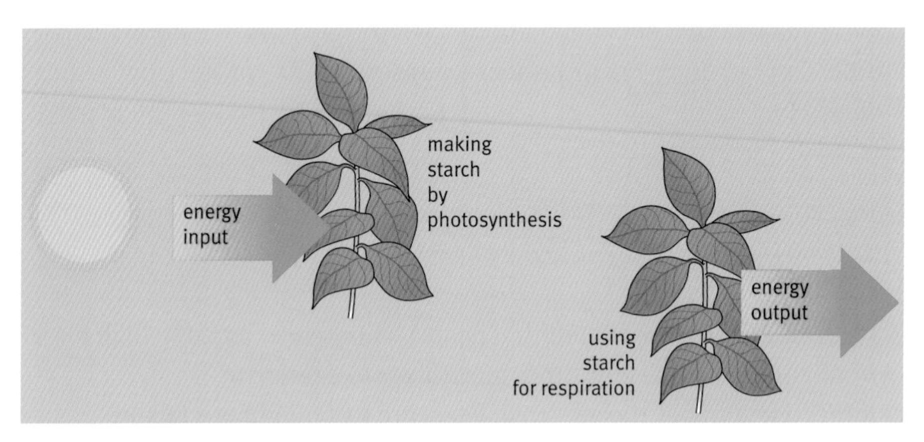

Photosynthesis needs an energy input. That comes from the light that leaves absorb. Respiration provides an energy output.

An absorbing atmosphere

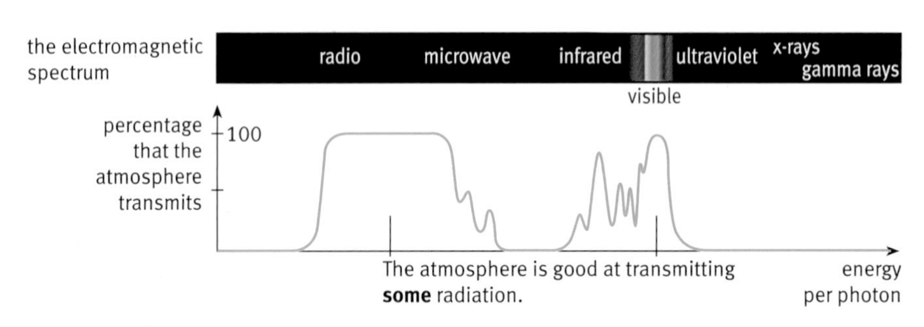

Radiation arrives at the top of the **atmosphere** from the Sun and other distant sources.

The atmosphere **transmits** some radiation, such as visible light and a lot of radio radiation. These radiations can reach the ground. But it absorbs other radiations like X-rays and most UV radiation.

Ozone protection

The atmosphere is a mixture of gases, including oxygen. In the upper atmosphere some of the oxygen is in the form of ozone. It makes an **ozone layer**.

The ozone layer is good at absorbing UV radiation. When UV radiation is absorbed its energy can

- break ozone molecules, to make oxygen molecules and free atoms of oxygen
- break oxygen molecules, to make free atoms of oxygen.

These chemical changes are reversible. Free atoms of oxygen in the ozone layer are constantly combining with oxygen molecules to make new ozone.

UV radiation is harmful to living things. Life on Earth depends on the ozone layer absorbing UV radiation.

altitude/km
ionosphere (extends up to 550 km)
stratosphere (15–50 km)
ozone layer
meteorological balloons
troposphere (up to 15 km)

O_2 – oxygen

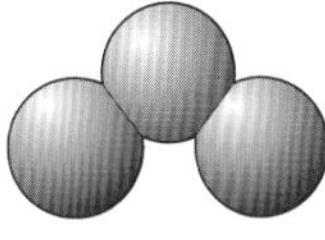

O_3 – ozone

The ozone layer is good at absorbing harmful UV radiation.

Ozone holes

Humans have created a problem. Some synthetic (man-made) chemicals, such as **CFCs** (chlorofluorocarbons) used in fridges, have been escaping into the atmosphere. They turn ozone back into ordinary oxygen. So more UV radiation reaches the Earth's surface. This happens strongly over the North and South Poles in Winter and Spring. Thin ozone in those places is called 'the hole in the ozone layer'.

The international community is now dealing with this problem. Aerosol cans once used CFCs, but this use has been stopped worldwide. At your local waste recycling centre fridges are stacked in a separate section. They must go to a specialist centre so that CFCs can be carefully removed from them.

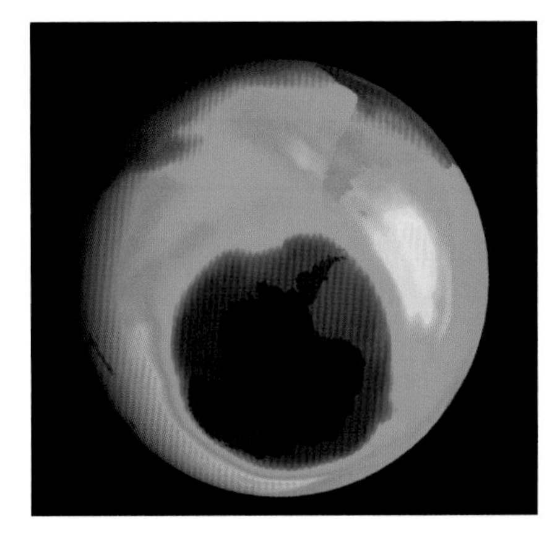

This image has been made by sensing ozone. Dark colours represent less dense ozone. There seems to be a 'hole' in the protective layer.

Old fridges waiting to have CFCs removed.

Key words
absorbs
reflect
photosynthesis
respiration
atmosphere
transmits
ozone layer
CFCs

Questions

4 Briefly describe one type of selective absorption that takes place in the atmosphere.

5 What are the names of the three main layers of the atmosphere? In which of these is the ozone layer?

6 What effect do CFCs have on ozone?

7 What action is being taken to reduce damage to the ozone layer?

Find out about:

- a family of radiations called the electromagnetic spectrum
- sources and detectors of radiation
- why some kinds of radiation are more dangerous than others

B Radiation models

A beautiful world

All radiation has a **source** that **emits** it. Then it has a journey. It spreads out, or 'radiates'. Radiation never stands still. Some radiation, at the end of its journey, causes chemical changes at the back of your eye. That radiation is visible light.

Some materials, like air, are good at transmitting light. They are clear, or transparent. On the way from the source to your eyes light can be reflected by other materials. The objects around you would be invisible if they did not reflect light.

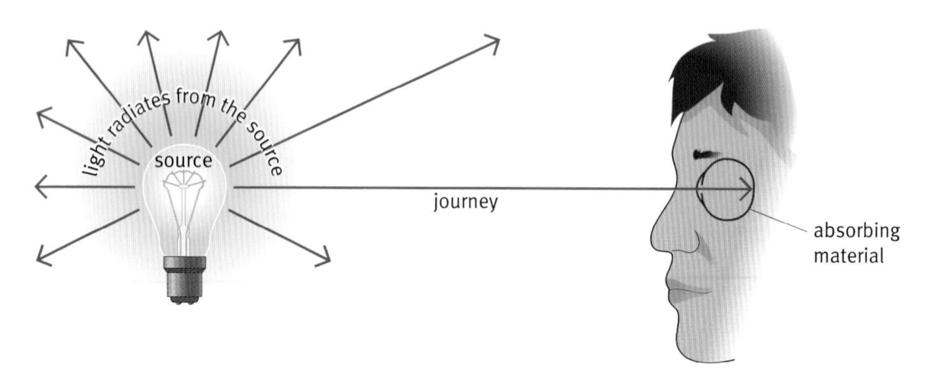

A journey of visible light from source to eye.

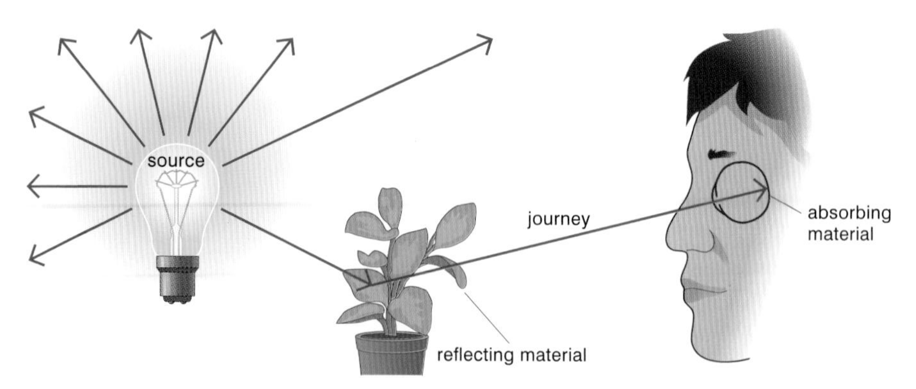

A journey of visible light, from source to reflector to eye.

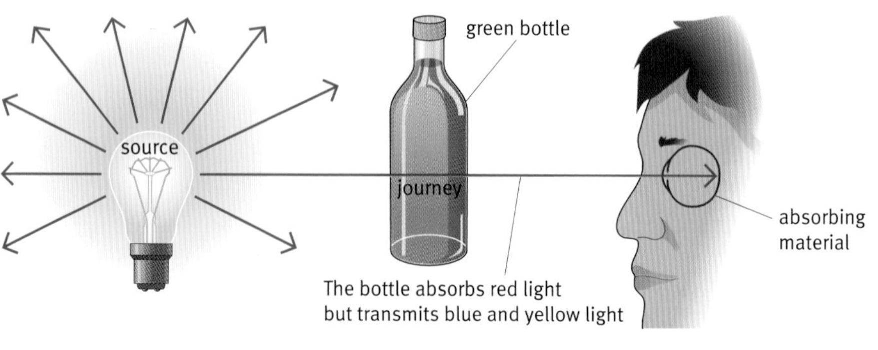

A journey from source to detector, but with absorption of light on the way.

Coloured materials added to glass can absorb some colours of light and transmit others.

A single source of light, the Sun, made this picture possible. But the light interacted with materials along the journey between the source and the detector (a camera).

- Air mostly transmitted the light, though there was some particle-by-particle reflection ('scattering') so that light arrives at the scene from all over the sky.
- Water surfaces are good reflectors, though some light also travels down into its depths.
- Tree leaves transmit some colours of light, and absorb others. But they also reflect light into the camera.

Transmission, reflection, and absorption of light make the world so beautiful.

Hidden messages

Detectors can make invisible radiation visible, as you will see later in this chapter. There are also ways of detecting radiation without producing pictures at all. For example:

- gamma radiation can make audible clicks in a speaker
- a bowl of soup in a microwave oven responds to radiation by getting hot
- the aerial of a radio detects radiation by making electrical signals in the radio's circuits

For all of these examples of electromagnetic radiation, there is a source, a journey, and a detector. The detector must absorb radiation for it to work.

Questions

1 Glass is a weak absorber of visible light. How would you show that it does absorb some light?

2 Can glass reflect light? Explain.

3 Materials can transmit, reflect, or absorb light. Which one of these is glass best at?

Communication

Radiation can carry information from a source to a detector by having coded patterns. The simplest way to do that is to turn the source on and off. Digital radio communication uses this idea, switching at incredibly high speeds. A TV 'remote' also uses different flickering patterns of infrared light to send information to a TV set.

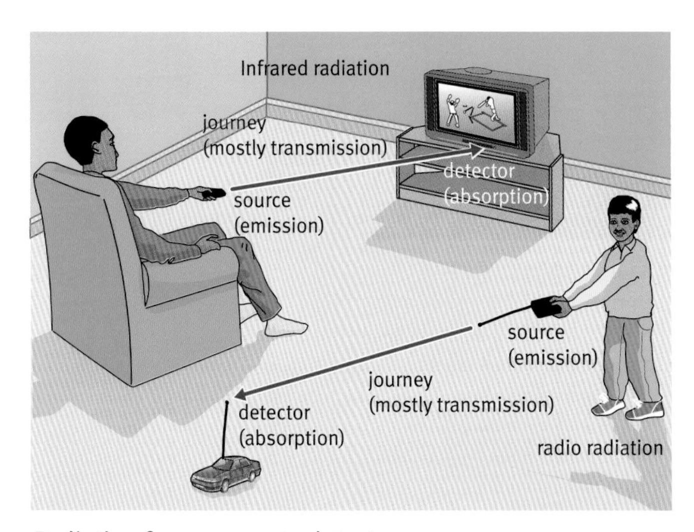

Radiation from source to detector.

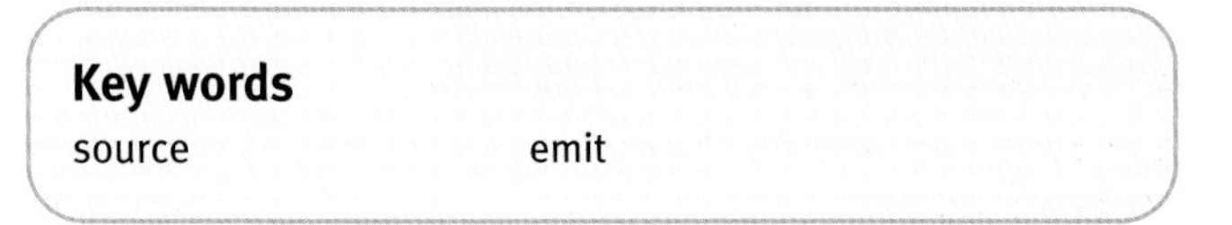

Key words
source
emit

Find out about:
- reasons for studying craters
- possible explanations for the extinction of dinosaurs

Absorbing electromagnetic radiation

When materials absorb electromagnetic radiation they gain energy. Exactly what happens depends on the energy absorbed.

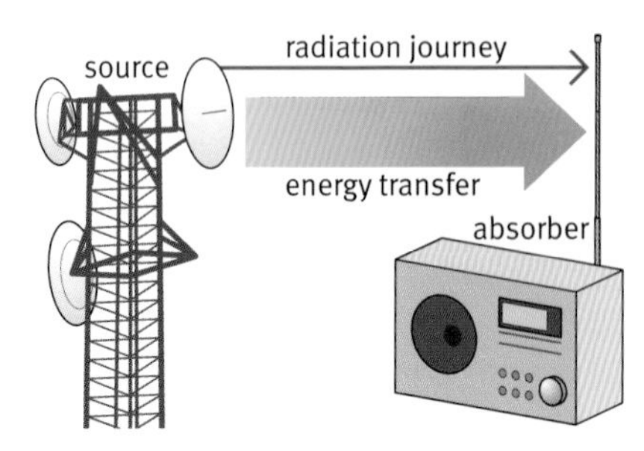

Metal aerials can absorb radio and microwave radiation. The process creates electrical vibrations inside the metal.

Radiation can make patterns of electric current in metals

Patterns of microwave and radio radiation can make patterns of electric current in radio aerials.

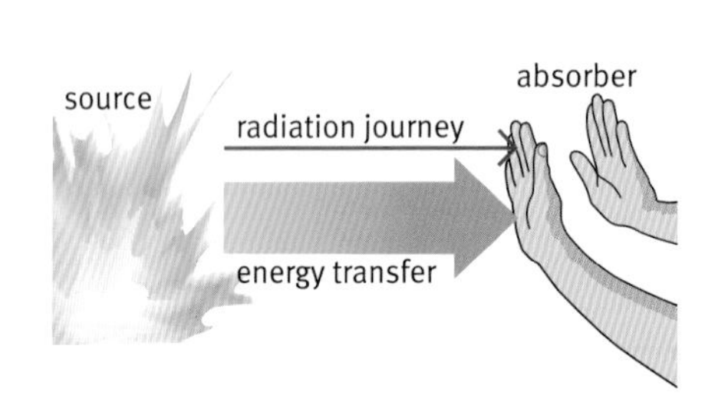

A fire transfers energy to the world around it. Surfaces in its surroundings, including people, absorb the radiation and gain the energy.

Radiation can have a heating effect

Radiation absorbed by a material may increase the vibration of its particles (atoms and molecules). The material gets warmer.

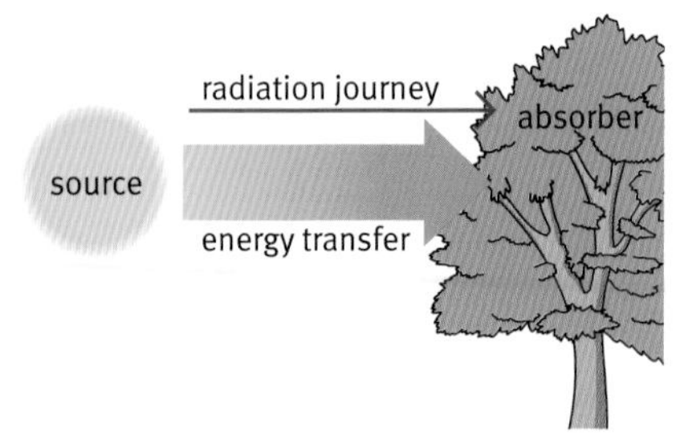

A leaf takes energy from the Sun's radiation so that photosynthesis can happen.

Radiation can cause chemical changes

If the radiation carries enough energy, the molecules that absorb it become more likely to react chemically. This is what happens, for example, in photosynthesis, and in the retinas of your eyes.

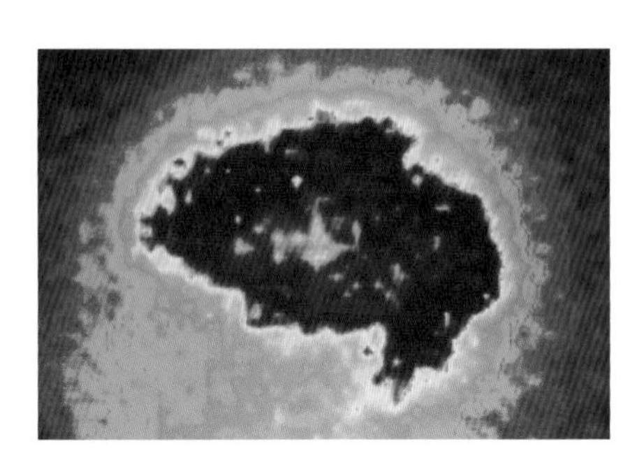

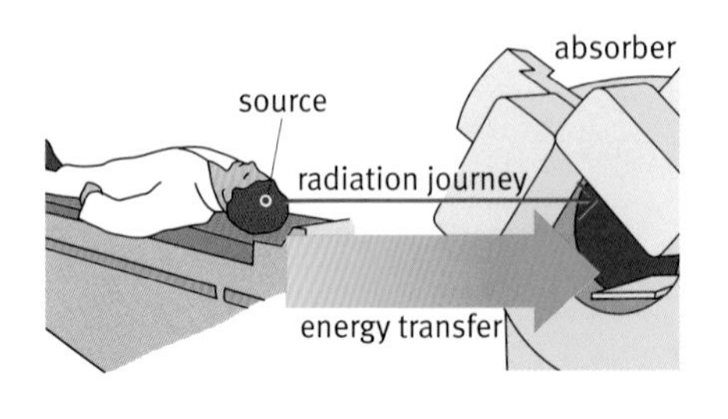

This medical image was made by a gamma camera. Each dot on the image was made by a single ionization event.

Ionization can damage living cells

If the radiation carries a large amount of energy, it can break up the molecules that absorb it into smaller 'bits', called **ions**. This process is called **ionization**. The ions then take part in other chemical reactions. Ionization can damage living cells.

Radiation arrives in energy packets

It is useful to think about radiation in terms of **photons**. A photon is an energy packet of radiation:

- sources emit energy photon by photon
- absorbers gain the energy photon by photon

The energy deposited by a beam of electromagnetic radiation depends on both:

- the number of photons arriving
- and the energy that each photon delivers.

Sitting in sunlight, infrared, and visible radiations have a warming effect; UV radiation ionizes and can cause a chemical change that could (though not very often) start skin cancer.

Ionizing and non-ionizing radiation

Sources of gamma radiation, X-rays, and UV radiation pack a lot of energy into each photon. So absorbers get a lot of energy from each photon. These photons have a strong local effect – they can ionize. Gamma radiation, X-rays, and UV radiation are called **ionizing radiation**.

Look back at pages 36–37.
The electromagnetic spectrum shows the order of the amount of energy each photon carries. X-ray photons and gamma ray photons carry most energy. Radio photons carry least energy. Visible, infrared, microwave, and radio radiation are all **non-ionizing radiation**. Their main effect is warming. The lower the photon energy, the smaller the heating effect.

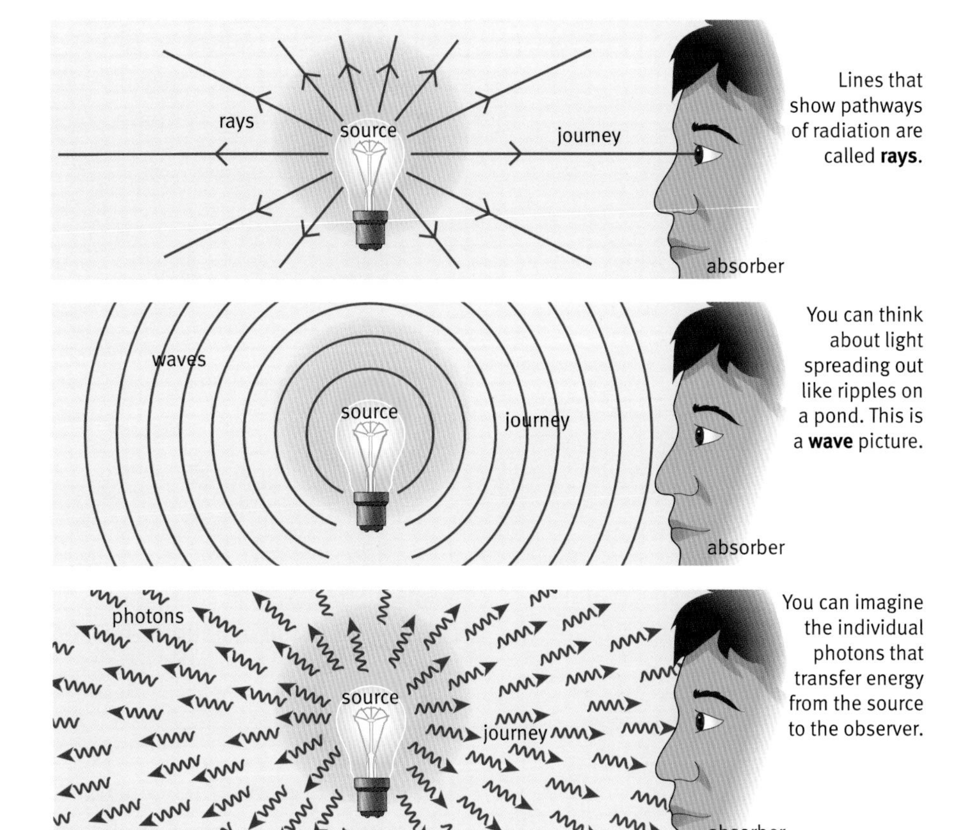

Radiation transfers energy. There are different ways of thinking about how it travels between source and absorber.

Questions

There is radio radiation passing through your body right now.

4 Where does the radio radiation come from?

5 Why does it not have any ionizing effect?

6 Does it have a heating effect? Explain.

Key words

ions
ionization
photons
ionizing radiation
non-ionizing radiation
ray
wave

Find out about:

- how microwaves cause heating
- design features that make microwave ovens safe to use

C Using radiation

Microwave ovens

In a microwave oven, microwave radiation transfers energy to absorbing materials. Once the radiation is absorbed it loses all of its energy, and it ceases to exist.

Molecules of water, fat, and sugar are good absorbers of microwave radiation. Microwaves make these molecules vibrate. Food containing them gets hot. A potato, for example, is made mostly of water, with carbohydrate and just a little fat.

Other particles, like the particles in glass or crockery, take very little energy from the radiation. It does not increase their vibrations at all. So the radiation in a microwave oven doesn't heat a bowl or a mug directly. The bowl or the mug is heated by the food or drink inside it.

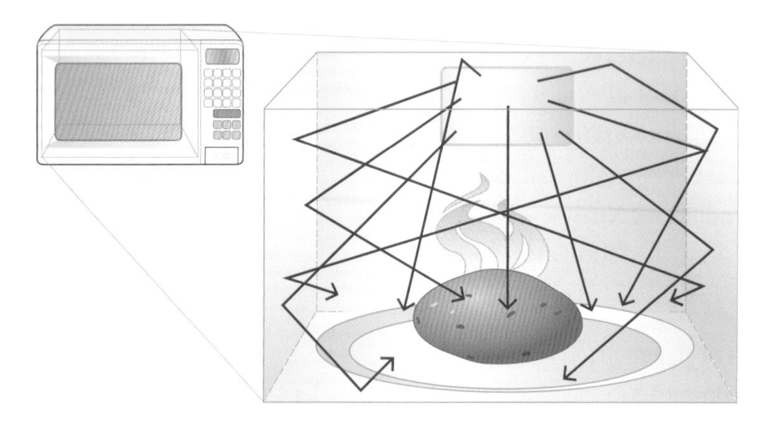

Inside a microwave oven, materials like glass and pottery are partially transparent to the radiation. The metal walls reflect it. Some substances, including water, absorb the energy.

How deep?

Absorption does not take place until the radiation enters the material. Water in a potato is good at absorbing microwave radiation. But it is not so good that the energy is all absorbed near the surface of the potato. Some energy is transferred quite deeply into the potato.

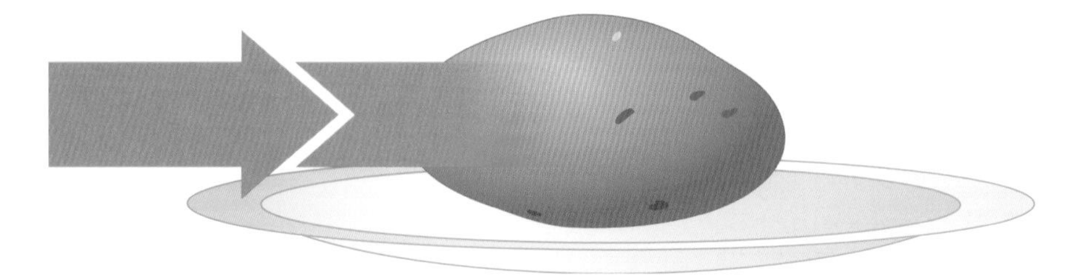

Absorption of energy by a potato. The plate absorbs very little energy from the microwave radiation.

Questions

1. What radiations are on either side of microwave radiation in the electromagnetic spectrum?
2. Why doesn't microwave radiation cause ionization?
3. (3) In a conventional oven, how does energy reach the centre of a potato to cook it?
4. A simple microwave oven is not very good for 'baking' potatoes. To get better, crispy skin you need a microwave oven that also has a 'grill' heater emitting infrared radiation. Why is infrared radiation good for producing a crispy skin, but not so good for rapid cooking of a potato all the way through?

How cooked?

Because most of the energy goes into heating the food, using microwaves is an energy-efficient way of cooking. Microwave ovens are typically rated at 600–800 watts.

The heating effect of non-ionizing radiation on an absorbing material always depends on its **intensity** (the energy arrives every second) and its **duration** (the exposure time).

You control the amount of cooking in a microwave oven by adjusting:

- the power setting
- the cooking time

Look inside any microwave oven. It will either have rotating metal blades near its top, or a rotating plate at the bottom. Without these devices, there would be 'hot spots' in the oven – places of higher intensity.

Safety features

People contain water, and fat. So a human body is a good absorber of microwave radiation.

Exposure to sufficient microwave radiation from an oven could cook you. The oven door has a metal grid to reflect the radiation back inside the oven. And a hidden switch prevents ovens from operating with their door open.

Key words
intensity
duration

Questions

5 How much larger is the power of a microwave oven than a 60-W lamp?

6 Why is it important that the walls and door of a microwave oven reflect the microwave radiation?

Find out about:
- radiation from mobile phones and masts
- how to judge whether a health study is reliable
- using X-rays

D Is there a health risk?

Mobile phones – Gt th msg?

A mobile phone stops radiating when you stop speaking. It also sends a weaker signal when you are close to the phone mast. That's to save the battery, but it also means that less radiation penetrates your head.

Cooked brain?

Mobile phones use microwave radiation. They receive microwaves from a nearby phone mast (or 'base station') and send microwaves back. Patterns in the radiation carry information. Phone masts radiate at powers up to 100 watt and mobile phones up to $\frac{1}{4}$ watt.

When you make a call, the fairly thick bones of your skull absorb some of this radiation. But some reaches your brain and warms it, ever so slightly. Vigorous physical exercise has a greater heating effect.

Distance from a radiation source is important. The intensity of microwaves decreases with distance, because the energy spreads out as it travels. Some people use a hands-free kit to keep the mobile phone away from their head.

Perfectly safe?

Nothing is completely safe. Even drinking a glass of water can be hazardous. You could choke on it. Or drop it and get a cut from the broken glass.

There are usually ways of reducing risks to an acceptable level.

Phone mast radiation

People have concerns about the radiation from phone masts. Fortunately, phone masts are designed so their radiation beam is shaped like the beam of light from a lighthouse. If you stand directly under one, its radiation is much weaker than the radiation from your phone.

Mobile phone mast
antenna
beam
15 – 50 m
ground
50 – 300 m

The microwave beam of a mobile phone mast.

Comparing risks

Some people might think that road travel is less risky than going by train. But over 3000 people die each year on the UK's roads, and only a few on the railways. This is an example of the difference between a **perceived risk** and an **actual risk**.

The precautionary principle

The health outcomes of some hazards are delayed. For example, skin cancer can develop many years after a person is over-exposed to UV radiation. Nobody knows yet whether mobile phones will have some long-term effect.

Are mobile phones safe?

At present, scientific evidence suggests that the microwave radiation produced by mobile phones is unlikely to harm the general population of the UK. But it is still too early to be certain. Health problems may take some time to develop.

Some people may be at higher risk because of genetic factors. Children may be more vulnerable because of their developing nervous system, the greater absorption of energy in the tissues of the head, and a longer lifetime of exposure.

Until we know more, it makes sense for mobile phone users to minimize their exposure to such radiation. This can be done in several ways, including making fewer and shorter calls.

In line with our precautionary approach, we believe that the widespread use of mobile phones by children under 16 for non-essential calls should be discouraged. Children under 8 should not use mobile phones.

UK report on mobile phone safety, January 2005

The **precautionary principle** can be stated like this: if the costs of some activity may turn out to be greater than any benefit, it makes sense to restrict the activity (or stop it). You could think of this as 'better safe than sorry'.

Questions

1. Getting dressed in the morning is an everyday activity. Explain why even this is not 'completely safe'.
2. Sal says, 'I will not eat GM food until it is proven safe.' Later he says, 'I'll continue using my mobile phone until it is proven unsafe.' Explain to him what is wrong with both statements.
3. You can choose whether or not to use a mobile phone. But you may have no choice about living near a phone mast. Does having a choice make a difference to your perceived risk?
4. Why does the report on mobile phone safety support the use of the 'precautionary principle'?

Key words

perceived risk
actual risk
precautionary principle

Researchers often collect data by sampling a whole population.

Health studies

Over 50 million people in the UK use mobile phones. Few people worry about unknown risks. People like the benefits they get from mobile phones. But research is underway to see if there are any harmful effects.

Scientists search for any harmful effects by comparing a sample of mobile phone users with a sample of non-users. But they need to be careful, because people are all different. They have different genes. Their homes, eating habits, and work are different. All of these are factors that affect health.

Are the results reliable?

The news often has reports of studies that compare samples from two groups – to see if a particular factor or treatment makes a difference. When you think about studies like these, there are two things worth checking:

What to check ...	**... and why**
Look at how the two samples were selected. Can you be sure that any differences in outcomes are really due to the factor claimed?	Suppose there was a study to see whether mobile phone use increases the risk of brain tumour (cancer). It would need to compare two groups – a sample of mobile phone users and a sample of non-users. People in both samples should be matched on as many *other* factors as possible. For example, each should have similar numbers of people of each age. Why? The development of brain tumours might be age-related The researchers could also select the samples randomly, so that other factors (e.g. genetic variation) are similar in both groups.
Were the numbers in each sample large enough to give confidence in the results?	With small samples, the results can be more easily affected by chance. Larger sample sizes can give a truer picture of the whole population. Why? With bigger samples, the effect of chance is more likely to average itself out. So, for example, if you toss a coin five times, it is not uncommon to get four or five heads. But if you tossed the coin 20 times, you would be very unlikely to get 19 or 20 heads.

Applying the precautionary principle

The UK government's Chief Medical Officer 'strongly advises' that children and young people use mobile phones for essential calls only and keep calls as short as possible. The Chief Medical Officer is using the precautionary principle. So far, there is no evidence of a problem. But nobody has proved that there is NOT a problem.

Government officers consider carefully before making public statements and giving advice. There are many new technologies, not just mobile phones, for which decisions like that have to be made. In most cases, the expected benefits of a new technology clearly outweigh any risks. New bridges and cable networks are built, or new vaccines and food products are introduced.

How great is the risk?

Generally, health outcomes are reported as relative risks. For example, 'people exposed to high levels of sunlight were four times more likely to develop eye cataracts'. What might this mean?

- If your risk was one in a million, it rises to 4 in a million - not a worry!
- If your risk was 5 in 100, it rises to 20 in 100 - worth thinking about!

Some people will be at higher risk because of their skin type, their diet, or their personal or family medical history. Your doctor can help you interpret information about health risks.

Questions

5 a Look at the first row of the table. What factor and what outcome are being studied?

b Describe a second way that the samples should be matched in the same study.

6 You are trying to answer the question 'Are girls better at maths than boys?' What sort of sample of people would you need to study this?

7 In a survey comparing mobile phone users and non-users, why would you need to know how much time each person spends on the phone?

8 Read the newspaper article (right). Using ideas in this module, write a letter to the council for or against the new phone mast. Give reasons for your view.

Storm over mobile phone mast

Kate Beach had used her mobile phone for years without giving it a thought. Earlier this month her phone company was given planning permission to install a new mast. The site agreed was a roof near her children's primary school. When she discovered the location she became concerned.

Ms Beach has started a campaign group to discuss with the Council where local phone masts are located. The group wants the Council to adopt a precautionary principle and not to grant permission for masts near schools.

Phone masts operate at a much lower power output than TV and radio broadcasting stations. Jane Wells, a company spokesperson, said: 'In a city you're going to have more masts because there are more people using mobile phones. But the exposure is hazardous only directly in front of a mast. Using a phone handset, the exposure is 10,000 times more than standing close to a mast.'

Some people blame the radiation from phone masts for symptoms such as insomnia, dizziness, nausea, migraines, eye damage, and cancers. Scientific studies are currently underway to find out whether any of these concerns are justified.

Ms Beach will present the group's views to the Full Council meeting on Monday.

X-ray safety

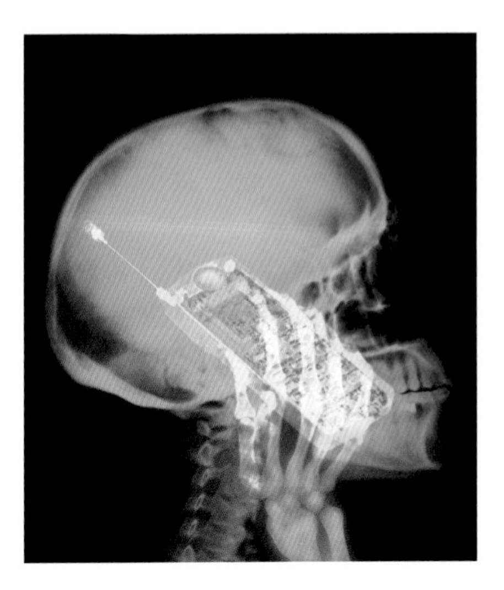

X-rays were discovered in the 1890s. They soon caught on as a useful medical tool, and they have saved many thousands of lives. But they are a form of ionizing radiation. Any benefits must be balanced against risks.

Both the health benefits and the risks of X-rays are well known. Using mobile phones has benefits but uncertain risks.

Shoe shops used to boast that they could check the fit of your shoes using an X-ray machine. The assistant and the customer peered down into the machine. They saw a shadow image of the bones of the foot and the outline of the shoe. By the late 1950s, people realized that this produced an unnecessary exposure to ionizing radiation, which could be damaging. The machines were banned.

Discovery of a correlation

Alice Stewart (see photo) and George Kneale carried out a survey on a large number of women and their children. They discovered a correlation between X-ray exposure of mothers during pregnancy and cancers in their children.

There is a plausible **mechanism** that could explain this correlation. X-ray photons can ionize molecules in your body. This can disrupt the chemistry of body cells, and cause cancer. So the link is more than just a correlation. X-rays can, in a few cases, cause cancer.

This study made doctors more cautious about using X-rays. The risks associated with X-rays for small children and pregnant women usually outweigh any benefit.

Obituaries

Alice Stewart

Alice Stewart was a British doctor. She collected and analysed information from women whose children had died of cancer between 1953 and 1955. Soon the answer was clear. On average, one medical X-ray for a pregnant woman was enough to double the risk of early cancer for her child.

ALARA

When a patient has an X-ray, the equipment and procedures keep the X-ray exposure to the minimum that still produces a good image. Because digital detectors work with a lower dose, they have now replaced X-ray film in many uses. This approach, making the patient's exposure as low as reasonably achievable, is called the ALARA principle.

There are three ways to achieve ALARA exposure to X-rays:

- **time**: the shorter the time of exposure, the less radiation is absorbed
- **distance**: as radiation spreads out from its source it becomes less intense due to the spreading
- **shielding**: lead is an extremely good absorber of X-rays. Lead screens provide excellent shielding

Questions

9 Why is the link between X-ray exposure during pregnancy and childhood cancer believed to be a 'cause' and not just a 'correlation'?

(10) Why are medical X-rays still used for patients other than pregnant women, despite this link?

(11) Make a list of ways to achieve ALARA exposure to microwave radiation from mobile phones.

Key words

mechanism
ALARA

Find out about:

- records of the Earth's past temperatures
- how the atmosphere keeps the Earth warm
- why the amount of CO_2 in the atmosphere is changing

E Global warming

Are summers now hotter and winters milder than they once were? This is a question about **climate**, or average weather in a region over many years. You cannot answer it from personal experience, because you can only be in one place at a time. And memory can be unreliable. Instead, you need to collect and analyse lots of data.

Past temperatures

Weather stations have been keeping temperatures records for over a century. Climate scientists study these records. They also study Nature's own records, going back thousands of years:

- growth rings in trees
- ocean sediments
- air trapped in ancient ice

There is a clear pattern. The Earth's average temperature has been increasing since 1800. The last decade was the hottest since temperature records began. Possibly the hottest in a thousand years.

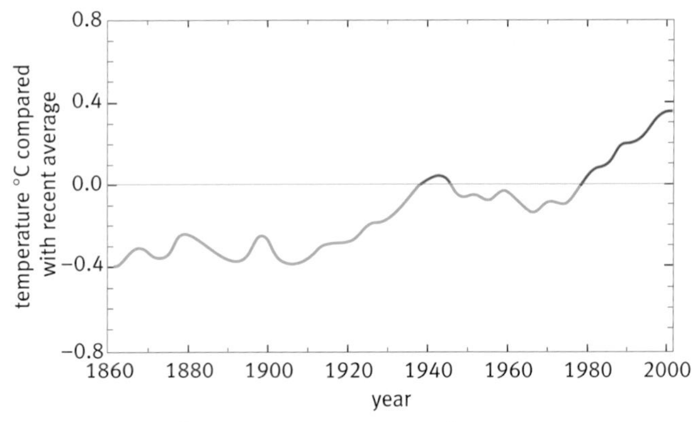

The Earth's surface temperature over the past 140 years (data from thermometers)

Questions

1 Personal experience does not provide reliable evidence of climate change. Why not?

2 All of the statements about CO_2 and the Earth's average temperature describe correlations. Which statement is also about cause and effect?

A correlation with CO_2?

Most scientists think that CO_2 in the atmosphere is causing the Earth's average temperature to rise. Why?

- Global temperatures and CO_2 level have increased recently.
- There is evidence from the distant past that temperature and CO_2 level change together.
- Scientists know how CO_2 in the atmosphere warms the Earth.
- Computer climate models show that global temperatures are related to CO_2 levels.

The greenhouse effect

Without its atmosphere, the Earth's average surface temperature would be −18 °C. That's how cold it is on the Moon. In fact the Earth's average temperature is 15 °C. This warming of the Earth by its atmosphere is called the **greenhouse effect**.

There is an energy balance between radiation coming in and going out of the atmosphere.

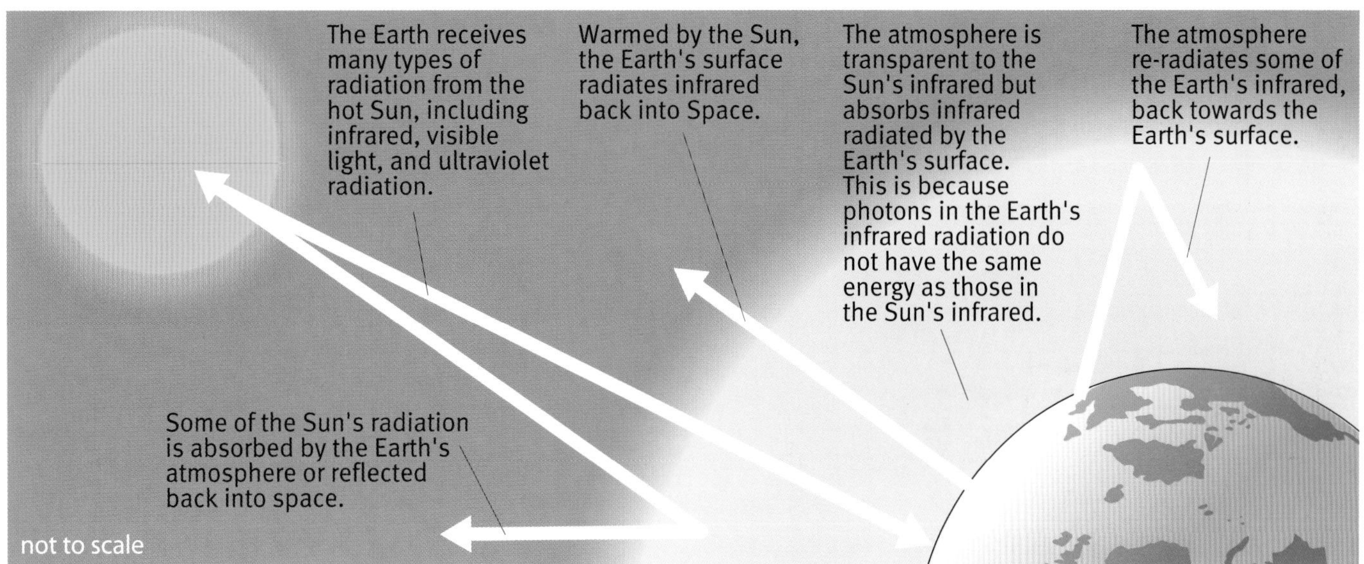

Life on Earth depends on the greenhouse effect. Without it, the Earth's water would be frozen. Water in its liquid form is essential to life.

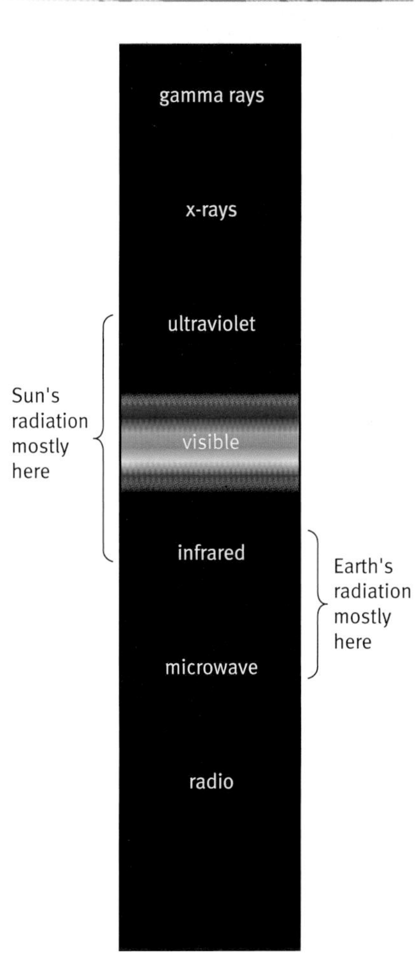

Greenhouse gases

Tiny amounts of a few gases in the atmosphere make all the difference. Carbon dioxide, methane, and water vapour absorb some of the Earth's infrared radiation. They are called **greenhouse gases**. The nitrogen and oxygen that make up 99% of the atmosphere do not absorb this radiation, and so they have no warming effect.

Key words

climate
greenhouse effect
greenhouse gases

Questions

3 a Which of the following gases are found in the Earth's atmosphere: nitrogen, methane, oxygen, carbon dioxide, water vapour, argon?

b Which of them are *not* greenhouse gases?

4 Explain why it gets cold at night, by describing the radiation arriving and leaving the Earth. Hint: it is often colder on a clear night than on a cloudy one.

The carbon cycle

Carbon dioxide is a greenhouse gas that plays a key role in global warming. Industrial societies produce CO_2 as never before.

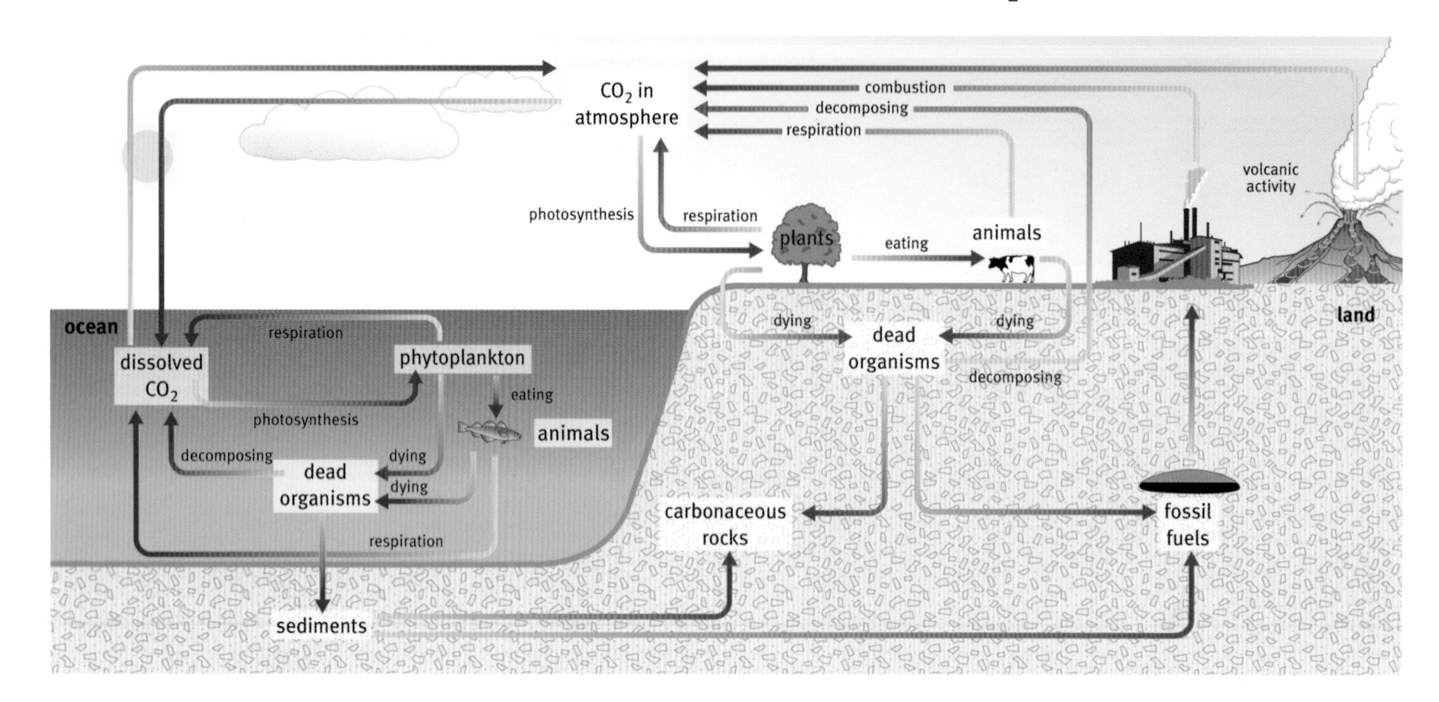

The Earth's crust, oceans, atmosphere, and living organisms all contain carbon. Carbon atoms are used over and over again in natural processes. The **carbon cycle** describes stores of carbon and processes that move carbon.

Carbon dioxide (CO_2) in the atmosphere

Hundreds of millions of years ago, the amount of CO_2 in the atmosphere was much higher than it is today. Green plants made use of CO_2 and released oxygen. This made life possible for animals. Eventually, lots of carbon was locked up underground in the form of fossil fuels, as well as carbonaceous rocks such as limestone and chalk.

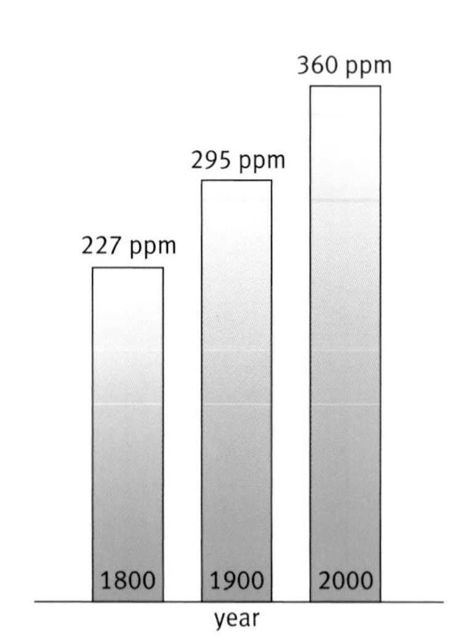

In 1800, the concentration of CO_2 in the atmosphere was only 277 parts per million (ppm). This means there were 277 molecules of CO_2 for every 1 000 000 molecules that make up dry air.

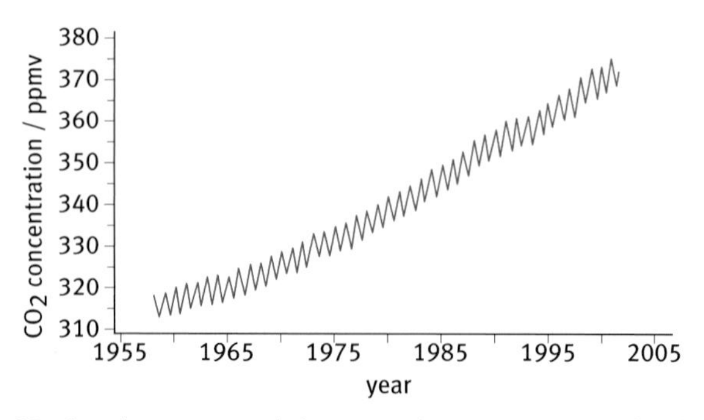

CO_2 levels go up and down each year, as a result of photosynthesis. They fall each summer and rise again each winter. The average is rising steadily by 2 ppm every year.

Human activities release carbon

People want to live comfortably. In some parts of the world, many feel they have a right to processed foods, unlimited clean water and electricity, refrigerators and other manufactured goods, bigger houses and flats. All of these things require energy.

But whenever fossil fuels – coal, oil, and gas – are burned, they increase the amount of carbon dioxide in the atmosphere. Methane, another greenhouse gas, is produced by grazing animals and from rice paddies.

Although methane is the more effective greenhouse gas, carbon dioxide produced by human activities has a bigger effect. This is because the amount of CO_2 is so huge – thousands of millions of tonnes each year. This is why there is talk of reducing 'carbon emissions'.

Motor vehicles are a major source of greenhouse gas emissions.

A power station like this supplies enough electricity for a major city. Every day it uses several trainloads of coal and sends thousands of tonnes of carbon dioxide into the atmosphere.

Air transport is a big user of fossil fuels. Aviation fuel is cheap because it is untaxed, unlike petrol for cars.

People in the UK use more energy on keeping buildings warm than on anything else.

Questions

5 Forest land can be cleared for farming by burning the trees. This is called **deforestation**. Why does tree-burning increase the amount of carbon dioxide in the atmosphere? Explain using a diagram.

(6) Look at the graph of CO_2 levels on page 56.

a Explain its shape – why does it go up and down every year, and why is the long-term trend upwards?

b The data was collected on the Hawaiian Islands, in the middle of the Pacific Ocean. Why is that a good place to make measurements?

(7) If aviation fuel were heavily taxed, what might happen to the amount of air travel? Explain your answer.

Key words

carbon cycle
deforestation

Find out about:

- possible effects of climate changes
- tways of reducing the amount of CO_2 released into the atmosphere

F Changing climates?

Nature's records

The polar ice caps are frozen records of the past. In parts of the Antarctic, ice made from annual layers of snow is four kilometres thick. That ice contains tiny bubbles of air, a record of the atmosphere over 740 000 years. It shows that climate has always changed. There have been ice ages and warm periods.

But never before have temperatures increased so fast as during the last 50 years.

Natural factors change climates

Over the long term, natural factors cause climate change. For example:

- the Earth's orbit changes the distance to the Sun by tiny amounts
- the amount of radiation from the Sun changes in cycles
- volcanic eruptions increase atmospheric CO_2 levels

Climate modelling

The atmosphere and oceans control climates. Climate scientists use computer models to predict the effects of increasing CO_2 levels.

What these models show is alarming.

- Human activities are now contributing more to climate change than natural factors.
- Future emissions of greenhouse gases are likely to raise global temperatures by between 1.4 and 5.8 °C during your lifetime.
- If CO_2 concentration rises above 500 ppm, climate change may become irreversible.
- To stabilize climates, carbon emissions would need to be reduced by 70% globally.

Climates change slowly. It may take 20 to 30 years for climates to react to the extra CO_2 already in the atmosphere. So global temperatures are guaranteed to rise by 2 °C. Ice will continue to melt, and sea levels continue to rise, for the next 300 years or so, even if humans today stopped producing any CO_2 at all.

Uncertain risks of climate change

Global warming is expected to produce a variety of effects in different parts of the world. To evaluate a risk, you need to consider both the chances of something happening and the consequences if it does. The risks associated with global warming are enormous.

We are already seeing its effects.

- Mountain glaciers are retreating everywhere. Mt Kilimanjaro, a famous snow-capped peak in Tanzania, may be bare of snow by 2015.
- Some polar regions are warming at a rate two to three times the global average.
- Many parts of the world are experiencing extreme weather – high winds, heavy rains, or heat-waves and droughts.

Climate models predict that winters will become wetter and summers may become drier across all of the UK.

In March 2002 a giant ice sheet broke away from Antarctica. Larsen B was the same size as Somerset.

Possible effects on people

Human societies depend on stable climates. Climate change may cause problems for:

- food and water resources
- coastal populations and industry affected by rising sea levels
- insurance companies and other financial services
- human health (for example, malaria will spread if mosquitoes can breed in more places)

Global warming sceptics

During the 1980s and 90s, scientists argued a lot about what is happening to climates. Now even the sceptics accept global warming, and the fact that human activities contribute to it. But the sceptics still argue that temperatures will only increase by about 1.4 °C. They say that global warming is harmless.

Questions

1 Make a list of the scientific uncertainties mentioned on these pages.

② Choose any two things from the list of 'possible effects on people'. For each one, explain exactly how climate change could produce a harmful effect.

Time for action?

The world's poorest countries will be least able to deal with the effects of climate change, so their people are most vulnerable. Even people in developed countries could be badly affected.

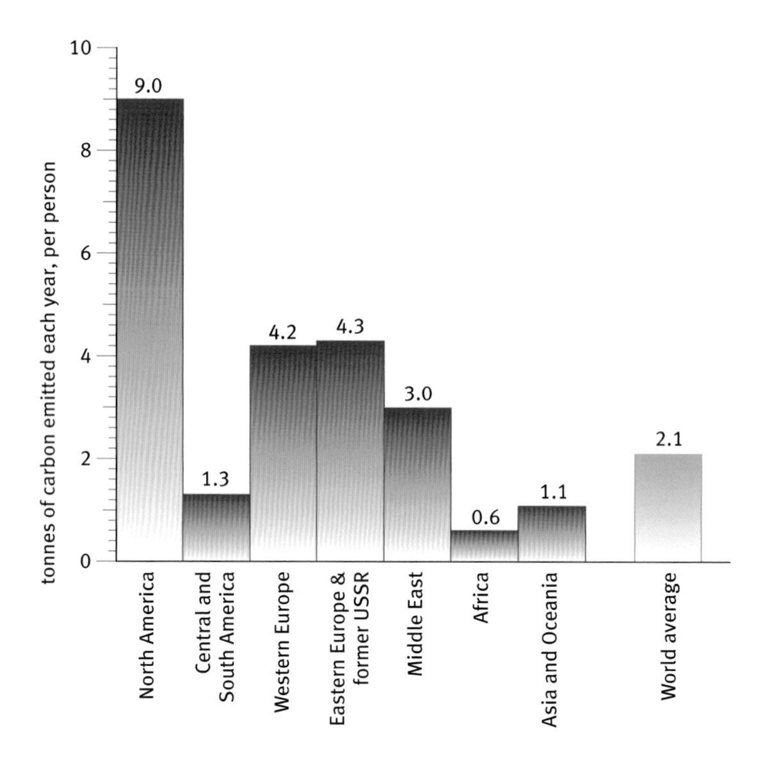

Europe and North America have just one-fifth of the world's population. But they account for more than 60% of carbon emissions.

The UK too is at risk

The UK climate is kept mild by the Gulf Stream, a warm current from the Caribbean that flows towards Europe across the North Atlantic.

The Earth has a giant 'conveyor belt' system of ocean currents. It helps to warm land in northern latitudes.

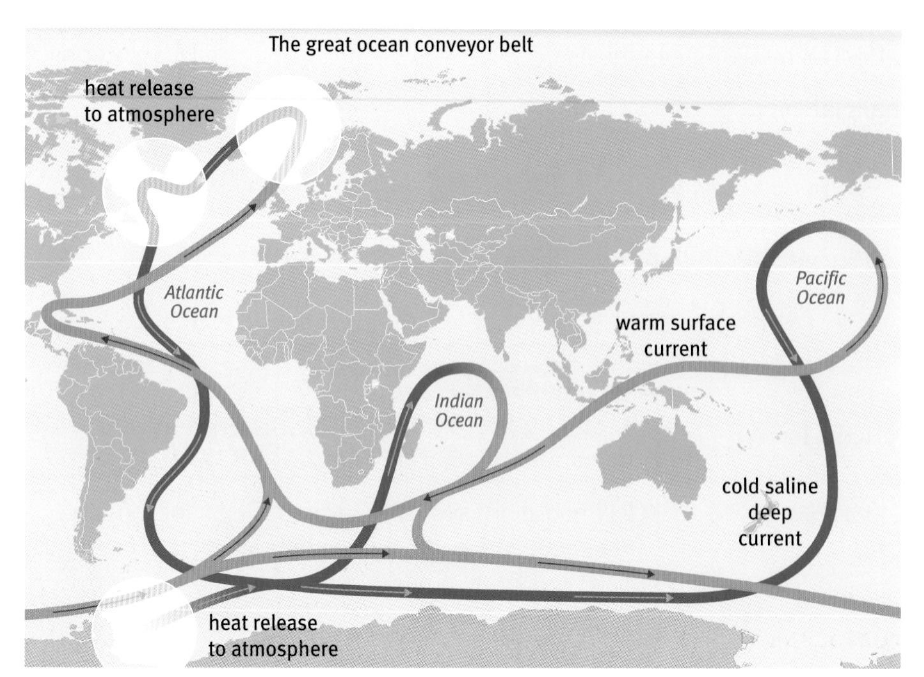

There is evidence that this current slowed down in the past, making the UK an icy place. There are signs that the Gulf Stream may be slowing again.

What can governments do?

The UK government aims to reduce greenhouse gas emissions:

- 20% by the year 2010
- 60% by 2050

The baseline is 600 million tonnes emitted in 1990. To reach those ambitious targets people's expectations and behaviour need to change.

The government can spend tax money in different ways. It can introduce new taxes, laws, and regulations.

But democratic governments are sensitive to public opinion, because they face election every few years. They find it difficult to do what's best for the long term. People may protest if they feel their freedoms are being taken away. Businesses may fight to protect their profits.

And nobody can predict climate futures accurately. As a result, some politicians are not convinced of the need for action. Joint international action is very hard to achieve.

In September 2000, protesters opposing higher fuel taxes staged a dramatic series of blockades at petrol depots.

Science to the rescue

Several solutions have been proposed that might get around the difficulties of reducing carbon emissions.

- Spread iron granules on the southern oceans. This would help the growth of plankton, which take dissolved CO_2 from the ocean. The oceans would remove more CO_2 from the atmosphere.
- Capture the CO_2 produced at power stations. Then compress it into a liquid and pump it into disused oil reservoirs beneath the sea-bed.
- Cement production counts for 5% of the greenhouse gases produced in Europe and America and more than 10% in China. A new type of 'eco-cement' absorbs CO_2 while setting and goes on absorbing CO_2 for years afterwards.

None of these have yet been tried and evaluated.

Extract from a popular science magazine

A global challenge

Most of the world's population is poor and would like to have a better standard of living. Can humanity find a way to reduce inequalities without at the same time wrecking the atmosphere, and with it climates, land, and oceans? This represents an enormous challenge.

What can you do?

Perhaps you will take action yourself, now and in the future. You could

- turn the heating down
- use a car less
- have fewer holidays involving air travel
- use electricity from non-fossil energy sources

Questions

3. Do you think action should be taken now to reduce carbon emissions? Justify your answer.
4. Look at the photos on page 57. For each one, suggest what the government could do to reduce carbon emissions.
5. Do you think people should rely on technical solutions, like those suggested in the science magazine? Justify your answer.

P2 Radiation and life

Science explanations

This chapter introduces the electromagnetic spectrum.

You should know:

- how to think about any form of radiation in terms of its source, its journey path and what happens when it is absorbed
- that a beam of electromagnetic radiation delivers energy in 'packets' called photons
- how to describe the electromagnetic spectrum, with its parts in order of their photon energies
- what different parts of the electromagnetic spectrum can be used for
- two factors that affect the energy deposited by a beam of electromagnetic radiation
- how the intensity of an electromagnetic beam changes with distance.
- why ionizing radiation is hazardous
- which parts of the electromagnetic spectrum are ionizing
- how people can be protected from ionizing radiation
- how microwaves heat materials, including living cells
- about features of microwave ovens that protect users
- that sunlight provides the energy for photosynthesis and warms the Earth's surface
- how photosynthesis affects what molecules are in the atmosphere
- what the greenhouse effect is (and be able to identify greenhouse gases)
- how to use the carbon cycle to explain several things about the atmosphere
- how the atmosphere's ozone layer protects living organisms
- what global warming means
- some possible effects of global warming
- how computer models provide evidence that human activities are causing global warming

Ideas about science

To make personal and social decisions about health or global warming, it can be important to assess the risks and benefits. For risks and benefits from different parts of the electromagnetic spectrum

You should be able to:

- explain why nothing is completely safe
- suggest why people will accept (or reject) the risk of a certain activity, e.g. sunbathing because they want a tan
- suggest ways of reducing particular risks
- interpret information on the size of risks, presented in different ways
- discuss a given risk, taking account of both the chances of it occurring and the consequences if it did
- explain that if it is not possible to be sure about the results of doing something, and if serious harm could result, then in makes sense to avoid it (the 'precautionary principle')
- explain the ALARA principle and how it applies to an issue
- correctly use the ideas of correlation and cause when discussing topical issues related to this chapter
- suggest factors that might increase the chance of an outcome
- explain that individual cases do not provide convincing evidence for or against a correlation
- evaluate a health study by commenting on sample size or sample matching
- explain why a correlation between a factor and an outcome does not necessarily mean that one causes the other, and give an example to illustrate this
- evaluate a claimed causal link by discussing the presence (or absence) of a plausible mechanism
- discuss personal and social choices in terms of actual risk and perceived risk

These ideas are illustrated through Case Studies, including: whether sunlight is good for you; UV and the ozone layer; microwave ovens, mobile phones, and X-ray scans; global warming.

Why study radioactive materials?

People make jokes about radioactivity. If you visit a nuclear power station, or if you have hospital treatment with radiation, they may say you will 'glow in the dark'. People may worry about radioactivity when they don't need to.

Most of us take electricity for granted. But today's power stations are becoming old and soon will need replacement. Should nuclear power stations be built as replacements?

The science

Radiation from radioactive materials comes from deep inside their atoms. To use radioactive materials safely you need to know about the different types of radiation.

Nuclear power stations produce nuclear waste. This waste can be dangerous for tens of thousands of years.

Ideas about science

Nothing can be completely safe. Before any medical procedure uses radioactive materials, doctors and their patients carefully weigh up the benefits against the risks.

Soon, decisions about getting rid of nuclear waste, or building new power stations will be made. Who will decide, and how can you have your say?

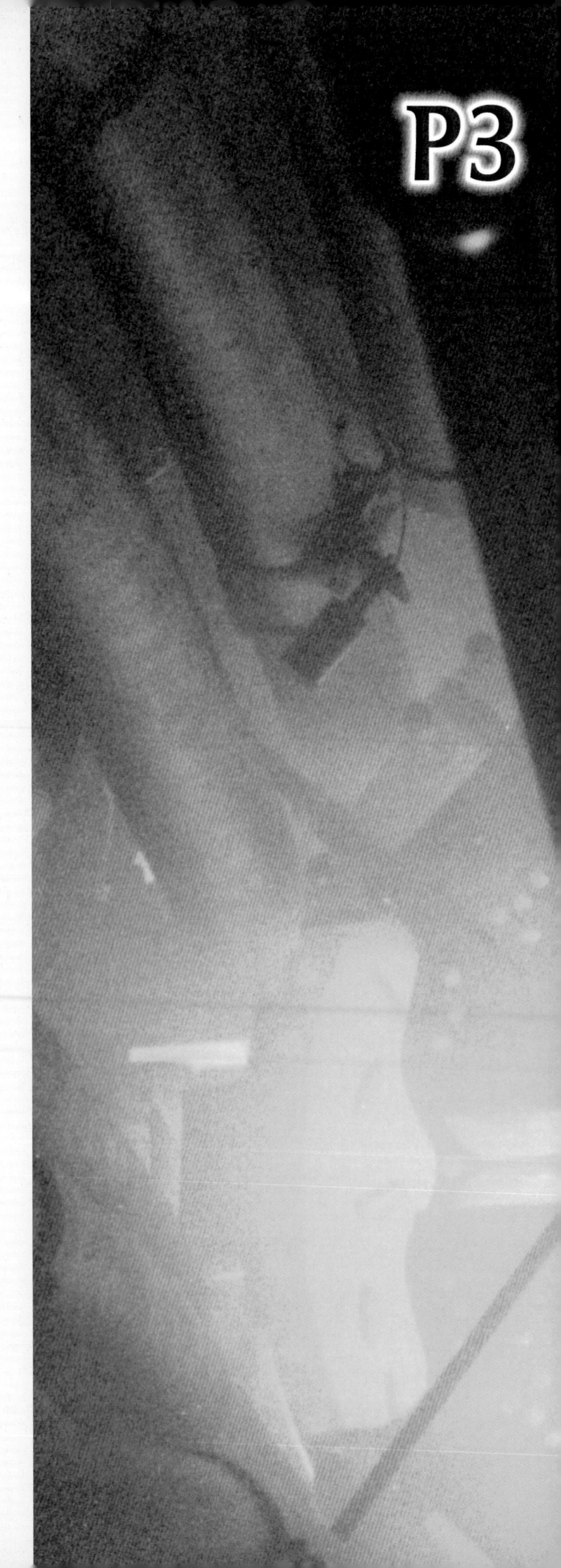

Radioactive materials

Find out about:

- what 'causes' radioactivity
- radioactive materials being used to treat cancer
- ways of reducing risks from radioactive materials
- different ways of generating electricity

Find out about:
- primary and secondary energy sources
- the importance of energy efficiency
- energy use around the world

A Energy patterns

Energy consumer

Every day you use energy sources. You blow dry your hair, listen to some music, or use a computer. In winter the heating goes on. The heating system may use natural gas; the other three rely on electricity. Natural gas is a **primary energy source**. Electricity is called a **secondary energy source** because it is generated from primary sources.

High voltage cables carry electricity from power stations to the National Grid.

Easy electricity

Electricity is convenient and clean. You just flick a switch. There are no flames and no fumes in your home. But there are flames and fumes hundreds of miles away – in a power station.

About a hundred power stations in the UK supply electricity to consumers through a network called the National Grid.

Electricity demand changes

Boiling a kettle makes a demand on the National Grid. A typical kettle demands a power of 1 kW. Every other mains electrical device makes a demand too, whenever it is switched on.

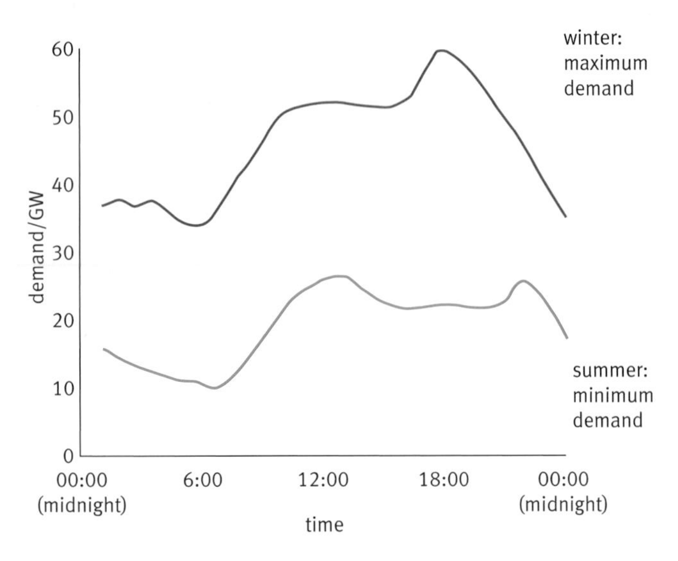

The peak demand is 60 gigawatts (GW). This is equivalent to each of the 60 million people in the UK switching on a kettle, all at the same time.

The total demand varies through the day and over the year. It rises to a peak at teatime on a winter's day. When demand rises, more power stations are brought 'on stream'.

Meeting demand

Minute by minute, the National Grid must be able to meet demand. Otherwise there will be a power blackout. The knock-on effects can be serious.

One evening in August 2003, south London was without power for just 40 minutes. But 250 000 people were affected. Buildings along the Thames were in darkness. Hundreds of traffic lights failed. Tens of thousands of commuters were stuck in tunnels on London's underground, for several hours.

Is electricity efficient?

Electricity is convenient. But it is also wasteful – especially when used for heating. A gas-fired power station wastes nearly half of the primary energy source. More energy is wasted in the cables and transformers of the National Grid. By contrast, a domestic gas water boiler is about 80% efficient.

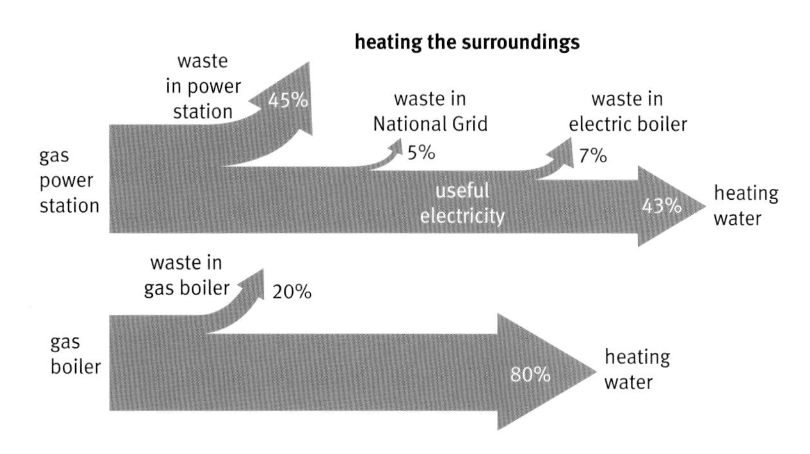

Using a primary source to heat water directly is much more efficient than using electricity.

Is electricity clean?

In 2005, three-quarters of the UK's electricity came from fossil fuels, like coal and natural gas. Burning fossil fuels releases carbon dioxide (CO_2) into the atmosphere. This contributes to climate change (see Module P2 *Radiation and life*).

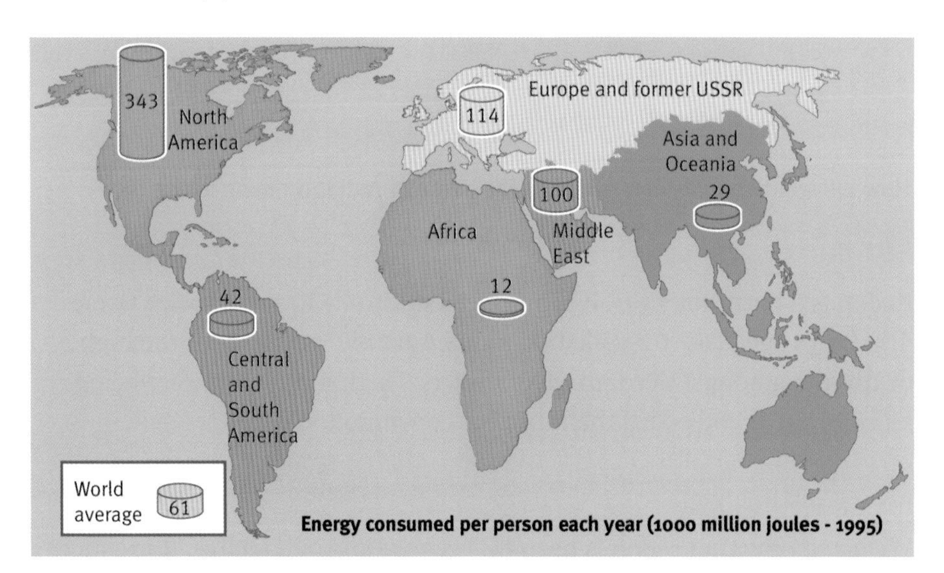

As countries become industrialized, living standards rise and energy use increases. Energy use in India and China is now growing especially fast.

Limiting climate change

Global energy demands are expected to grow by 60% between 2005 and 2030. This has the potential to cause a significant increase in greenhouse gas emissions associated with climate change.

In 1997, government ministers from around the world met in Kyoto, Japan. They produced targets to reduce CO_2 emissions. This is difficult when global demand is increasing.

Some people think building new nuclear power stations could help, because they produce practically no CO_2 when operating. But nuclear power too has problems.

Key words

primary energy source
secondary energy source

Questions

1 Write down three things you do during a day that directly use

  a a primary energy source

  b a secondary energy source

2 Look at the graph showing electricity demand through the day. Describe and explain how the demand changes.

(3) Has electricity reduced or increased the pollution in

  a towns

  b the world?

  Explain your answers.

Find out about:

- background radiation
- a radioactive gas called radon
- radiation dose and risk

B Radiation all around

Radiation sources

If you switch on a Geiger counter, you will hear it click. It is picking up **background radiation**, which is all around you. Most background radiation comes from natural sources.

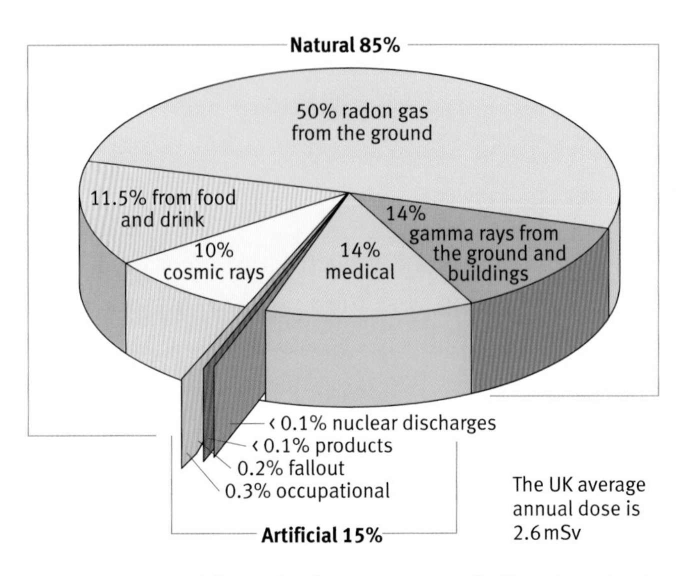

How different sources contribute to the average **radiation dose** in the UK.

Silver mines were contaminated with radon gas. The miners breathed it in and suffered.

Radon

Radon is a hazardous gas. It is produced naturally in some rocks. Over 400 years ago, a doctor called Georgius Agricola wrote about the high death rate amongst German silver miners. He thought they were being killed by dust, and called their disease 'consumption'.

Radioactive gas

We now know that radon is harmful because it is **radioactive**. It produces **ionizing radiation** that can damage cells. The silver miners were dying of lung cancer.

Health effects of radiation

If radiation passes through a living organism, any one of these things may happen to a cell.

- The cell is not damaged as the radiation passes through it.
- The cell is damaged but repairs itself.
- The cell is killed.
- The cell's DNA is damaged, and the cell may develop out of control – a cancer has been started.
- If a sex cell is hit, the radiation may change a gene (cause a mutation).

Damage happens for this reason.

- When ionizing radiation strikes molecules, it makes them more likely to react chemically.

Radiation dose

The risk to miners was high for two reasons:

- Radon can build up in enclosed spaces, such as mines.
  In the atmosphere, what little radon there is spreads out. It's a different story in enclosed spaces like mines. The rocks keep producing the gas and it cannot escape. So the radon concentration can be 30 000 times higher than in the atmosphere.
- The miners breathed in the radon.
  The miners became ill because the radon gave off its harmful radiation *inside their lungs*. Lung tissue is easily damaged.

Both of these reasons led to the miners getting a large radiation dose.

Dose summary (1)
Radiation dose is affected by
- amount of radiation
- type of exposed tissue

Measuring dose

Radiation dose is measured in millisieverts (mSv). The UK average annual dose is 2.6 mSv. For comparison, with a dose of 1000 mSv (400 times larger) three out of a hundred people, on average, develop a cancer.

Ionizing radiation from outer space is called cosmic radiation. Flying to Australia gives you a dose of 0.1 mSv, from cosmic rays. That's not much if you go on holiday, but it soon adds up for flight crews making repeated journeys.

It is difficult to be sure about the harm that low doses of radiation can cause. In the 1970s, Alice Stewart (see p. 53) studied the health of people working in the American nuclear industry. Her early results suggested that radiation is more harmful to children and to elderly people. She was attacked for her ideas, and the employers prevented any further access to medical records.

Is there a safe dose?

There is no such thing as a safe dose. Just one radon atom might cause a cancer. Just as a person might get knocked down by a bus the first time they cross a road. The chance of it happening is low, but it still exists. The lower the dose, the lower the risk. But the risk is never zero.

Questions

1 Explain fully how the silver miners developed lung cancers.

2 a In what units is radiation dose measured?

b What two factors increased the dose for a silver miner?

c Make a reasoned estimate for the annual dose of a long-haul airline pilot.

3 On what two factors does radiation dose depend?

Key words
background radiation
radiation dose
radioactive
ionizing radiation

A hazard at home

Radon gas builds up in enclosed spaces. In some parts of the UK, it seeps into houses.

Living with radon

Government Information Leaflet

There is radon all around you. It is radioactive and can be hazardous – especially in high doses.

Radon gives out a type of ionizing radiation called **alpha radiation**. Like all ionizing radiations, alpha radiation can damage cells and might start a cancerous growth.

Radon is a gas that can build up in enclosed spaces. Some homes are more likely to be contaminated with radon.

What about my home?

You and your family are at risk if you inhale radon-contaminated air. The map shows the areas where there is most contamination.

If you live in one of these areas, get your house tested for radon gas.

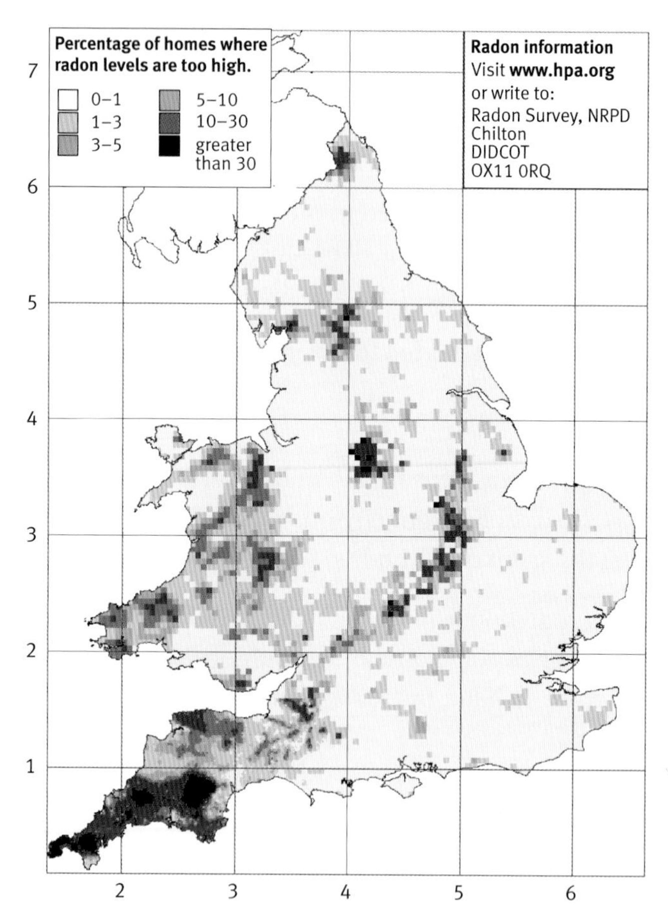

Radon-affected areas in England and Wales. Based on measurements made in over 400 000 homes.

What if the test shows radon?

Radon comes from the rocks underneath some buildings. It seeps into unprotected houses through the floorboards. If your house is contaminated, get it protected. An approved builder will put in

- a concrete seal to keep the radon under your floorboards and
- a pump to remove it safely

The risk is real: put in a seal.

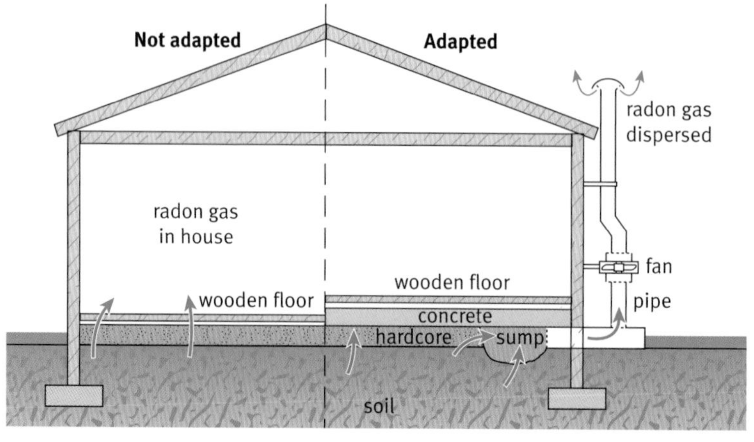

Radon gas can build up inside your home. Sealing the floor and pumping out the gas is an effective cure.

Irradiation and contamination

Radon in the air exposes you to alpha radiation. Exposure to a radiation source outside your body is called **irradiation**. Radon irradiation presents a very low risk because alpha radiation:

- only travels a few centimetres in air
- is easily absorbed

Your clothes will stop alpha radiation. So will the outer layer of dead cells on your skin.

But if a radiation source enters your body, or gets on skin or clothes, it is called **contamination**. You become contaminated. If you swallow or breathe in any radioactive material, your vital organs have no protection. They will absorb its radiation. Breathing in radon gas is dangerous.

Cause of death	Average number of deaths per year
cancer caused by radon	2500
cancer among workers caused by asbestos	3000
skin cancer caused by ultraviolet radiation	1500
road deaths	3400
cancer caused by smoking	40 000
CJD	82
House fire	570
All causes	500 000

Estimated deaths per year in the UK population of 60 million (2002)

Radon and risk

On average, radon makes up half the UK annual radiation dose. About 2500 people die each year from its effects, or about 1 in every 20 000 people. But radon is only one of the hazards that people face every day. There are risks associated with driving to school, sunbathing, swimming, catching a plane, and even eating.

Many risky activities have a benefit. You need to decide whether the risk is worth taking.

Alpha radiation

- highly ionizing
- short range in air
- easily absorbed (e.g. by paper, clothes or dead skin cells).

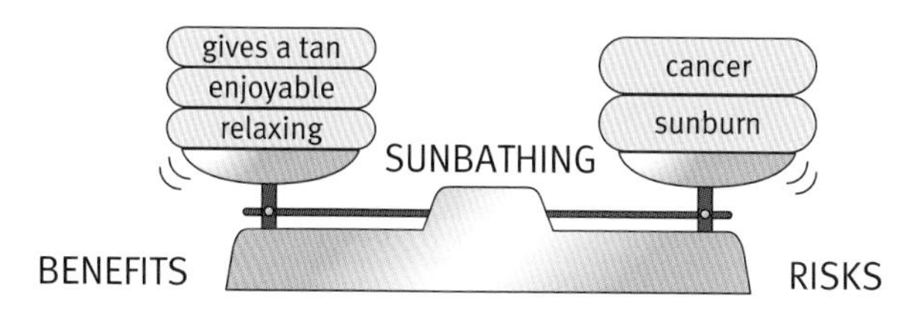

Many people sunbathe. They reckon the benefits outweigh the risks.

Key words

alpha radiation irradiation contamination

Questions

4 Explain the difference between irradiation and contamination.

5 **a** How big a dose of radiation do you get by catching a flight to Australia?

b Where do cosmic rays come from?

c Is this irradiation or contamination?

6 There is a risk from radon gas building up in houses. Which of these are good ways to reduce the risk?

- stop breathing
- wear a special gas mask
- move house
- adapt the house

7 Choose three causes of death from the table on the left. Write down two ways of reducing the risk from each chosen cause on the left (for example walk to school).

8 Imagine that alpha radiation damages a cell on the outside of your body. Why is this less risky than internal damage? Give two reasons.

Find out about:

- different uses of radiation
- types of radiation
- benefits and risks of using radioactive materials
- limiting radiation dose

C Radiation and health

Radioactive materials can cause cancer. But they can also be used to diagnose and cure many health problems.

Medical imaging

Jo has been feeling unusually tired for some time. Her doctors decide to investigate whether an infection may have damaged her kidneys when she was younger.

They plan to give her an injection of DMSA. This is a chemical that is taken up by normal kidney cells. Before doing this, they need to be sure that she is not pregnant.

The DMSA has been labelled as radioactive. This means its molecules contain an atom of technetium-99m (Tc-99m), which is radioactive. The kidneys cannot tell the difference between normal DMSA and labelled DMSA. They absorb both types.

The Tc-99m gives out its **gamma radiation** from within the kidneys. Gamma radiation is very penetrating. So nearly all of it escapes from Jo's body and is picked up by a special gamma camera. Parts of the kidney, which are working normally will appear to glow. Any dark or blank areas show where the kidney isn't working properly.

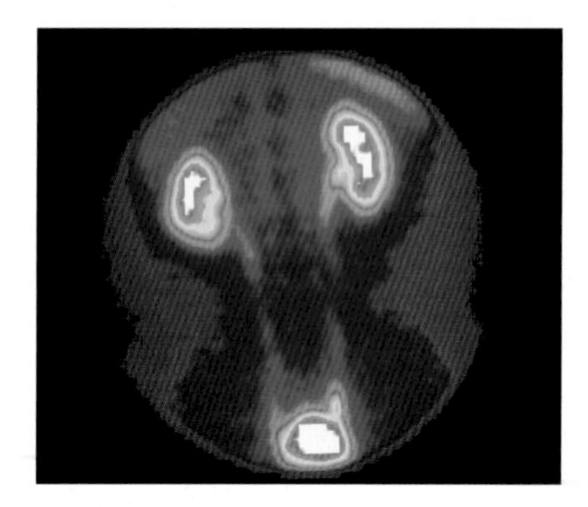

This gamma scan shows correctly functioning kidneys – the top two white areas.

Jo's scan shows that she has only a small area of damage. The doctors will take no further action.

Glowing in the dark

Jo was temporarily contaminated by the radioactive Tc-99m. For the next few hours, until her body got rid of the technetium, she was told to:

- flush the toilet a few times after using it
- wash her hands thoroughly
- avoid close physical contact with friends and family

Is it worth it?

There was a small chance that some gamma radiation would damage Jo's healthy cells. Before the treatment, her mum had to sign a consent form.

Jo's mum said 'We felt the risk was very small. And it was worth it to find out what was wrong. Even with ordinary medicines, there can be risks. You have to weigh these things up. Nothing is completely safe.'

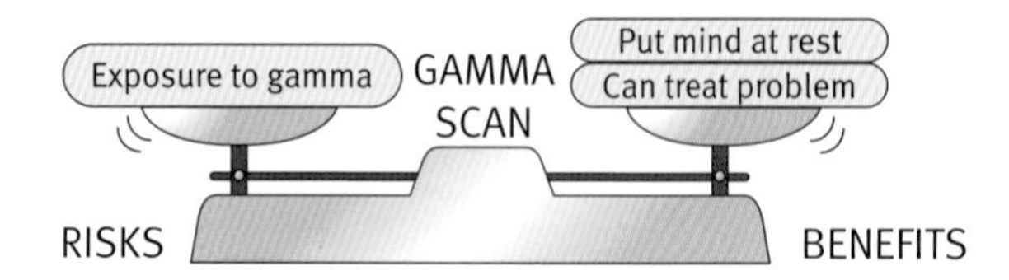

Jo's mum weighed the risk against the benefit.

Treatment for thyroid cancer

Alf has thyroid cancer. First he will have surgery, to remove the tumour. Then he must have **radiotherapy**, to kill any cancer cells that may remain.

A hospital leaflet describes what will happen.

Radioiodine treatment

You will have to come in to hospital for a few days. You will stay in a single room.

You will be given a capsule to swallow, which contains iodine-131. This form of iodine is radioactive. You cannot eat or drink anything else for a couple of hours.

- The radioiodine is absorbed in your body.
- Radioiodine naturally collects in your thyroid, because this gland uses iodine to make its hormone.
- The radioiodine gives out **beta radiation**, which is absorbed in the thyroid.
- Any remaining cancer cells should be killed by the radiation.

You will have to stay in your room and take some precautions for the safety of visitors and staff. You will remain in hospital for a few days, until the amount of radioactivity in your body has fallen sufficiently.

Many other conditions can be treated with radiation too.

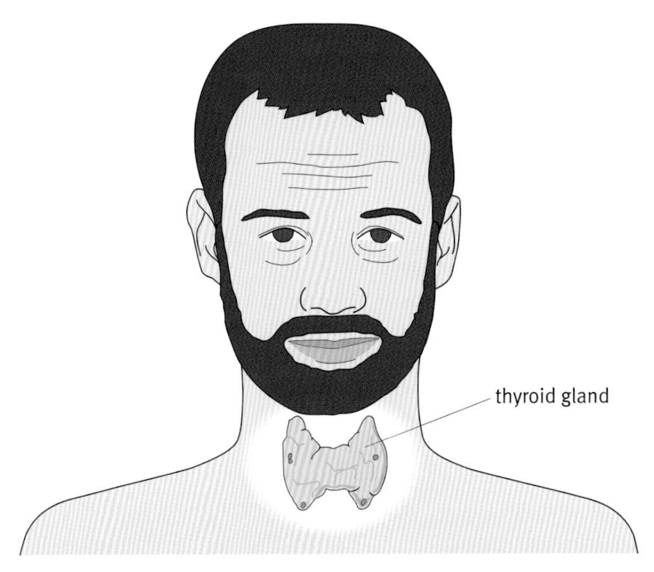

The thyroid gland is located in the front of the neck, below the voice box.

Diagnosis using radioactive materials takes place in the nuclear medicine department of a hospital.

Questions

1. Look at the precautions that Jo has to take after the scan. Write a few sentences explaining to Jo why she has to do each of them.
2. Look at the paragraph headed 'Medical imaging'. Write out the key steps as a flow diagram or bullet points.
3. It would be safe to stand next to Jo but not to kiss her. Use the words 'irradiation' and 'contamination' to explain why.

Key words

radiotherapy
beta radiation
gamma radiation

Regulating radiation dose

The Health Protection Agency (HPA) studies radiation hazards and gives advice to protect against them. It also keeps a close eye on the many people who regularly work with radioactive materials – in hospitals, industry, and nuclear installations. They are called 'radiation workers'.

Hospital radiologists wear badges like this to monitor their radiation dose.

The ALARA principle

Employers must ensure that radiation workers receive a radiation dose 'as low as reasonably achievable'.

The ALARA principle applies when better equipment or procedures can reduce the risks of an activity. Any extra cost this involves must be balanced against the amount by which the risk is reduced.

To reduce their dose, medical staff take a number of precautions. They:

- use protective clothing and screens
- wear gloves and aprons
- wear special badges to monitor their dose

The ALARA principle applies equally to hospital patients who receive radiation treatment. If the HPA finds that one hospital uses smaller doses but is just as effective as any other, then all hospitals are encouraged to copy them.

What affects radiation dose?

The dose measures the potential harm done by the radiation. On page 69, you saw that it depends on the amount of radiation and the type of tissue that is exposed. It also depends on the type of radiation.

Alpha is the most ionizing of the three radiations. Therefore it can cause the most damage to a cell. The same amount of alpha radiation gives a bigger dose than beta or gamma radiation.

Dose summary (2)

Radiation dose is affected by

- amount of radiation
- type of exposed tissue
- type of radiation

Properties of three types of radiation

Radiation	Range in air	Stopped by	Ionization	Dose factor
alpha	a few cm	paper/dead skin cells	strong	20
beta	10 to 15 cm	thin aluminium	weak	1
gamma	metres	thick lead	very weak	1

Amount of radiation

Radiation is all around you. At any time, there is a tiny chance that it might collide with something crucial within one of your cells. It's a bit like a game of dodge ball, with tennis balls bouncing around a court. The more moving tennis balls there are, the higher the risk of being hit.

A gamma scan is similar. Increasing the intensity of gamma radiation increases the dose. Time, too, is important. Having a scan is not too risky, because the patient is only exposed for a short period of time.

The chance of being hit goes up with the number of tennis balls in play.

Sterilization

Ionising radiations can kill bacteria. Gamma radiation is used for sterilizing surgical instruments and some hygiene products such as tampons. The products are first sealed from the air and then exposed to the radiation. This passes through the sealed packet and kills the bacteria inside.

Food can be treated in the same way. Irradiating food kills bacteria and prevents spoilage. As of 2005, irradiation is permitted in the UK only for herbs and spices. But the label must show that they have been treated with ionizing radiation. This is a useful alternative to heating or drying, because it does not affect the taste.

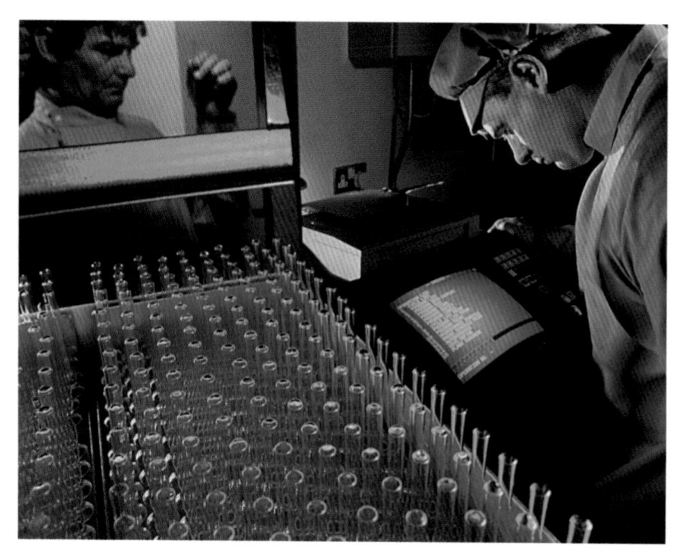

Gamma rays kill the bacteria on and inside these test tubes.

Questions

4 Look at the paragraph headed 'The ALARA principle'. For each of the bullet points, describe how the precaution prevents contamination and irradiation.

(5) a Why seal the packets of surgical instruments before sterilizing them?

b Does the gamma radiation make them radioactive? Explain your answer.

(6) a Why is alpha radiation more harmful to cells than beta or gamma radiation?

b Would alpha radiation be a suitable source for **i** scanning a patient **ii** treating cancer? Say why.

(7) Write down three uses of radioactive materials mentioned in this section. Choose one of these and write the key points on how radiation is used.

Find out about:

- radioactive decay
- what makes an atom radioactive

D Changes inside the atom

A cut diamond sitting on a lump of coal. Each of these is made of carbon atoms. Some of the atoms will be radioactive.

Many elements have more than one type of atom. For example, there are carbon-12 and carbon-11 atoms. In most ways they are identical. Both can:

- be part of coal, diamond, or graphite
- burn to form carbon dioxide
- be a part of complex molecules

Radioactive decay

The main difference is that carbon-12 atoms do not change. They are stable.

Carbon-11 atoms are radioactive. Randomly, these atoms give out energetic radiation. Each carbon-11 atom does it only once. And what is left afterwards is not carbon, but a different element – boron. The process is called **radioactive decay**. It is not a chemical change; it is a change *inside* the atom.

Inside the atom

Atoms are small – about a ten millionth of a millimetre across. Their outer layer is made of electrons. Most of their mass is concentrated in a tiny core, called a **nucleus**.

The nucleus itself contains two types of particle: **protons** and **neutrons**. All atoms of any element have the same number of protons. For example, carbon atoms always have six protons. But they can have different numbers of neutrons and still be carbon. The word **isotope** is used to describe different atoms of the same element. Carbon-11 and carbon-12 are different isotopes of carbon.

Compared to the whole atom, the tiny nucleus is like a pinhead in a stadium.

Carbon-11 will give out its radiation whether it is in diamond, coal, or graphite. You can burn it or vaporize it and it will still be radioactive.

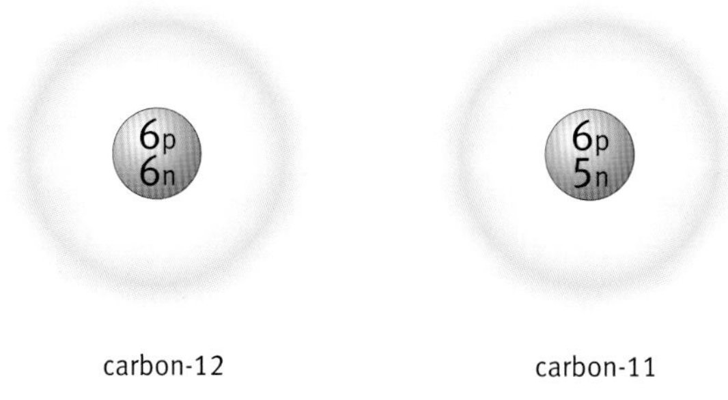

Carbon-11 has 11 particles in its nucleus: 6 protons and 5 neutrons. The nucleus of carbon-12 has 6 protons and 6 neutrons.

What makes an atom radioactive?

Some atoms, with particular combinations of protons and neutrons in the nucleus, are **unstable**. The atom decays to become more stable. It emits energetic radiation and the nucleus changes. This is why the word 'nuclear' appears in *nuclear reactor*, *nuclear medicine*, and *nuclear weapon*.

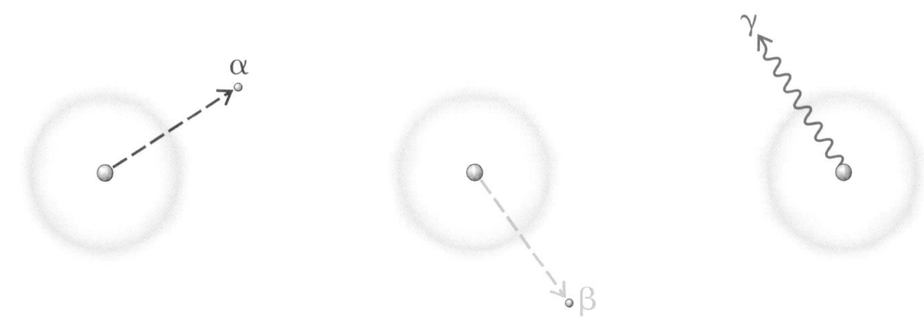

It is the nucleus of an atom that makes it radioactive and emits the radiation.

Radiation	What it is
alpha α	helium nucleus
beta β	high-speed electron
gamma γ	electromagnetic radiation

Making gold

When platinum-197 decays it turns into a new element – gold. A good way to make money? No. The price of gold is only half the price of platinum.

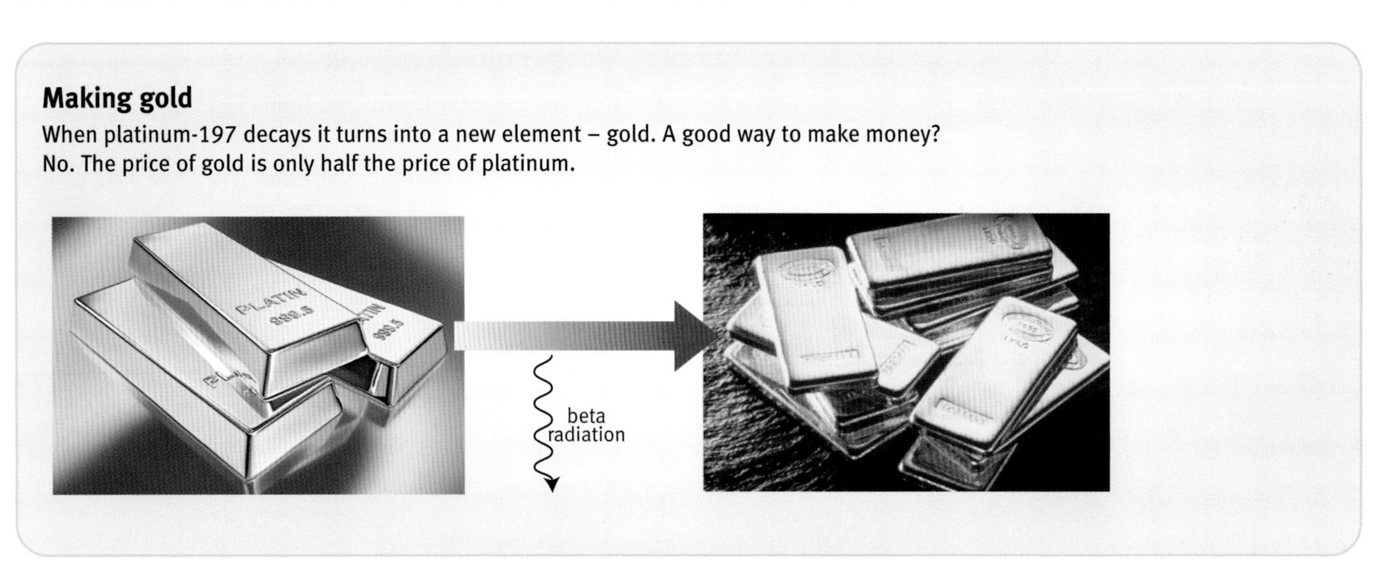

Radioactive changes

The emission of either an alpha or a beta particle from any nucleus produces an atom of a different element, called a 'daughter product' or 'decay product'. The daughter product may itself be unstable. There may be a series of changes, but eventually a stable end-element is formed.

After emitting an alpha or beta particle, protons and neutrons remaining in the new nucleus sometimes rearrange themselves to a lower energy state. When this happens, the nucleus emits gamma radiation. This does not cause a change of element.

Medical isotopes

Many elements occur in a form where the nucleus is unstable. These different forms are called radioactive isotopes.

Radioactive isotopes are quite rare in Nature – because most of them have decayed. But hospitals need a regular supply of several isotopes for diagnosis and treatment. These are made in nuclear reactors, or in accelerators, and are prepared in laboratories and hospitals around the country.

Carbon-11 atoms can be put into molecules of carbon methionone. This is a chemical that is absorbed by the brain. Doctors use the radioactive form of this chemical to produce brain scans.

Questions

1 Look at these isotopes:

A carbon-11
B boron-11
C carbon-12
D nitrogen-12

a Which two are isotopes of the same element?

b Which ones have the same number of particles in the nucleus?

c Do any of them have identical nuclei?

d A nucleus of carbon-14 has **i** how many protons? **ii** how many neutrons?

2 Which of these will test whether something is radioactive?

A look at it just with your eye
B burn it
C put in acid
D put it by a Geiger counter
E look at it through a microscope

(3) Alpha radiation is the most ionizing type of radiation. Explain why, in terms of what it is.

Key words

radioactive decay
nucleus
isotope
protons
neutrons
unstable

E Nuclear power

Find out about:
- energy from nuclear fission
- nuclear power stations

Nuclear fission

Radioactive atoms have an unstable nucleus. Other nuclei can be made so unstable that they split in two. This process is called **nuclear fission**.

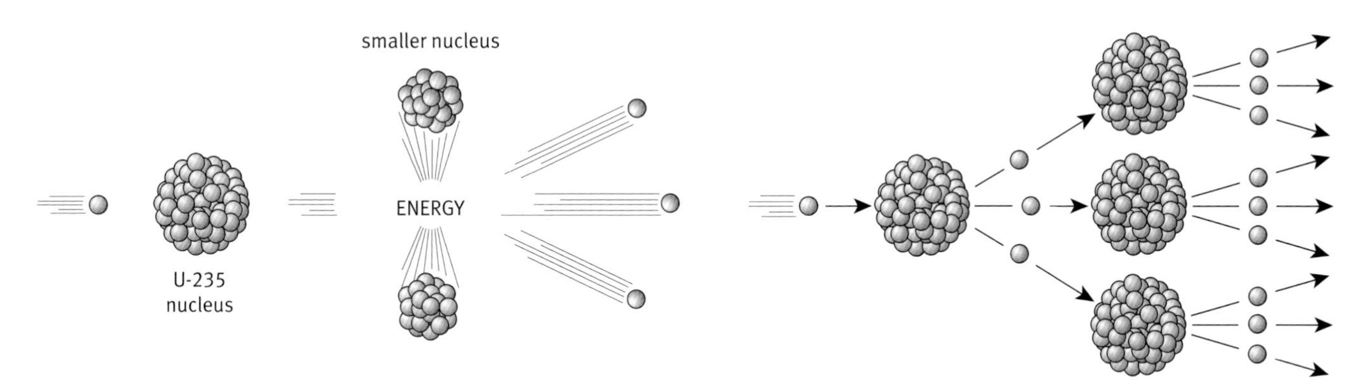

Splitting the nucleus of an atom

A chain reaction

For example, the nucleus of a uranium-235 atom breaks apart when it absorbs a neutron. And the products of nuclear fission all have kinetic energy.

In the 1930s, scientists realized that they could use nuclear fission to release a huge amount of energy. During World War II, there was a race to 'split the atom' and harness the energy in a bomb.

The fission of one atom can set off several more, because each fission reaction releases a few neutrons. If there are enough U-235 atoms close together, there will be a **chain reaction**, involving more and more atoms.

Each fission reaction produces roughly a million times more energy than changing a molecule does, in any chemical reaction.

Key words
nuclear fission
chain reaction

Nuclear weapons

On 16 July 1945, in the deserts of New Mexico, a group of scientists waited tensely as they tested 'the gadget'. Some thought it would be a flop. Others worried that it might destroy the atmosphere.

It had taken years of research to isolate enough uranium-235 to make the first atom bomb. At 5.29 a.m., it was detonated and filled the skies with light. The bomb vaporized the metal tower supporting it. All desert sand within a distance of 700 m was turned into glass.

Some of the scientists were worried about the power of the bomb and wanted the project stopped. A few weeks later, the Americans dropped two nuclear bombs on Japan, at Hiroshima and Nagasaki.

The devastating power of a nuclear weapon.

Key words
fuel rods
control rods
coolant

The Nuclear Installations Inspectorate monitors the design and operation of nuclear reactors. Reactor cores are sealed and shielded. Very little radiation gets out.

Controlling the chain

At the heart of a nuclear power station is a reactor. It is designed to release the energy of uranium at a slow and steady rate, by controlling a chain reaction.

- The fission takes place in **fuel rods** that contain uranium-235. This makes them extremely hot.
- **Control rods**, which contain the element boron, absorb neutrons. Moving control rods in or out of the reactor decreases or increases the reaction rate.

Once fuel rods are in service, decay products build up in them. They become more radioactive.

Generating electricity

A fluid, called a **coolant**, is pumped through the reactor. The hot fuel rods heat the coolant to around 500 °C. It then flows through a heat exchanger in the boiler, turning water into steam.
The steam drives turbines that, in turn, drive generators.

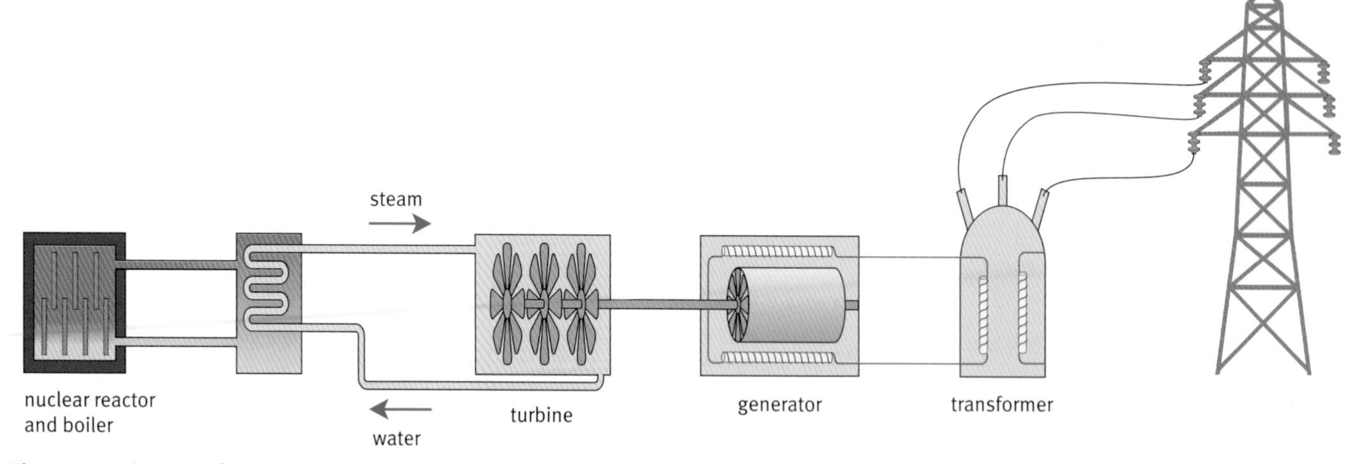

The stages in a nuclear power station.

In the 1950s, many countries started building nuclear reactors. They hoped that nuclear power would:

- produce cheap electricity
- reduce the need to import fossil fuels

But the building of nuclear power stations in Europe and North America stopped in 1986. The year of ...

The Chernobyl disaster

Chernobyl is a small town in the Ukraine. It is now deserted: a ghost town. In 1986 its nuclear reactor overheated, as a result of a mistaken test experiment. The reactor was not designed to be fail-safe. It produced too much steam, and the reactor's top blew off – like steam lifting a saucepan lid. Winds carried radioactive dust as far as Wales, where some fields are still contaminated.

Fortunately, major accidents at nuclear power stations are rare. Many safety systems operate in nuclear power stations, to prevent accidents. But when they do happen, they can be very serious.

An American team, led by the Italian immigrant, Enrico Fermi, built the first nuclear reactor in a squash court at Chicago University. Fermi was born in Rome. In 1938 he was awarded a Nobel Prize. Italy then had a Fascist government, and Fermi feared for his Jewish wife's safety. She was permitted to accompany him to the Nobel awards ceremony, in Sweden. They went straight from the ceremony to America.

In the news

Nuclear reactors produce new elements. One of these is plutonium, which can be used to make bombs. It is difficult to keep track of all the world's plutonium.

UN Nuclear Weapons Inspectors try to ensure that nuclear power stations:

- are very secure
- account for all their waste
- are not operated in unstable countries

Questions

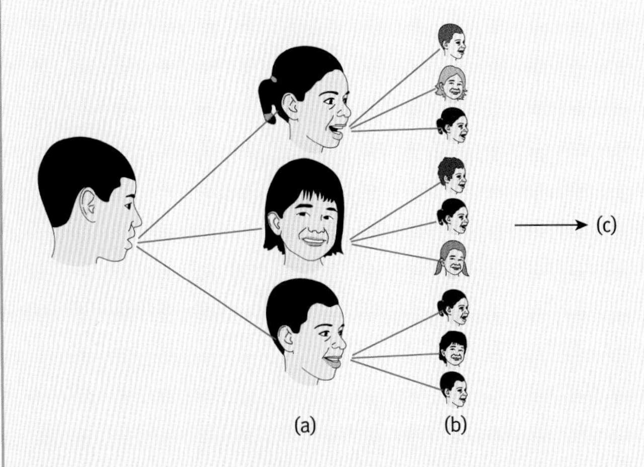

1 A rumour is a bit like a chain reaction. Alex tells a story to three friends. Each of them tells three friends and so on. How many people have heard the story when it is told

  a the first time (from Alex)

  b the second time (from his three friends)

  c the third time

  d the tenth time?

2 Which part of a nuclear power station

  a produces steam

  b produces electricity

  c contains the energy source

  d uses the steam to turn a shaft?

③ Why do nuclear reactors use coolants? In other words, why not circulate ordinary water and boil the steam directly?

④ Look at the box called *In the news*. Explain what each bullet point means and why it is important.

Find out about:
- the UK's nuclear legacy
- the half-life of radioactive materials
- possible methods of disposal

F Nuclear waste

A legacy of nuclear waste

In 2004, the government set up the Nuclear Decommissioning Authority (NDA) to clean up hazardous nuclear waste at sites around the UK. More than 95% of the radioactive waste comes from nuclear power stations. The rest comes from medical uses, industry, and scientific research. Nuclear waste is a cocktail made of different radioisotopes. They call it UK's 'nuclear legacy'.

Initially the total NDA budget was expected to be £500 000 million (£1 000 every year, for 50 years), but the projected costs increase every year. Before the NDA can start disposing of nuclear waste, it must find some method that is acceptable to the public

Nuclear waste is hazardous

Radioactive waste has very little effect on the UK's average background radiation. But it is still hazardous. This is because of contamination. Imagine that some waste leaks into the water supply. This could be taken up by a carrot, which you eat. The radioactive material is now in your stomach, where it can irradiate your internal organs. This is dangerous – it is like the radon and the silver miners on page 68.

Some radioactive materials last for tens of thousands of years. The NDA must dispose of nuclear waste in ways that are safe and secure for many, many generations.

Key words
activity
half-life
High Level Waste
Intermediate Level Waste
Low Level Waste

The pattern of radioactive decay

The amount of radiation from a radioactive material is called its **activity**. This decreases with time.

- At first there are a lot of radioactive atoms.
- Each atom gives out radiation as it decays to become more stable.
- The activity of the material falls because fewer and fewer radioactive atoms remain.

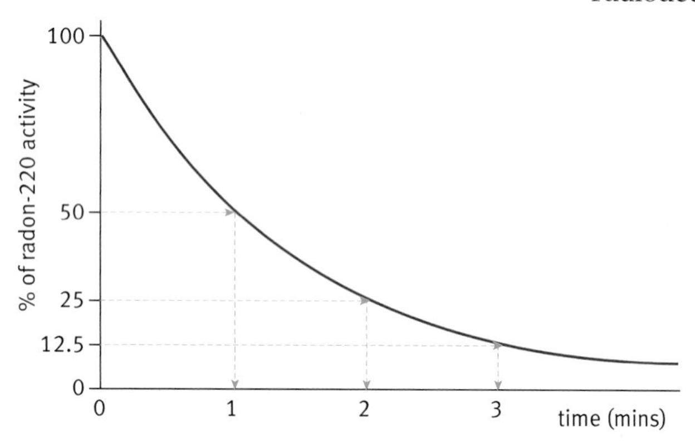

The decay curve for radon-220

The graph shows the pattern of radioactive decay for radon.

Notice that the amount of radiation halves every minute. This is the **half-life** of radon-220. The half-life is the time it takes for the activity to drop by half.

All radioactive materials follow the same pattern of decay. But they can have different half-lives. For example:

- iridium-192 has a half-life of 74 days
- strontium-81 has a half-life of 22 minutes
- radon-220 has a half-life of about a minute

There is no way of slowing down or speeding up the rate at which radioactive materials decay. Some decay slowly over thousands of millions of years. Others decay in milliseconds – less than the blink of an eye.

The shorter the half-life, the greater the activity for the same amount of material. Of the three radioactive isotopes listed above, radon-220 is the most active.

High level radioactive waste is hot, so it is stored underwater.

The control room at a nuclear waste storage plant enables people to monitor the waste continuously.

Types of waste

The nuclear industry deals with three types of nuclear waste.

- **High Level Waste** (HLW). This is 'spent' fuel rods. HLW gets hot because it is so radioactive. It has to be stored carefully but it doesn't last long. And there isn't very much of it: all the UK's HLW is kept in a pool of water at Sellafield.
- **Intermediate Level Waste** (ILW). This is less radioactive than HLW. But the amount of ILW is increasing, as HLW decays to become ILW.
- **Low Level Waste** (LLW). Protective clothing and medical equipment can be slightly radioactive. It is packed in drums and dumped in a landfill site that has been lined to prevent leaks.

Type of waste	Volume (m^3)	Radioactivity	% of radioactivity
LLW	15 000	weak	1 millionth
ILW	75 000	strong	10
HLW	2 000	extremely strong	90

The amount of nuclear waste in store (2001). The problem of what to do with it remains unsolved.

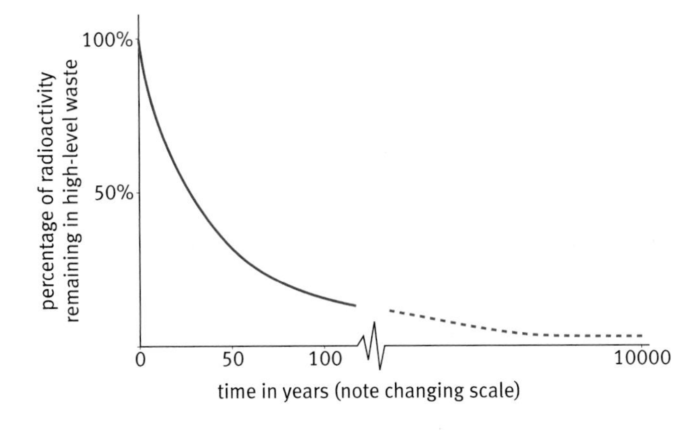

High Level Waste decays quickly at first. When its activity falls, it becomes Intermediate Level Waste. ILW stays radioactive for thousands of years.

Questions

1 Carbon-14 has a half-life of 5700 years. What fraction of its original activity will a sample have after 11 400 years?

(2) Iodine-132 is used to investigate problems with the thyroid gland, which absorbs iodine. It is a gamma emitter.

  a Explain why the element iodine is chosen.

  b Explain why it is useful that iodine-132 gives out gamma radiation.

  c Iodine-132 has a half-life of 13 hours. Why would it be a problem if the half-life was:

    i A lot shorter

    ii a lot longer?

Sellafield

A government-owned company runs the biggest nuclear site in the UK – Sellafield, in Cumbria. Thousands of workers – professional, skilled and unskilled – contribute to its important work. Sellafield reprocesses nuclear waste produced in the UK and abroad. It also prepares and stores nuclear waste for permanent disposal.

Risk management is a major concern at Sellafield. They must plan in advance how to maintain production and safety in the event of any possible problem.

Intermediate Level Waste presents the biggest technical challenge, because it is very long-lived. Currently it is chopped up, mixed with concrete, and stored in thousands of large stainless-steel containers. This is secure but not permanent. The long-term solution has to be secure and permanent.

Permanent disposal?

Years ago the UK dumped nuclear waste at sea, to be dispersed. Later, people suggested burying it in Arctic ice, or firing it into space. But these options are too risky. Current possibilities include:

- keeping it on the surface, in storage containers
- burying it deep in rock

The UK first investigated deep disposal for nuclear waste in the late 1980s. In secret, a shortlist of 12 possible sites was decided.

In the news

Preventing nuclear materials from falling into the wrong hands is a real problem, according to the UN's International Atomic Energy Agency. Their records show a dramatic rise, since the 1990s, in the level of smuggling of radiological material.

A 'dirty bomb' is a conventional explosive designed to spread radioactive material. A terrorist attack using a dirty bomb is 'a nightmare waiting to happen', said one expert. It could contaminate a large urban area.

The precautionary principle

The **precautionary principle** is relevant here.

> If ... you are not sure about the possible results of doing something
>
> And if ... serious and irreversible harm could result from it
>
> ... then it makes sense to avoid it.

As some people say, 'Better safe than sorry!'

In other words, be careful. Only proceed once you are sure that you have minimized the risks involved, and that the benefits outweigh those risks.

Questions

③ Disposing of ILW needs to be both secure and permanent.

a Explain why both criteria are important.

b Five disposal methods are mentioned in the text. Choose two of them and, for each one, describe how well it meets these requirements.

④ Choose one of the views expressed by the people above. Write a letter to that person to try to change their view.

⑤ Some people say that waste should be stored above ground until a safe method is found for permanent storage. This illustrates the precautionary principle. Explain how.

Key words
precautionary principle

Find out about:
- different ways of generating electricity
- their benefits and risks
- how to make your choices known

Oil and gas are fossil fuels that formed over millions of years. They are extracted from underground reserves through wells like this.

G Energy futures

Who decides?

Electricity is a secondary energy source. Energy companies, operating under government regulation, generate and distribute it.

Energy companies also make decisions on your behalf. When you boil a kettle, the electricity may have come from any type of primary source.

Primary sources of energy

Fossil fuels like coal, oil, and gas are finite. One day they will run out. Power stations burn fossil fuels and release waste, including carbon dioxide, into the atmosphere.

Nuclear fuel comes from uranium mines. There are large but finite reserves. It produces solid radioactive waste that has to be handled carefully.

Renewable energy sources like wind, geothermal, and solar power produce very little waste. They are sustainable primary sources, because they should last forever.

Primary source	Estimated generating cost in 2020 (pence per unit)	CO_2 produced (per unit, estimated over full lifetime of power station)	Typical power output (MW)	Other issues
coal	3.0–3.5	1000–1300	1000	CO_2
gas	2.0–2.5	440–690	600	CO_2
nuclear	3.4–8.3	9–21	1000	radioactive waste long build
wind	1.5–2.5 onshore 2.0–3.0 offshore	10–48	2 (per turbine)	not constant
solar (photovoltaic)	15–20 ? (70 in 2005)	100–280	Peak 1 kW per m^2	small scale only

Different ways of generating electricity (source: IAEA (2000))

Generating electricity

Fossil and nuclear fuels are used to boil water and make steam. The high-pressure steam passes through a steam turbine.

This turbine has lots of small blades that drive it round.

Regular maintenance keeps the generators running smoothly.

Key words

fossil fuels
nuclear fuel
renewable energy sources
decommissioning

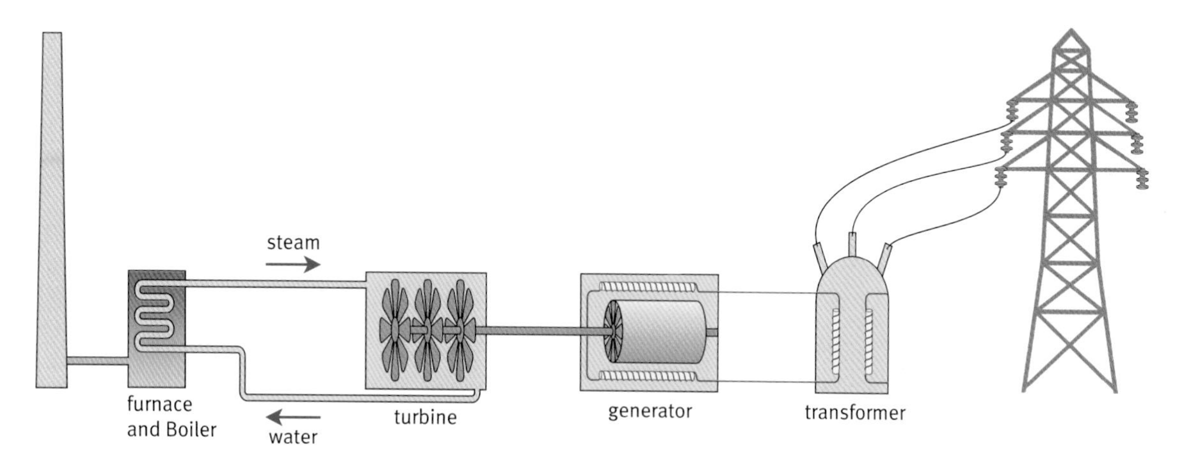

How a thermal power station works

Power stations burning natural gas have an extra turbine that harnesses the flow of hot exhaust gases. This makes them the most efficient type of power station.

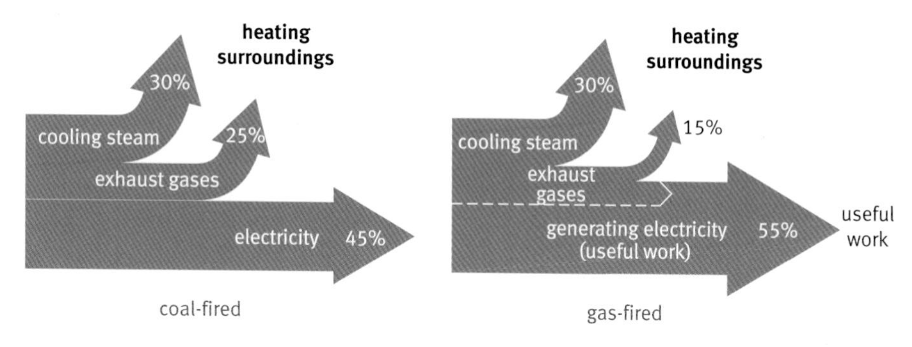

The Sankey diagrams show where energy the energy goes. Less is wasted in a gas-fired power station.

Reducing CO_2 emissions

Using more efficient gas-fired power stations is one way of reducing the amount of CO_2 produced. Others include:

- using nuclear power
- using renewable energy sources
- reducing total electricity consumption

None of these is the perfect answer. Each one presents challenges.

And you have to assess the whole life of the power station to get the full story. At the end of their lifetime, power stations must be dismantled. This is called **decommissioning**.

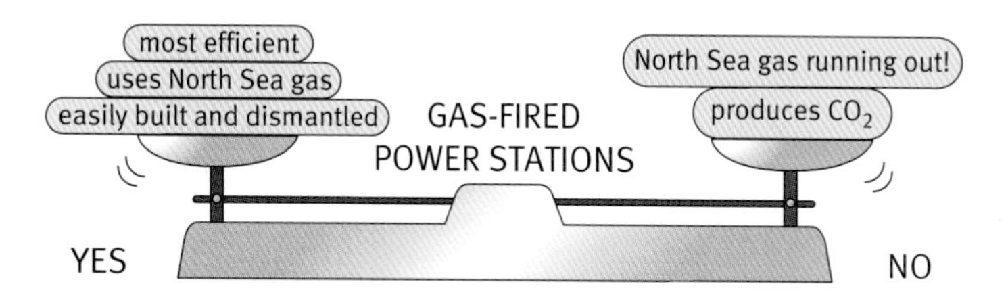

Build gas-fired power stations?

The dismantling of Berkeley nuclear power station, in Gloucestershire. Electricity costs take in the whole life of a power station, from start to finish.

The Energy Debate

Nuclear

YES

Nuclear power is the only energy source that can meet a substantial electricity demand. It releases no damaging carbon dioxide.

The best way to use world stockpiles of uranium and plutonium is as fuel in civilian reactors to generate electricity. Otherwise they remain available for making nuclear weapons.

UK nuclear power stations use tried and tested technology. Safety systems meet high standards. Waste disposal is a problem that can be solved.

NO

Nuclear power stations may release little CO_2 while operating. But large amounts of CO_2 are associated with materials and energy used during construction and decommissioning. Most importantly, they produce radioactive waste.

A new generation of reactors would take about a decade to build and cost roughly £2 billion each. No insurance company will cover their risks, during operation or decommissioning. The public will have to pay if anything goes wrong.

Renewable energy sources

NO

Renewable energy is unreliable. Winds don't always blow. The Sun doesn't always shine.

No wind farms should be built where people live and work. Each wind turbine is a huge and noisy machine, as high as an office block.

- Their noise carries long distances.
- They can interfere with TV reception.
- If a turbine blade breaks, it becomes a very dangerous projectile.

In remote but beautiful areas, wind turbines are a blot on the landscape.

YES

The UK should exploit its own energy sources and not rely on imports. Recent studies suggest that renewable energy sources could provide the UK with a reliable supply of electricity.

What we need is a full range of generators – very big to very small – at sites all around the country. A decentralized power system would be based on micro-generation. Installing wind generators and solar cells on the rooftops of many offices and homes will be relatively cheap and easy.

A life cycle assessment shows that power from the Sun and winds releases little CO_2.

Use less energy – for and against

YES

Energy consumption rises year by year. In your lifetime, you are likely to use as much energy as all four of your grandparents put together. Every energy saving you can make will help.

The government can help by:

- requiring new buildings to use less energy for heating and lighting
- providing grants to help householders install domestic combined heat and power systems
- ensuring that new appliances are energy efficient
- taxing fossil fuels more highly

NO

It's all very well to dream of using less energy. But energy makes the world go round. It's essential to education, business, and pleasure. And everyone has a right to a good standard of living at home.

Making decisions

In the coming decade, older power stations will be closed down as they come to the end of their useful lives. New power stations must be built to replace them.

Energy companies will continue to use a variety of primary energy sources. You can influence the amount of each type that they use. Make your views heard. One thing is certain: there's no easy answer.

Year	Percentage of electricity generated				
	gas	coal	renewables	nuclear	other
2002	38	32	3	23	4
2010	56	16	10	16	2
2020	?	?	?20	?8	?

Future energy sources. These figures show that many decisions have yet to be made.

Questions

1 What does 'a sustainable supply of energy' mean?

(2) The cost of decommissioning contributes to the price of electricity. It is much larger for nuclear power stations than for stations burning fossil fuels. Explain why.

(3) Look at the leaflets for and against each energy source.

a Draw balance diagrams for each option, listing statements on each side.

b Distinguish statements of fact from opinion, by putting a tick next to facts.

c Using the information tables on these pages, and any other sources, add further statements to your diagrams.

(4) Write a letter to your Member of Parliament expressing your views about future power stations. Use your answer to question **3** to make your letter persuasive and show that you have considered the issues.

P3 Radioactive materials

Science explanations

This Module is about radioactive materials and how electricity is generated.

You should know:

- that radioactive materials randomly emit ionizing radiation all the time
- about three kinds of radiation and their different properties
- the difference between contamination and irradiation
- what radiation dose measures, and what factors affect it
- how ionizing radiation can damage living cells
- that atoms have an outer shell of electrons and a nucleus, made of protons and neutrons
- that all atoms of any element have the same number of protons, but they can have different numbers of neutrons
- how the nucleus changes in radioactive decay
- that the activity of radioactive sources decreases over time
- that radioactive elements have a wide range of half-life values
- about uses of ionizing radiation from radioactive materials
- that there are three categories of radioactive waste, each with different methods of disposal
- why electricity is called a secondary energy source
- what renewable energy sources are used for generating electricity
- that burning carbon fuels in power stations produces carbon dioxide
- what nuclear fission means
- how nuclear power stations work and what waste they produce
- how to label a block diagram showing the main parts of a power station
- how to interpret a Sankey diagram
- how to evaluate information about different types of power station

Ideas about science

To make personal and social decisions about health, it can be important to assess the risks and benefits.

For risks and benefits about the use of radioactive materials you should be able to:

- explain why nothing is completely safe
- suggest ways of reducing particular risks
- interpret information on the size of risks
- suggest why people will accept (or reject) the risk of a certain activity
- discuss a given risk, taking account of both the chance of it occurring and the consequences if it did
- identify, or propose, an argument based on the 'precautionary principle'
- discuss personal and social choices in terms of actual risk and perceived risk
- explain the ALARA principle and use it in context

Where there are health risks associated with radioactive materials, you should be able to:

- identify the groups affected, and the main benefits and costs of a course of action for each group
- explain and use the idea of sustainable development
- distinguish what can be done from what should be done
- explain why different courses of action may be taken in different social and economic situations
- show you know that regulations and laws control scientific research and applications

These ideas are illustrated by: radon in homes; medical imaging and treatment; debates about the disposal of nuclear waste; and possible energy futures.

Why study motion?

Humans have always been interested in how things move and why they move the way they do. Motion is such an obvious part of our everyday lives that we cannot really claim to know very much about the natural world if we cannot explain and predict how objects move.

The science

One tantalizing thought has always driven people who have studied motion – is it possible that every example of motion we observe can be explained by a few simple rules (or laws) that apply to everything? Remarkably the answer is 'yes'. And these laws are so exact and precise that they can be used to predict the motion of an object very precisely. A key idea is force. A force acting on an object, unless it is cancelled out by another force, causes a change in its motion.

Physics in action

Understanding forces and motion has enabled scientists and engineers to design and build more efficient cars and trains, to develop aircraft, and to fly spacecraft with enormous accuracy to the furthest reaches of the Solar System. The same science ideas are used in testing new materials. Although our understanding of motion was developed by Isaac Newton in the 17th century, it is still at the heart of many innovations and developments in the 21st century.

Explaining motion

Find out about:

- how forces always arise from an interaction between two objects
- friction and reaction of surfaces
- instantaneous and average speed, and velocity
- the idea of momentum, and how the momentum of an object changes when a force acts on it
- everyday examples of motion, including the principles on which traffic safety measures are based
- gravitational potential energy and kinetic energy

Find out about:

- how forces arise when two objects interact
- contact and action-at-a-distance forces

The chemical reaction inside the firework shell produces forces that send the burning fragments out in all directions, producing a sphere of shooting stars.

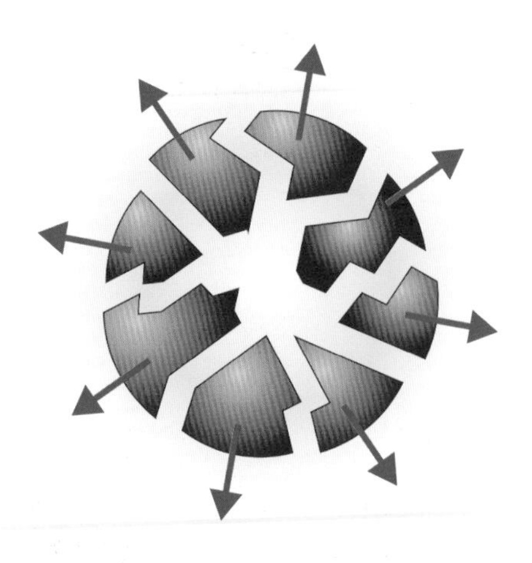

The size of the force on each fragment is shown by the length of the arrow.

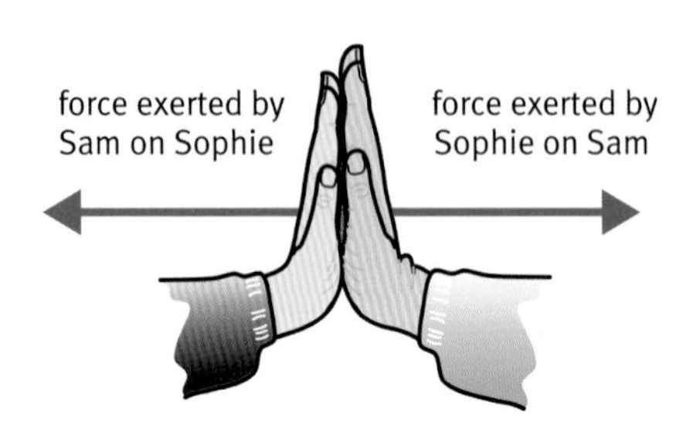

Forces always arise in pairs. Here Sophie pushes Sam and experiences a force in return.

A Forces in all directions

To start anything moving requires a **force**. A firework explodes because of a chemical reaction inside. The forces on the sparks in the starburst are a result of the chemical reaction. The firework fragments in the photograph on the left are being pushed out in all directions. The force on each fragment can be represented by an arrow, as shown below the photograph. In this case, there are lots of fragments and lots of forces.

How do forces arise?

To help understand how forces arise, it is easier to start with a simpler situation where there are only two objects moving apart.

Sophie and Sam have gone out sailing on the lake, but the wind has dropped. Their boats are together but neither is moving. They have no oars. They do not know what to do to get back to land.

Two stationary boats and no oars: how can Sophie or Sam get moving?

Perhaps one of them could push the other. But when they do this, they both move. However hard they try, Sam and Sophie are unable to make only one of the boats move. It is not possible for them to push so that one experiences a force but not the other.

You can tell something very important about forces from this:

- Forces always arise from an **interaction** between two objects.

So forces always come in pairs. The two forces in an **interaction pair** are

- equal in size
- opposite in direction

This is always true. And it does not depend on the size or strength of the two people involved. Another important thing to notice is that

- the two forces act on different objects

In this example, one force of the pair acts on Sam and the other on Sophie.

'Things' can push (and pull) too!

You are used to the idea that people and animals can push and pull. You also know that machines and motors can exert forces. But in fact anything can exert a force if it is involved in an interaction.

The diagram on the right shows Deborah, a roller-skater. She pushes against the wall and immediately starts to move backwards. When she pushes the wall, the wall pushes back on her. It pushes back with an equal force in the opposite direction. This force makes Deborah start to move.

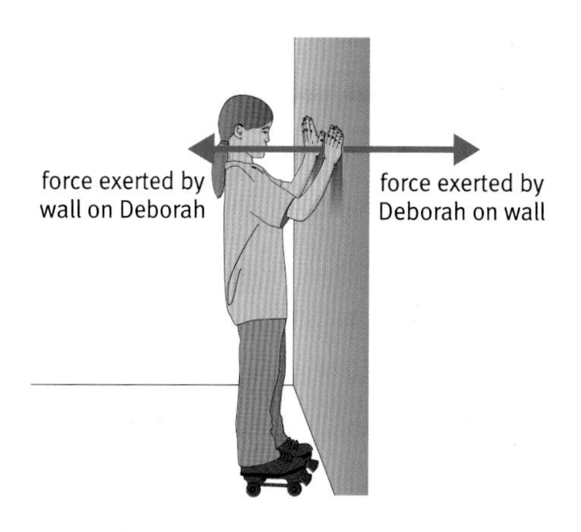

Deborah pushes against the wall. The other force in the interaction pair then starts her moving on her roller-skates.

Action at a distance

Where the interacting objects touch each other, the forces are known as contact forces. There are also forces that act at a distance. The forces caused by gravity and magnetism are examples. However, *all* forces arise from interactions.

An apple falls from a tree because of the force pulling it downwards, towards the centre of the Earth. But gravity is an attraction between two objects. So the apple also exerts an equal and opposite force on the Earth! This does not have much effect, however, because the Earth is so large.

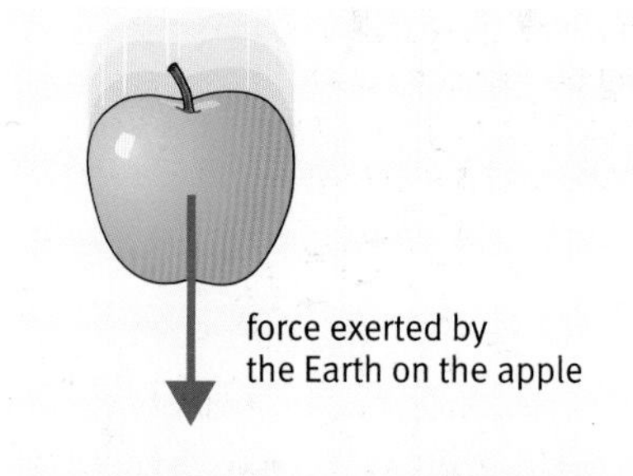

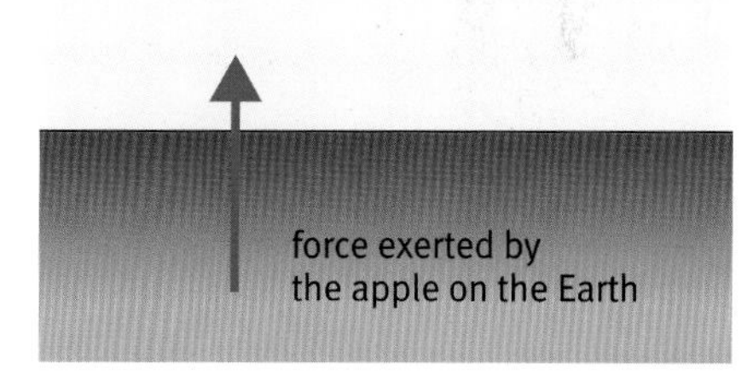

A gravitational interaction – again forces always arise in pairs.

Questions

1. List four examples of interaction pairs of forces mentioned on these pages.
2. What three things are always true about interaction pairs?
3. (circled) These two ring magnets are repelling each other. Notice that both magnets are being pushed aside.

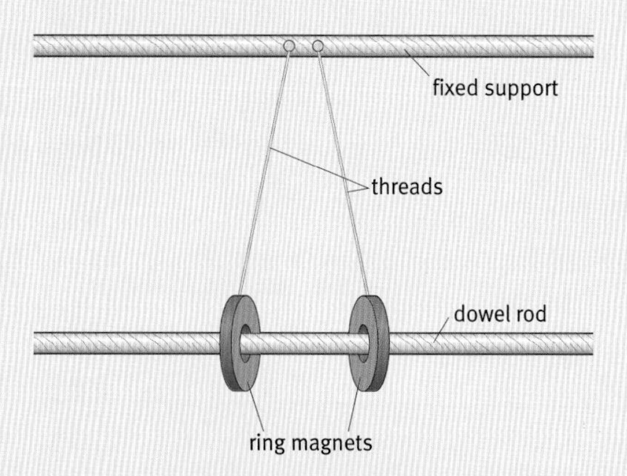

How could you modify this apparatus to show that *attraction* forces between magnets also arise in pairs? Sketch how you would set it up, and write down what you would expect to see.

Key words

force
interaction
interaction pair

Find out about:
- the forces that enable people and vehicles to get moving
- rockets and jet engines

B How things start moving

Rockets

When objects explode, the pieces usually travel outwards in all directions. But if an object is designed so that it does not break up, and everything that comes out of it goes in one direction, then you have a rocket.

The photograph on the left shows one of the most famous rocket launches: the *Apollo 11* mission to land the first people on the Moon. The interaction pair of forces is shown on the photo. Burning hot gases are pushed out of the base of the rocket, and the rocket is pushed in the opposite direction.

Rockets carry with them everything they need to make the burning gases they push against. This means that they can work in space as well as in air.

The start of the longest journey humans have made so far – to the Moon. A huge force is needed to push a rocket like this upwards. It is provided by the hot exhaust gases, which are formed by burning the fuel.

Jet engines

Jet engines use the same basic idea as rockets. Air is drawn into the engine and pushed out at the back. The other force of the pair pushes the engine forward. Jet engines need to draw air in, so they cannot work in space.

How does a car get moving?

To make a car move, the engine has to make the wheels turn. This causes a forward force on the car. To understand how, think first about a car trying to start on ice. If the ice is very slippery, the wheel will just spin. The car will not move at all. The spinning wheel produces no forward force on the car. Now imagine a car on a muddy track. The rally car below is throwing up a shower of mud as it tries to get going.

As it rotates, the wheel exerts a force backwards on the ground – with dramatic results in this case!

You can see that there is an interaction between the wheel and the ground. The wheel is causing a backwards force on the ground surface. This makes the mud fly backwards. Mud, however, moves when the force is quite small. The other force of the interaction pair is the forward force on the car. It is equal in size. So it is also small – and not big enough to get the car moving.

Now imagine a good surface and good tyres, which do not slip. Again, the engine makes the wheel turn. It pushes back on the road. The wheel cannot slip, and it exerts a very large force backwards on the road surface. So the other force of the interaction pair is the same size. It is this large forward force which gets the car moving.

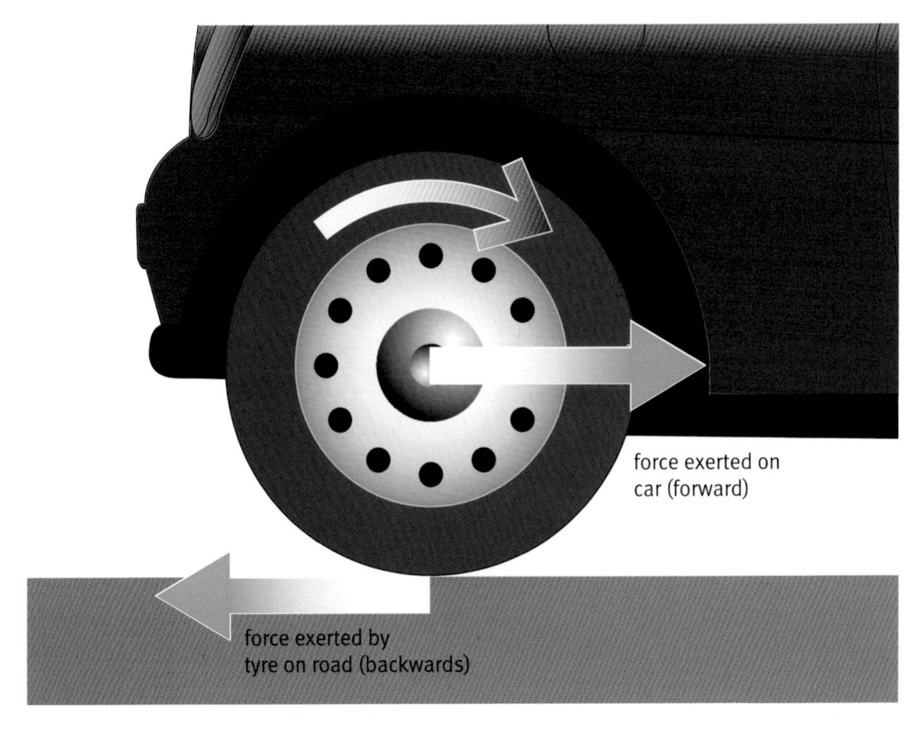

If the tyre grips the road and does not spin, the second force of the interaction pair results in a forward force on the axle. This pushes the car forward.

Walking

When you walk, you push back on the ground with each foot in turn. The ground then pushes you forward. You are not usually aware of this. If you tiptoe carefully across a floor, it does not feel as though you are pushing backwards on it. You only become aware of the importance of this interaction when the surface is slippery – when you try to walk on an icy surface, for example. Because you cannot push it back, it is unable to move you forward.

Questions

1 Jet engines are suitable for aircraft but not for travel in space. Explain why. How do rockets overcome this problem?

(2) A boat propeller pushes water backwards when it spins round. Use the ideas on these pages to write a short paragraph explaining how this makes the boat move forward. Draw a diagram and label the main forces involved, to illustrate your explanation.

(3) Sketch a matchstick figure walking. Mark and label the interaction pair of forces on the foot in contact with the ground.

Icy surfaces are difficult to walk on. Your foot cannot get a grip to push back on the surface, so the surface does not push you forwards.

Find out about:

- friction and what causes it

C Friction – a responsive force

Friction often seems a nuisance. But without it, you could not start walking. Cars could not get moving.

What is friction?

Jeff is a workman. He is trying to push a large box along a level floor. Think about the forces involved:

1 Jeff pushes the box with a force of 25 N, to try to slide it along. It does not move. The friction force exerted by the floor on the box is 25 N. This exactly balances Jeff's push.

2 Jeff then pushes harder, with a force of 50 N. The box still does not move. The friction force exerted by the floor on the box is now 50 N. Again, this balances Jeff's push.

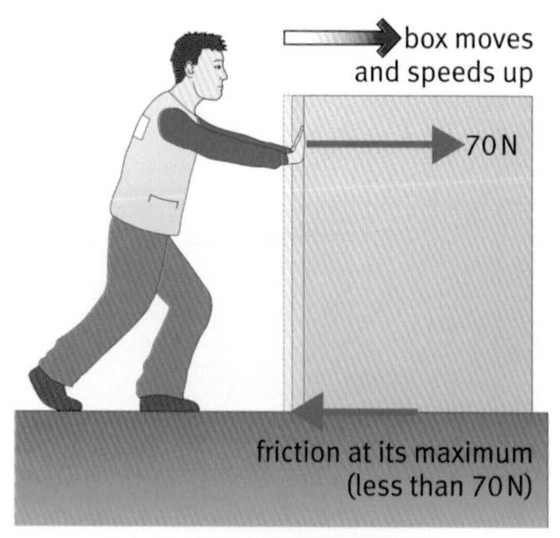

3 Jeff pushes harder still, exerting a force of 70 N. The box starts to move. It keeps speeding up while Jeff pushes. 70 N is bigger than the maximum possible friction force for this box and floor surface.

So friction is an unusual force. It adjusts its size in response to the situation – up to a limit. This limit depends on the objects and surfaces involved.

What causes friction?

Friction is a common type of force. But surprisingly, scientists do not yet agree on an explanation of the friction force between two sliding surfaces. Some things about friction are, of course, understood. It has to do with the roughness of the surfaces. Even surfaces that seem smooth have quite large humps and hollows if you look at them under a microscope.

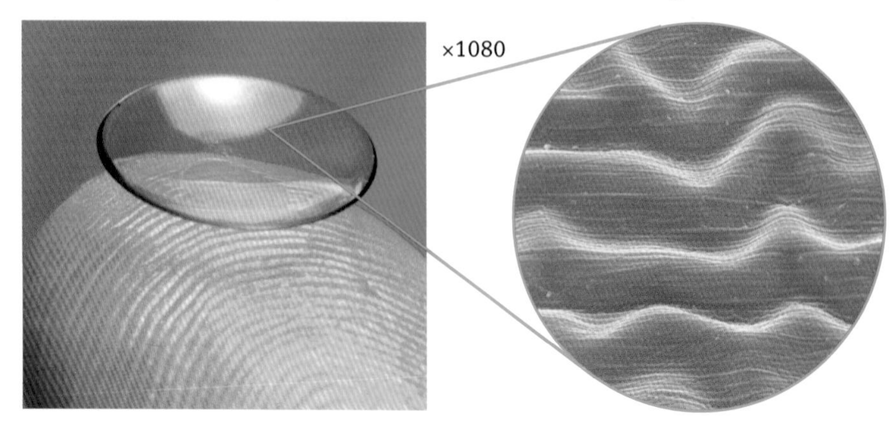

Even the smoothest surface is really quite rough. At the microscopic level, it has humps and hollows. The photograph on the right shows the surface of a contact lens magnified 1080 times.

When two surfaces are put together, the bumps on one can fit into the hollows of the other. When one object slides over another, it has to ride up and down over these bumps. To see why this requires a force, think about trying to slide two brushes past each other. The bristles on each brush will exert a sideways force on the other one.

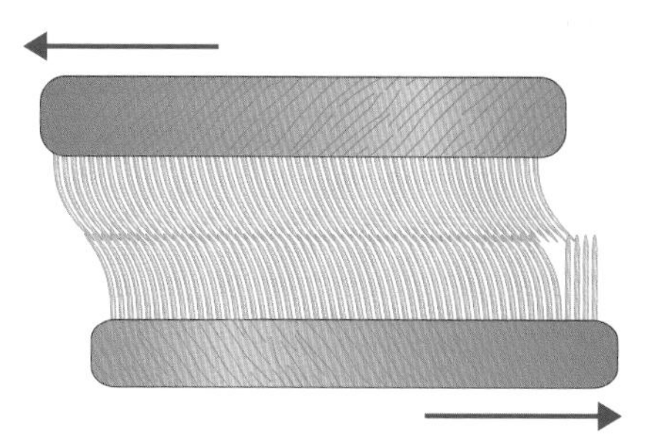

You can see how the friction force arises when you try to slide two brushes past each other.

But there is more to it than this. Because all surfaces are really quite rough, they only touch at a few points. They touch where a bump on one meets a bump on the other. So there are only a few real points of contact. As a result, the pressure at these points is very large. It is large enough to 'cold weld' them together. So when you slide one object over another, you have to keep breaking these tiny welds. And this needs a force.

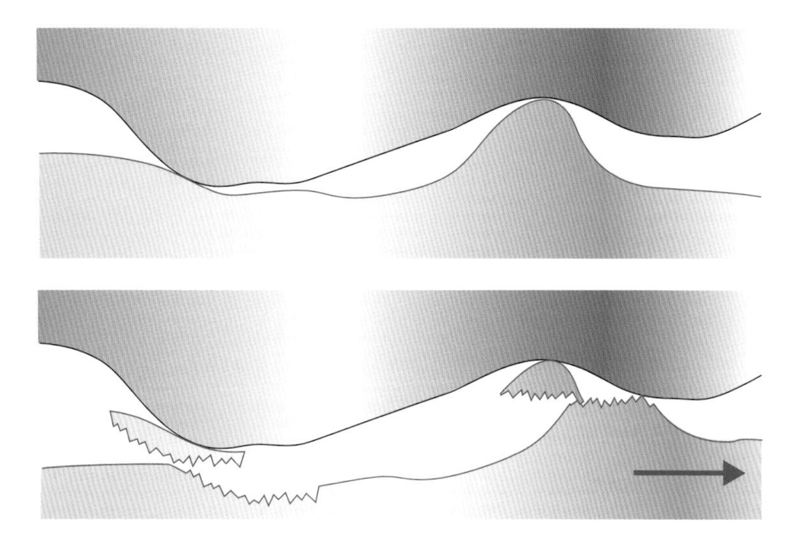

The force of friction arises because lots of tiny welds have to be broken while the objects slide past each other. With modern methods of detection, it is possible to show that tiny bits of one surface stick to the other, after they have been slid across each other.

Key words

friction

Questions

1. List three everyday situations in which we try to reduce friction, and three where we try to make friction as large as possible.
2. Use the ideas on these pages to write short explanations of the following observations:
   - **a** We can reduce the friction between two surfaces by putting oil on them.
   - **b** It is easier to push a box across the floor when it is empty than when it is full.
   - **c** Sometimes, when you polish two surfaces, the friction force between them gets bigger.
3. Scientists think that tiny 'spot welds' occur when two surfaces are in contact. What is the evidence for this?
4. Sketch three diagrams of the workman, Jeff, pushing the box on page 98. Mark and label the forces acting ***on Jeff*** as he pushes. Use the length of each force arrow to indicate the size of the force.

Find out about:

- how a surface exerts a reaction force on any object that presses on it

D Reaction of surfaces

If you hold a tennis ball at arm's length and let it go, it immediately starts to move downwards. There is a force acting on the tennis ball. This force is the pull exerted on it by the Earth. It is due to the interaction known as gravity.

But if you put a tennis ball on a table so that it does not roll about, it does not fall. The force of gravity has not suddenly stopped or been switched off. There must be another force that cancels it out. The only thing that can be causing this is the table. The table must exert an upward force on the ball that balances the downward force of gravity.

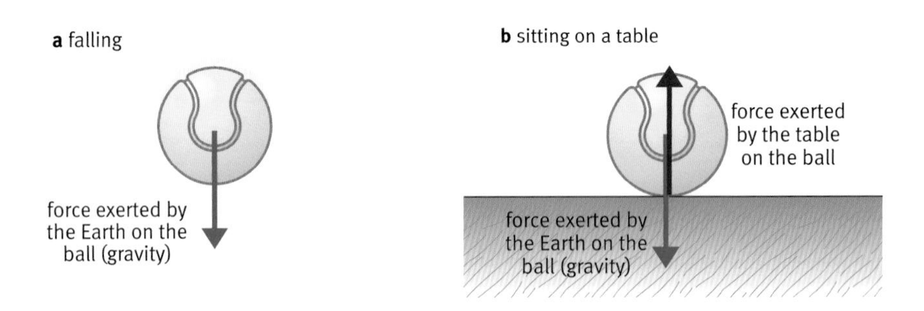

The forces acting on a tennis ball **a** falling and **b** sitting on a table.

How can a table exert a force?

Although it may seem strange, tables can and do exert forces.
To understand how, imagine an object, like a school bag, sitting on the foam cushion of a sofa. The bag presses down on the foam, squashing it a bit. Because foam is springy, it then pushes upwards on the bag, just like a spring. Like a spring, the more it is squeezed, the harder it pushes back. So the bag sinks into the foam until it reaches the point where the push of the foam on it exactly balances the downward pull of gravity on it.

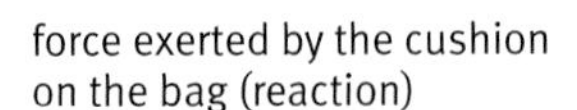

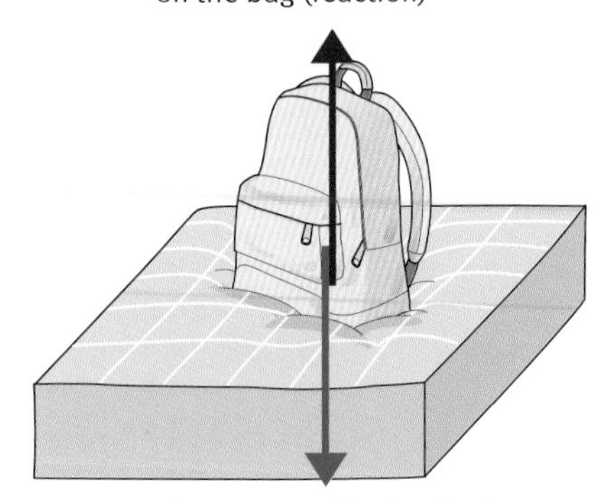

The bag squeezes the foam until the upward force of the springy foam on the bag exactly balances the downward gravity force on the bag.

The same thing happens, though on a much smaller scale, when the bag sits on a table top. A table top is not so easily squeezed as a foam cushion. But it ***can*** be squashed. This may not be visible to the naked eye, however. We call this upward force which a hard surface exerts when something presses on it the **reaction** of the surface.

Of course, a table cannot always exert an upward force on an object to balance the downward force on it. There is a limit. This limit depends on the material the table is made from. If the force exerted on the table top gets bigger than this, it is distorted beyond the point where it can spring back. It then breaks. Up to this point, however, it exerts an upward force that exactly matches the downward force exerted on it.

E Adding forces

Find out about:
- how to add the forces acting on an object

The discussion on pages 98-100 about friction and reaction of surfaces used an idea that may seem obvious:

- If there is a force acting on an object, but it is not moving, then there must be another force balancing (or cancelling out) the first one.

For Jeff, the workman pushing the box (page 98), the other force is friction. For a bag sitting on a table, the other force is the reaction of the table surface.

If the forces acting on an object balance each other, we say they add to zero. Adding several forces that act on the same object is straightforward. But you must take the direction of each force into account. The sum of all the forces acting on an object is called the **resultant force**. The diagrams below show some examples.

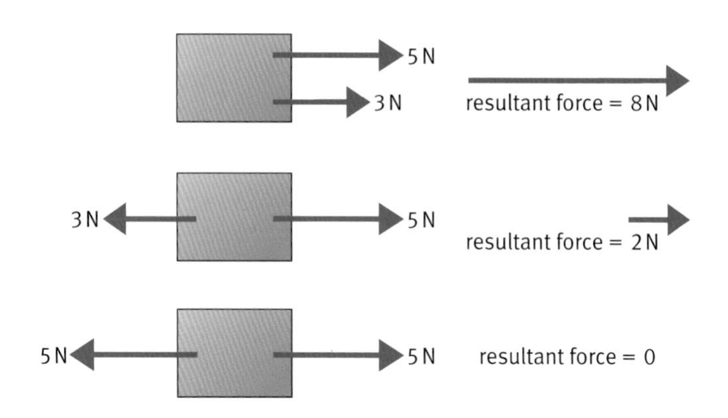

To find the resultant force acting on an object, you add the separate forces. You must take account of their directions.

Key words
reaction (of a surface)
resultant force

Questions

1 What happens to a 'hard' surface when something sits on it? Is it really as hard as it seems?

② Look back at the diagram on page 95 of Deborah pushing against a wall. She moves because the wall exerts a force on her. Use the ideas on these pages to write a paragraph explaining what happens to the wall at the point where Deborah pushes – and how the wall is therefore able to exert a force on her.

③ Imagine the bag in the diagram on page 100 hanging from a string. The string must be exerting an upward force on the bag, equal to the downward force of gravity on it. How does the string exert this upwards force? Use the ideas on these pages to suggest an explanation.

F How fast are you going?

Find out about:

- how to calculate the speed of a moving object
- catching speeding motorists

Cars have speedometers. But how can you tell how fast you are going on your bike?

Most bicycles do not have speedometers, so you cannot measure your speed directly. But you can time how long it takes to cycle between two places – two lamp-posts, for example. And you can measure how far you have cycled – the distance between the two lamp-posts. Then you can work out the **average speed**. Here's how.

Key words
average speed
instantaneous speed

Average speed

To calculate average speed is quite easy. You use the equation

$$\text{average speed} = \frac{\text{distance travelled}}{\text{time taken}}$$

However, knowing the average speed is sometimes not very useful. For most journeys, the speed is not always the same. It varies. For instance, imagine you are going to drive to a friend's house:

Your journey

1 Town traffic

During the first part of your journey, you drive from home to the motorway. This takes you through busy city streets. The speed limit is 30 mph (miles per hour), but often you are travelling slower than this.

2 Motorway

You travel on the motorway for 1 hour. In that time, you go 60 miles. So your average speed on the motorway is

$$\frac{60 \text{ miles}}{1 \text{ hour}} = 60 \text{ mph}$$

3 Country lane

You turn off the motorway. You have 6 more miles to go. But you get held up on a narrow country road. It takes you 30 minutes (0.5 hour)! So your average speed on this part of the journey is

$$\frac{6 \text{ miles}}{0.5 \text{ hour}} = 12 \text{ mph}$$

4 End of journey

You look at your watch and the mileometer when you arrive. The whole journey of 76 miles has taken you exactly 2 hours. So, for the whole journey, your average speed was

$$\frac{76 \text{ miles}}{2 \text{ hours}} = 38 \text{ mph}$$

So, you may know the average speed for the whole journey. But you cannot tell anything about how fast the car was going at any particular moment. If you did drive steadily at 38 mph (the average speed) for the whole journey, it would take you exactly 2 hours. But in practice, you did not. Your speed kept changing.

Instantaneous speed

The speed at a particular moment is called the **instantaneous speed**. If you were able to calculate average speeds over shorter and shorter time intervals, these would get closer and closer to the instantaneous speed of the car. In practice, to estimate the instantanous speed, you measure the average speed over a very short time interval.

The speedometer in a car shows the driver the instantaneous speed.

Questions

1 a Your whole journey above was 76 miles. How far was it from home to the motorway?

b Altogether, you drove for 2 hours. How long did it take you to reach the motorway?

c So, what was your average speed in miles per hour from home to the motorway?

Catching the speeders

The police have several methods they can use to measure a vehicle's speed:

1 **Gatso speed cameras** use radar (see method 3 below) to detect vehicles that are above the speed limit. The camera then takes two photographs of the vehicle, half a second (0.5 s) apart, to provide evidence. Distance markers on the road (they are 1.5 metres apart here) show how far the car has travelled in this time.

2 **Truvelo speed cameras** are triggered by detector cables in the road. Pressure sensors in the cables detect when a car is passing over. A computer in the camera measures the speed of passing cars by recording the time the car takes to travel from one cable to the next. If it is going faster than the speed limit, a picture is taken. These cables are 10 cm (0.1 m) apart.

3 **Police radar guns** bounce microwaves off approaching cars. Microwaves reflected off an oncoming car have a higher frequency than the original waves. These are picked up again by the radar gun. The gun uses the change in frequency to calculate the instantaneous speed of the car.

Questions

2 Look at the car moving away from you in the top two photographs.

a Estimate the distance that it moves between the top two photographs.

b Estimate the speed of this car in metres per second.

c Is this its average speed or its instantaneous speed? Explain your answer.

3 The detector cables in the bottom left photograph record a time of 0.008 s for this white van. The speed limit is 13 metres per second (30 mph).

a Is the van above the speed limit?

b Does this method measure the van's average speed or its instantaneous speed? Explain your answer.

4 You may have seen signs that display a car's speed as it enters a 30 mph zone – with a warning to 'Slow down!'. Which of these methods of measuring speed do you think these 'shame displays' are most likely to use? Explain why.

Find out about:
- how graphs can be used to summarize and analyse the motion of an object

G Picturing motion

You can use a graph to describe a journey more easily than with words:

- a distance–time graph shows how far a moving object is from its starting point at every instant during its journey
- a speed–time graph shows the speed of the moving object at every instant during its journey

Graphs do not only summarize information about the motion. They also help analyse it. This is how.

Distance–time graphs

On the left is a distance–time graph for the motorway section of the car journey described on page 102. Try taking readings from this graph of the distance the car has travelled after a quarter of an hour, half an hour, and one hour. The distance increases steadily as time goes by. The constant slope of the distance–time graph indicates a steady speed.

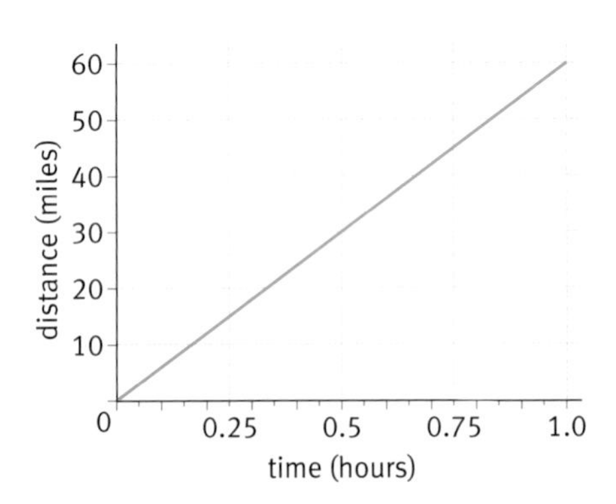

Distance–time graph for a car journey along the motorway

The distance–time graph below shows a cycle ride that Vijay did during his school holidays. The graph has four sections. In each section, the slope of the graph is constant. This means that his speed is constant during that section of the ride.

- In the first hour, Vijay travels 15 miles. His speed is 15 mph.
- In the second hour, he only travels 5 miles because of the strong wind. The shallower slope indicates a lower steady speed.
- In the third section (from 2.0 to 2.5 hours), his distance from home does not change. He is stopped. This is what a horizontal section of a distance–time graph means.

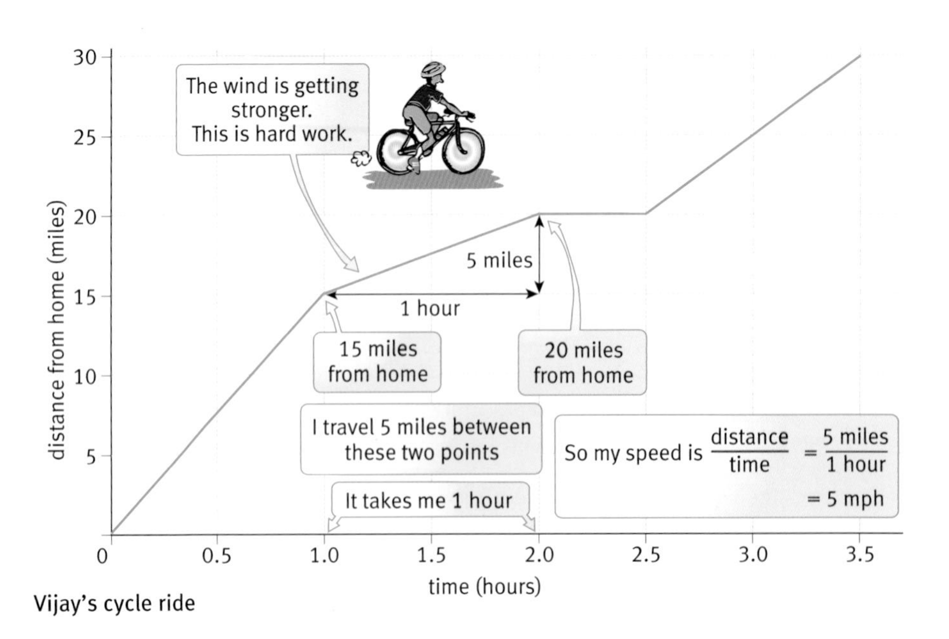

Vijay's cycle ride

Questions

1 a By looking at the slope of the graph on the right, say how Vijay's speed in the final section of his ride compares with his speed in the first and second sections.

b How far does Vijay travel during the final section of his journey (from 2.5 to 3.5 hours)? So what is his speed during this section?

Real journeys, however, do not consist of sections at steady speed. Instead the speed is always changing. And these changes may be gradual. More realistic graphs have slopes that change smoothly.

Look at the graph on the right. It shows a car journey. The slope tells you the speed of the car. So if the graph goes up ever more steeply, then the car must be speeding up, or accelerating. And if the slope is decreasing, then so is the speed. The final part of this distance–time graph slopes downwards. This shows that the distance from the starting point is getting less. The car is moving in the opposite direction, back towards the start.

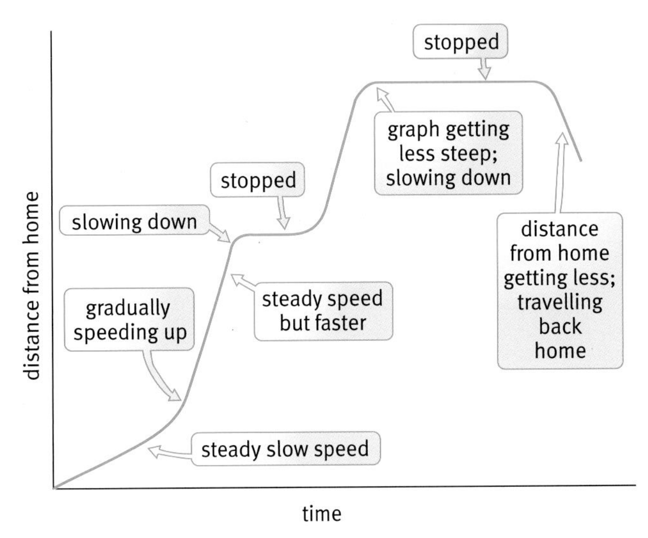

A more realistic distance–time graph

Speed–time graphs

A speed–time graph shows the speed of a moving object at every instant during its journey. The speed–time graph for Vijay's trip on page 104 would then look like the graph on the right. A steady speed is now shown by a straight horizontal line.

Again, the sudden changes of speed shown on this graph are not realistic. It would take time for the speed of a moving object to change. A more realistic speed–time graph would have smoother, more gradual changes from one speed to another.

The second speed–time graph on the right is for a tennis ball being dropped to the ground. It has no horizontal sections, so the speed of the ball is changing all the time. The constant slope of the speed–time graph shows that the speed is changing at a steady rate. It has steady (or uniform) acceleration downwards.

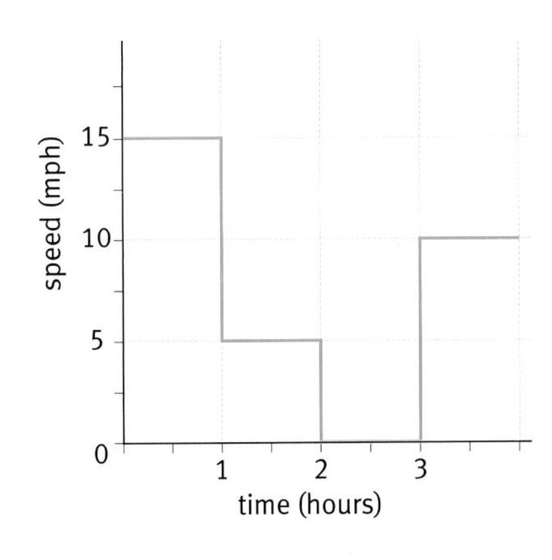

Speed–time graph for Vijay's cycle trip

Questions

2 What is the meaning of:

i a horizontal section
ii a section with a steady upward slope
iii a section with a steady downward slope

a on a distance–time graph?

b on a speed–time graph?

3 Roberta, an athlete, trains by jogging 20 m at a steady speed, then sprinting 20 m. She repeats this five times. Sketch a speed–time graph of her motion during a training session.

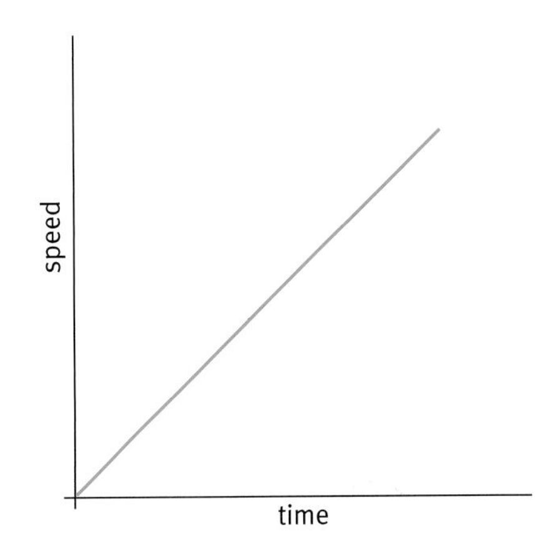

Speed–time graph for a falling tennis ball

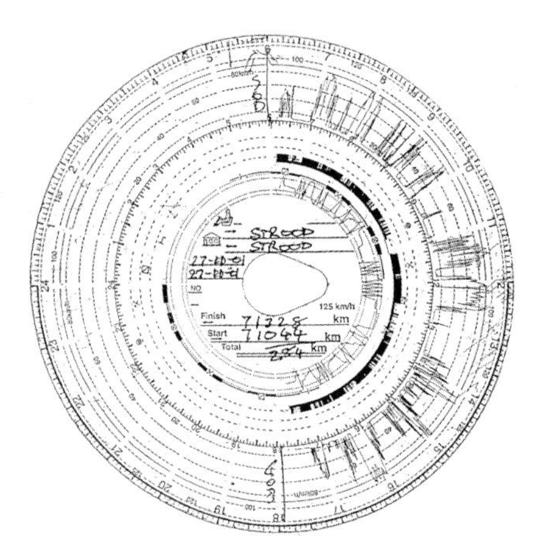

A tachograph trace – a speed–time graph of the vehicle's motion during a 24-hour period

Tachographs

According to EU regulations, lorry drivers are only allowed to drive for 9 hours per day. And they must take a break of at least 45 minutes every 4.5 hours. Lorries are also subject to speed limits on the road. So haulage companies have to keep a check on what their drivers are doing. They do this by installing a tachograph on each of their lorries. The tachograph monitors the lorry's distance and speed. It draws a graph of the lorry's speed against time. A tachograph trace is shown on the left.

An enlarged (and simplified) section of a tachograph trace might look like the speed–time graph below left. In the first section, the graph is horizontal. This indicates a constant speed. It then slopes upwards: the speed is increasing. The lorry travels for a while at a higher steady speed (another horizontal section). It then slows down again to its original speed.

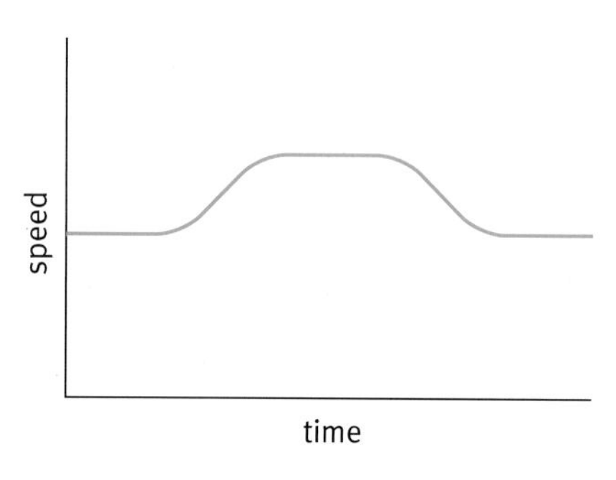

A speed–time graph for part of a lorry journey.

Velocity–time graphs

Look at the graph below. It shows the motion of Karl's skateboard up and then down a slope. At first Karl was travelling at 10 mph. He gradually slows down until his speed is zero. But then the line of the graph keeps on going down. The speed becomes negative. This may seem strange, but it is used to show that the skateboard's direction of motion has changed. Karl is now travelling in the opposite direction.

When you want to talk about speed in a certain direction, you use the term **velocity**. Velocity simply means speed in a certain direction.

Key words

distance–time graph
velocity
velocity–time graph

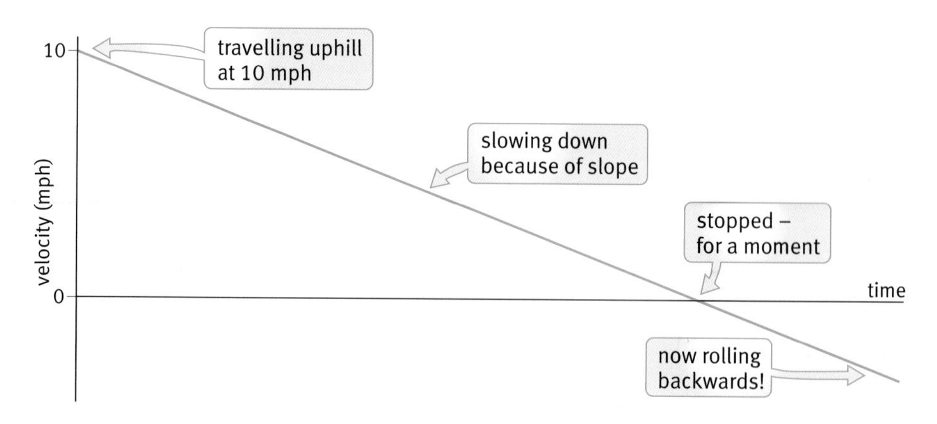

A velocity–time graph. The negative velocity means that Karl on his skateboard is travelling in the opposite direction to his original motion.

Questions

4 Sketch a distance–time graph for the part of the lorry's journey shown in the speed–time graph above left.

5 How would the speed–time graph of Karl on his skateboard be different from the velocity–time graph? Sketch a speed–time graph.

6 Think about the motion of a ball thrown upwards. Its speed gets steadily less on the way up and increases steadily again on the way down. Draw its

a speed–time graph

(b) velocity–time graph

from the moment it leaves your hand until the moment it lands back in your hand.

H Force, interaction, and momentum

Find out about:

- momentum
- the link between change of momentum, force, and time

Forces and motion

The key idea for explaining motion is **force**. If you know the forces acting on an object, you can predict how it will move. Look in more detail at what happens when two objects interact. Think about the three situations shown in the diagrams below:

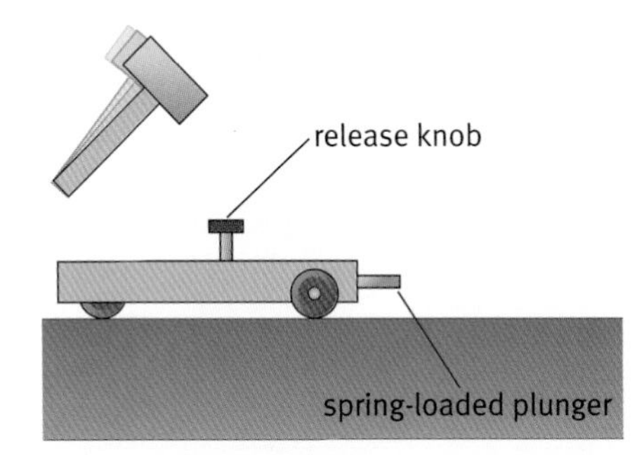

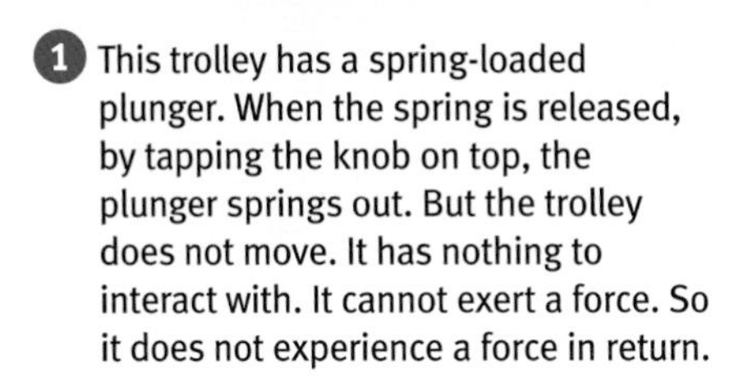

1 This trolley has a spring-loaded plunger. When the spring is released, by tapping the knob on top, the plunger springs out. But the trolley does not move. It has nothing to interact with. It cannot exert a force. So it does not experience a force in return.

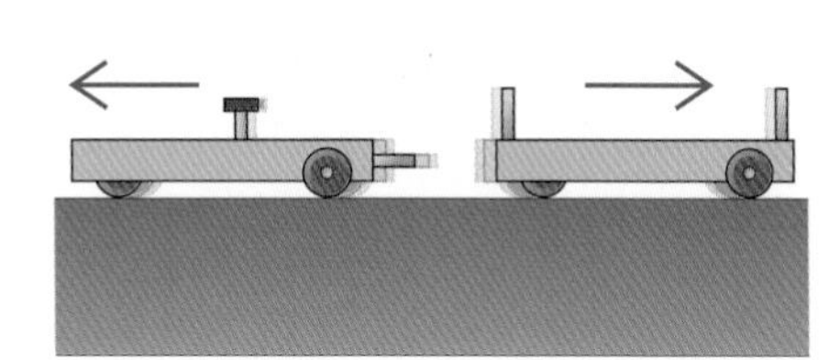

2 If you put a second trolley in front of the first one and then release the spring, both move, in opposite directions. The interaction causes two forces, one on each trolley. If the trolleys are identical, they both move with the same speed.

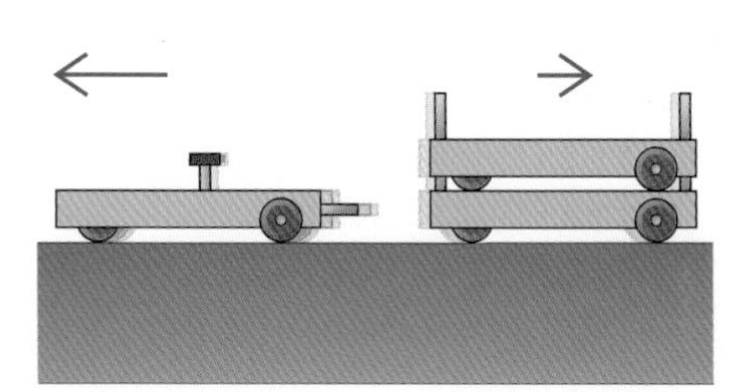

3 Here one trolley is twice as heavy as the other. When the spring is released, both move. But the heavier one has only half the speed of the lighter one.

When there is no interaction, there is no force on the trolley. So it does not move. When there is an interaction, both objects move. If the objects have different masses, the heavy one moves more slowly than the light one. In fact, the number you get if you calculate 'mass × speed' is the same for both. This seems to be an important quantity, so it is called the **momentum** of the object. Because the direction matters, momentum is defined by the equation

$$\underset{\text{(kilogram metre per second, kg m/s)}}{\text{momentum}} = \underset{\text{(kilogram, kg)}}{\text{mass}} \times \underset{\text{(metre per second, m/s)}}{\text{velocity}}$$

So, if an object is moving in one direction, its momentum is positive. And if it is moving in the other direction, its momentum is negative. You can choose which direction to call 'positive' in any situation.

Questions

1 What would you expect to happen if you carried out the exploding trolley investigation above with a stack of three trolleys on the right? Explain why.

2 What is the momentum of:

a a skier of mass 50 kg moving at 5 m/s?

b a netball of mass 0.5 kg moving at 3 m/s?

c a whale of mass 5000 kg swimming at 2 m/s?

3 A snooker ball has a momentum of 1 kg m/s just before it hits a cushion head-on. It bounces straight back with the same speed. What is its momentum now? Has its momentum changed?

Taking a free kick. The interaction between the footballer's foot and the ball causes a change of momentum.

Force and change of momentum

When a footballer takes a free kick, there is an interaction between his foot and the ball. His foot exerts a force on the ball. And the ball exerts a force on his foot. (That is why it can hurt to kick a ball with bare feet!)

This force lasts for only a very short time, the time for which the foot and the ball are actually in contact. After that, the player's foot can no longer affect the motion of the ball. The ball is on its own. So it is wrong to think of a kick 'giving the ball some force' or 'putting some force into the ball'. But the kick does give the ball some momentum. It causes a *change in momentum*.

Causing a change of momentum

Imagine pushing an object hard enough to make it start moving. If you keep pushing, it will get faster and faster. Its momentum is increasing. The change of momentum depends on two things:

- the size of the force you push with
- the time for which you keep pushing

This can be written in the form of an equation:

change of momentum = force × time for which it acts
(kilogram metre per second, kg m/s) (newton, N) (second, s)

All this is consistent with what you have seen in the interactions between two trolleys on page 107. After the interaction, the momentum of each of the moving objects is the same size. But in any interaction, the two objects involved always experience equal and opposite forces. These forces last for exactly the same length of time: the duration of the interaction. So 'force × time for which it acts' is also the same for both objects.

Questions

4 When a force makes an object move, which two factors determine the change of momentum of the object?

5 Which of the following will cause:

i the largest change in momentum?
ii the smallest change of momentum?

a a force of 40 N acting for 3 s
b a force of 200 N acting for 0.5 s
c a force of 3 N acting for 50 s?

H

Example

A football has a mass of around 1 kg. A free kick gives it a speed of 20 m/s. What is its momentum?

The football's momentum is given by the equation

momentum = mass × velocity
= 1 kg × 20 m/s
= 20 kg m/s

As it started with speed zero (when it was not moving), the ball's change of momentum during the kick is 20 kg m/s.

Using high-speed photography, it is possible to measure the contact time when a football is kicked. It is around 0.05 s (or one-twentieth of a second).

Estimate the force exerted on the ball during the kick.

Use the equation

change of momentum = force × time for which it acts

20 kg m/s = force × 0.05 s

Dividing both sides by 0.05 s, you get

$$\frac{20 \text{ kg m/s}}{0.05 \text{ s}} = \text{force}$$

So the force on the ball during the kick is 400 N. This is equal to the weight of a 40 kg object. No wonder it can hurt your foot! In fact, this is the *average* force during the kick. The maximum force will be even bigger.

Conservation of momentum

When there is an interaction between two objects, the change of momentum of one is equal in size to the change of momentum of the other but is opposite in direction. Another way to say this is:

- When two objects interact, the total change in momentum of the two objects (taking direction into account) is zero.

So the total momentum of the two objects is the same after the interaction as it was before.

This is true for any interaction. So it is a useful and important result. Scientists call it the principle of conservation of momentum.

You can use the idea of conservation of momentum to predict the speed of objects after an interaction. Look at the two skaters in the diagrams on the right. Zelda pushes on Jake's hands, and both move apart. Jake's mass is 60 kg and he is moving at 2 m/s. So his momentum is 120 kg m/s. You therefore know that Zelda's momentum must also be 120 kg m/s, in the opposite direction. As her mass is 40 kg, you can work out that she is moving at 3 m/s:

$$\text{momentum} = \text{mass} \times \text{velocity}$$

$$\text{velocity} = \frac{\text{momentum}}{\text{mass}}$$

$$= \frac{120\ \text{kg m/s}}{40\ \text{kg}} = 3\ \text{m/s}$$

Notice that the lighter person has the higher speed. This is just like the interacting trolleys on page 107.

The interactions you have been looking at so far are 'explosions', where two objects push apart. Collisions are another type of interaction. In any collision, momentum is also conserved.

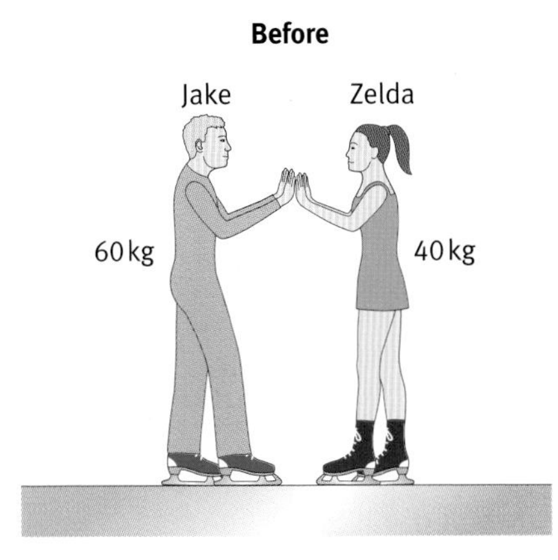

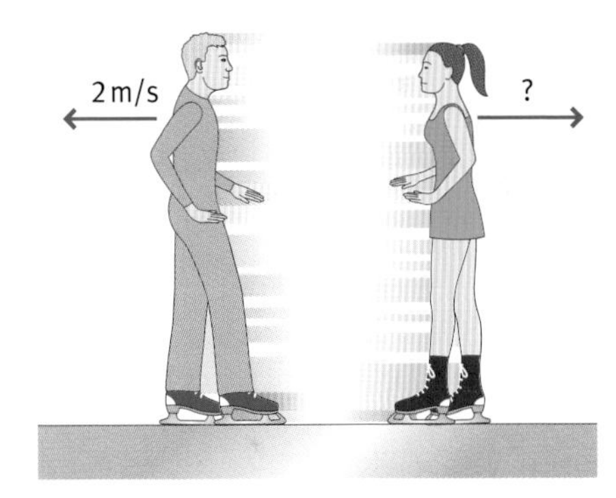

Jake pushes Zelda and they both move apart.

Key words
momentum

Questions

6 Look at this sequence of high-speed photographs of a tennis shot.

a If the frames are 0.01 s apart, estimate how many seconds the interaction between the ball and the racket lasts.

b The mass of a tennis ball is 0.06 kg. If it is moving at 12 m/s after this shot, estimate the force exerted on the ball by the racket.

Find out about:

- safety features in cars

1 Car safety

Cars today are much safer to travel in than cars ten or twenty years ago. As a result of crash tests like the one shown on the left, designs have changed and are still changing.

If a car is travelling at 70 mph, the driver and passengers are also travelling at that speed. If the car comes to a very sudden stop, owing to a collision, the occupants will experience a very sudden change in their momentum. This could cause serious injury.

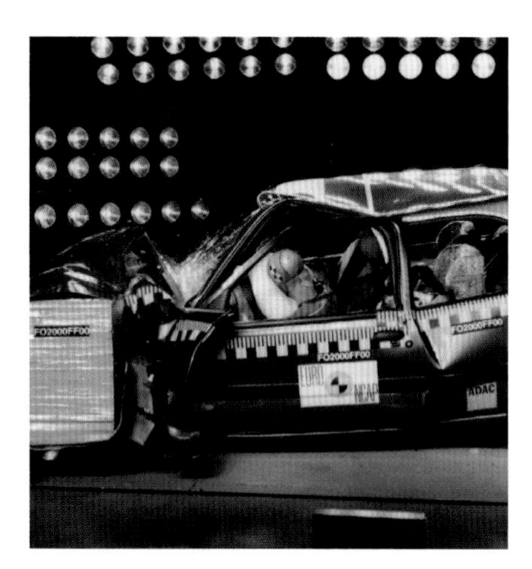

This driver, Hybrid 111, has experienced many crashes. With a steel skeleton and rubber skin it is packed with sensing equipment to record forces on different areas, like the head, the chest, and the neck. Each of these dummies costs more than £100,000 to build.

Crumple zones

Look at the diagram below. Which car would be safer in a collision? The answer may not be so obvious. You need to think about the change of momentum during the collision and the time the collision lasts.

a

b

Would you be safer in car **a** or car **b**?

The momentum of a moving car depends on

- its mass
- its velocity

In a collision, the car is suddenly brought to a stop. Its momentum is then zero. The size of the force exerted on the car during the collision depends on the time the collision lasts:

$$\text{change of momentum} = \text{force} \times \text{time for which it acts}$$

The bigger the time, the smaller the force – for the same change of momentum. This is why cars are fitted with front and rear crumple zones, with a rigid box in the middle. They are designed to crumple gradually in a collision. This makes the duration of the collision (the time it lasts) longer. This then makes the force exerted on the car less.

The passengers inside the car also experience a sudden change of momentum. A force exerted on their bodies (by whatever they come into contact with) causes this change. The longer it takes to change the passengers' speed to zero, the smaller the force they experience.

Questions

1. When you jump down from a wall or ledge, it is almost automatic to bend your knees as you land. Use the ideas on this page to explain why this reduces the risk of injury.
2. In railway stations, there are buffers at the end of the track. These are a safety measure – designed to stop the train if its brakes failed. Use the ideas on this page to explain how buffers would reduce the forces acting on the train and on the passengers, in the event of an accident of this sort.

Seat belts and air bags

Some people think that seat belts work by stopping you moving in a crash. In fact, to work, a seat belt actually has to stretch. Seat belts work on the same principle as crumple zones. They make the change of momentum take longer. So the force that causes the change is less.

With a seat belt, the top half of your body will still move forward, and you may hit yourself against parts of the car. Air bags can help to cushion the impact. Again, they reduce your momentum more slowly so that the force you experience is less.

Could you save yourself?

Some people think they could survive a car accident without a seat belt, especially if they are travelling in the back seats. The diagram below shows the position of a dummy driver at different times after an impact. The car was originally travelling at just 30 mph (or roughly 14 m/s). Without a seat belt, the dummy hits the steering wheel and windscreen about 0.07 s after the impact. Back-seat passengers would hit the back of the front seats at roughly the same time. As your reaction time is typically about 0.14 s, this would all happen before you even have time to react. Even if you could react in time, the force needed to change your speed from 14 m/s to zero in 0.1 s is larger than your arms or legs could possibly exert.

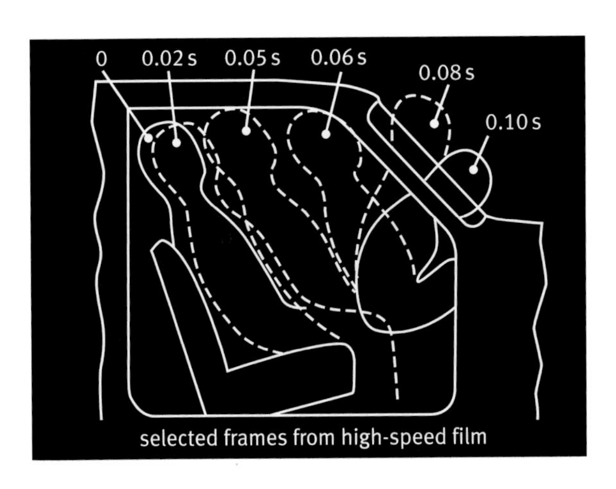

The position of a dummy driver at a series of instants after a crash.

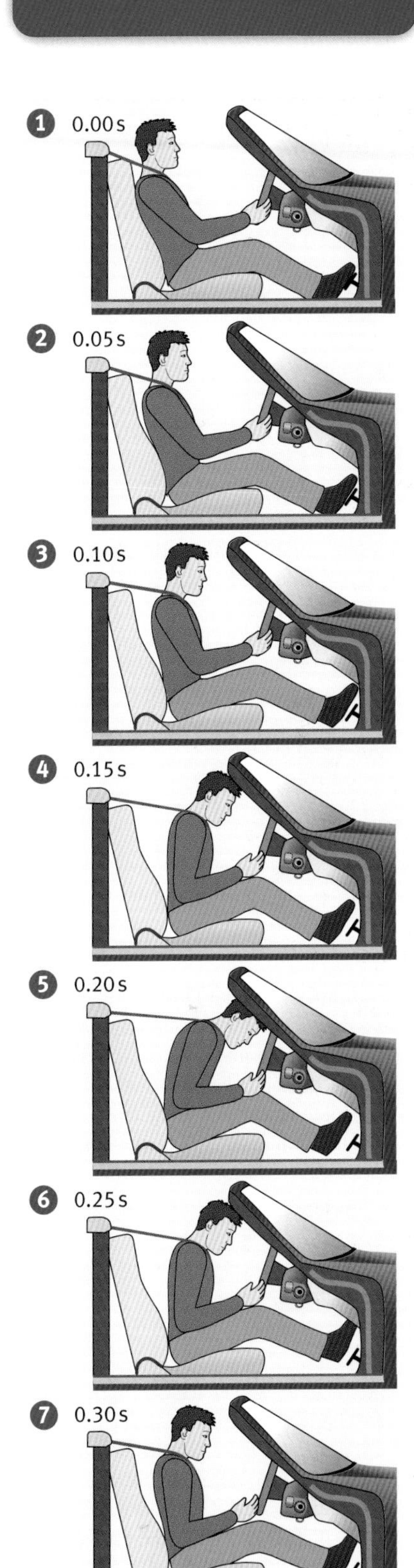

How seatbelts work. Notice how the seatbelt stretches during the collision. This 'spreads' the change of the driver's momentum over a longer period, making the force he experiences smaller.

Questions

3 Using the diagrams on the right, estimate how long it takes for the seat belt to bring the driver's body to a stop. If his mass is 70 kg, what is the average force that the belt has to exert to do this?

4 If the belt were made of material that did not stretch as much, would the force exerted on the driver be larger or smaller? Explain why. What if it were made of a material that stretched more?

Find out about:

- the laws (or rules) that apply to every example of motion
- how a resultant force is needed to change an object's motion

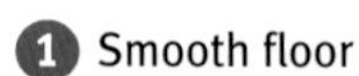

Imagine pushing this curling stone across a smooth floor. It will keep going after it leaves your hand, because of the momentum you have given it during the interaction with your hand. But it immediately begins to slow down because of friction, and soon it will stop.

2 Ice

Now think what would happen if you gave the same stone exactly the same push, but this time on ice. It would not slow down as quickly. But it would slow down all the same, because there is still some friction. Eventually it would stop.

3 'Perfect' ice

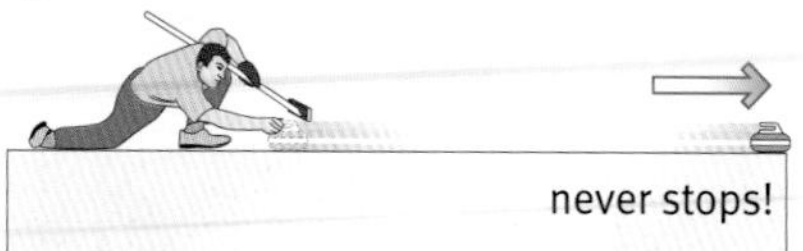

Now imagine 'perfect' ice, so slippery that there is no friction force between it and the stone. If you were able to give the stone the same push as before, it would not slow down after it left your hand. There is no friction force, so it just keeps on going, at the same speed. for ever.

Key words

driving force
counter-force

J Laws of motion

In everyday situations, there are always several forces acting on each object involved. To find out what will happen to the object, you need to find out what the combined effect of these forces is. To do this, you add the forces, taking their directions into account. This gives you the resultant force acting on the object (see page 101). You then apply the following laws (or rules) of motion:

- **Law 1:** If the resultant force acting on an object is zero, the momentum of the object does not change.
- **Law 2:** If there is a resultant force acting on an object, the momentum of the object will change. The change of momentum is given by (change of momentum = resultant force × time for which it acts) and is in the same direction as the resultant force.

These laws are completely general. They apply to every example of motion. That is why they are so useful.

When the resultant force is zero

Stationary objects are one example of law 1. If an object is stationary, its momentum is not changing. It is zero all the time. The resultant force acting on it is also zero. The forces acting on the object are balanced. They cancel each other out.

Now think about about an object travelling with constant velocity. Its momentum is not changing. The first law above says that the resultant force acting on it is zero. But you might be thinking that there must be a resultant force in the direction the object is going, to keep it moving. *In fact, this is not correct.* To see why, look at the diagrams on the left.

In the real world, there is *always* friction. So a **driving force** is needed to keep an object moving. But this just has to balance the friction force so that the resultant force *is* zero. If the driving force were bigger than the friction force. The object would not move at a steady speed. There would now be a resultant force. Its momentum would increase and it would speed up. Look now at how this works in an everyday situation: riding a bicycle.

Forces acting on a cyclist

When you press on the pedals of a bike, the chain makes the back wheel turn. The tyre pushes back along the ground. The other force in this interaction pair is the force exerted by the ground on the tyre. This pushes the bike forward. It is therefore called the driving force. As you move, air resistance and friction at the axles cause a **counter-force**. This is in the opposite direction to your motion.

A cyclist riding along at a steady speed. Is the resultant force on the cyclist zero? Or is there a resultant force forwards, in the direction she is going?

1 When you are starting off, the counter-force is very small. Your driving force is bigger. So you move forward, and your speed increases.

2 As you go faster, the air resistance force on you gets bigger. So the counter-force increases. You are still getting faster, but not as quickly as before.

3 Eventually you reach a speed where the counter-force exactly balances your driving force. Now your speed stops increasing. You carry on travelling, at a steady speed.

Questions

1 List three examples from everyday life of a situation where the resultant force on an object is zero. Explain how these are in agreement with the first law of motion on page 112.

2 List three examples from everyday life of a situation where there is a resultant force acting on an object. Explain how these are in agreement with the second law of motion on page 112.

3 Draw a fourth diagram in the series on this page to show the forces acting when the cyclist stops pedalling and freewheels. Write a caption, like those for the first three diagrams, to explain the motion.

Find out about:
- how to calculate the work done by a force
- the link between work done on an object and the energy transferred
- how to use energy ideas to predict the motion of objects

Pushing a car along is hard work!

K Work and energy

So far, you have seen how the ideas of force and momentum explain motion. However, it is sometimes hard to use these ideas to make an exact prediction, even though they apply to all situations. For example, to predict the speed something will reach when it slides down a slope, it is easier to use energy ideas. But first you need to make the connection between force and energy. The link is the idea of **work.**

Doing work

Imagine that you are out in your car and you break down. Luckily, there is a garage just down the road. You ask your passenger to steer the car while you push it to the garage. To do this, you have to transfer energy from your store of chemical energy (in your muscles) to the car. We say that you have to do work.

If the garage is a long way down the road, you are going to have to do more work than if it is nearby. You will also do more work the harder the car is to push. So the amount of work depends on

- the force you have to exert
- the distance moved in the direction of the force

The amount of work done by a force is defined by this equation;

work done by a force (joule, J) = force (newton, N) × distance moved in direction of force (metre, m)

Example

Calculate how much work is needed to push a car 50 m along a road.

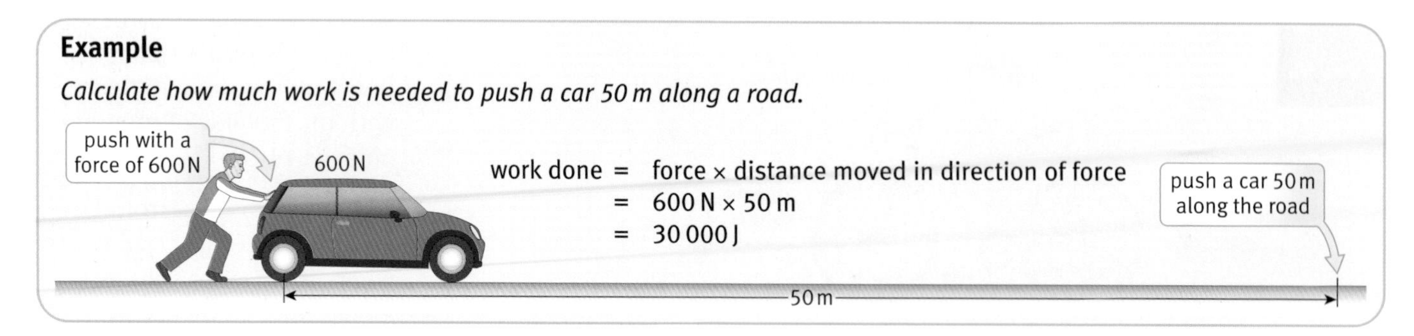

The amount of work that you do is equal to the amount of energy you transfer:

amount of work done = amount of energy transferred

Work, like energy, is measured in joules (J).

Lifting things: changing their gravitational potential energy

Lifting luggage into the boot of the car also involves doing work. You are transferring energy from your body's store of chemical energy. The **gravitational potential energy** of the luggage increases. The increase is equal to the amount of work you have done.

Suppose you have a suitcase that weighs 300 N. To lift it up, you have to exert an upward force of 300 N. If you lift it 1 metre into the boot of the car, then

$$\begin{aligned}\text{work done} &= \text{force} \times \text{distance moved in the direction of the force}\\ &= 300\,\text{N} \times 1\,\text{m}\\ &= 300\,\text{J}\end{aligned}$$

So the suitcase gains 300 J of gravitational potential energy. In general, when anything is lifted up, you can calculate its change in gravitational potential energy from the equation

$$\underset{\text{(joule, J)}}{\text{gravitational potential energy}} = \underset{\text{(newton, N)}}{\text{weight}} \times \underset{\text{(metre, m)}}{\text{vertical height difference}}$$

Notice that it is only the vertical height difference that matters. If you slide a suitcase up a ramp, the gain in gravitational potential energy is the same as if you lift it vertically. However, you would have to do more work because some energy is wasted on heating the ramp and suitcase, owing to the friction between them.

Doing work by lifting: increasing gravitational potential energy

Making things speed up: changing their kinetic energy

Imagine pushing a well-oiled supermarket trolley along a level floor. As you push, you are doing work. The trolley keeps speeding up. Its speed increases as long as you keep pushing. You are transferring energy from your body's store of chemical energy to the trolley, where it is stored as kinetic energy. If the trolley is absolutely smooth running, the amount of work you do pushing it is equal to the change in the trolley's kinetic energy.

Example

Calculate the change in kinetic energy of a trolley pushed with a force of 6 N over a distance of 5 m (assume there are no frictional forces acting).

$$\begin{aligned}\text{change in kinetic energy of trolley} &= \text{work done by pushing force}\\ &= \text{force} \times \text{distance}\\ &= 6\,\text{N} \times 5\,\text{m}\\ &= 30\,\text{J}\end{aligned}$$

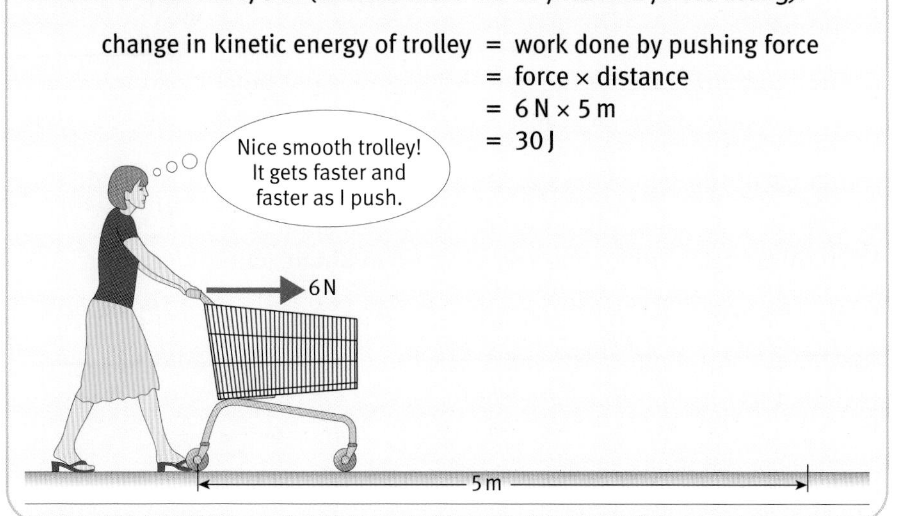

A real trolley will always have some friction, so its change in kinetic energy will be less than this. Some work is wasted in causing unwanted heating (and sound).

Questions

1. It takes a force of 1200 N to push a large car along the road. How much work would you have to do to push it 40 m?
2. The equation for work done by a force (page 114) is similar to the equation for the change of momentum caused by a force (page 108). But it has one important difference. What is it?
3. If your mass is 40 kg, then your weight is roughly 400 N. How much work do you have to do each time you go upstairs – a vertical height gain of 2.5 m?
4. (circled) A mother is pushing a child along in a buggy. She is doing work. So the amount of energy stored in her muscles is getting less. Where is this energy being transferred to? (Careful! The buggy is going at a steady speed.)

The equation for calculating the **kinetic energy** of a moving object is:

$$\underset{\text{(joule, J)}}{\text{kinetic energy}} = \frac{1}{2} \times \underset{\text{(kilogram, kg)}}{\text{mass}} \times \underset{\text{(metre per second, m/s)}^2}{(\text{velocity})^2}$$

Notice that the amount of kinetic energy depends on the velocity squared. So small changes in velocity mean quite big changes in kinetic energy.

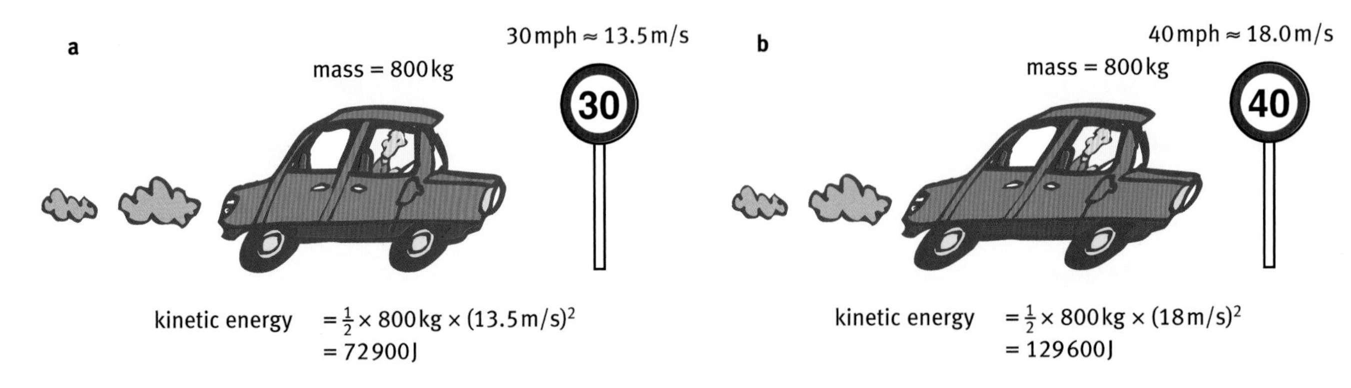

A car travelling at 40 mph (**b**) has nearly twice as much kinetic energy as the same car at 30 mph (**a**). This explains why there is a much greater risk of injury to pedestrians at at speeds greater than 30 mph.

The 'Oblivion' vertical roller coaster ride at Alton Towers has one section where the drop is vertical!

All the fun of the fair

When a roller coaster runs down a slope, it

- loses gravitational potential energy
- gains kinetic energy

If the slope has a complicated shape, it would be very difficult, maybe even impossible, to work out how fast the roller coaster is going at the bottom of the slope using the ideas of force and momentum. It is easier to use the principle of **conservation of energy**. If friction is small enough to ignore, then:

$$\text{amount of gravitational potential energy lost} = \text{amount of kinetic energy gained}$$

The shape of the slope does not matter at all. Only the vertical height difference is important in working out the change in gravitational potential energy – and hence the increase in kinetic energy.

And you do not need to know the direction in which the roller coaster is moving at the bottom of the slope. The final speed is the same, whatever the direction.

Key words
work
gravitational potential energy
kinetic energy
conservation of energy

To see how this works out in practice, look at the following example:

Example

Calculate the speed of the roller coaster shown below at the bottom of the ride (assuming there are no friction forces).

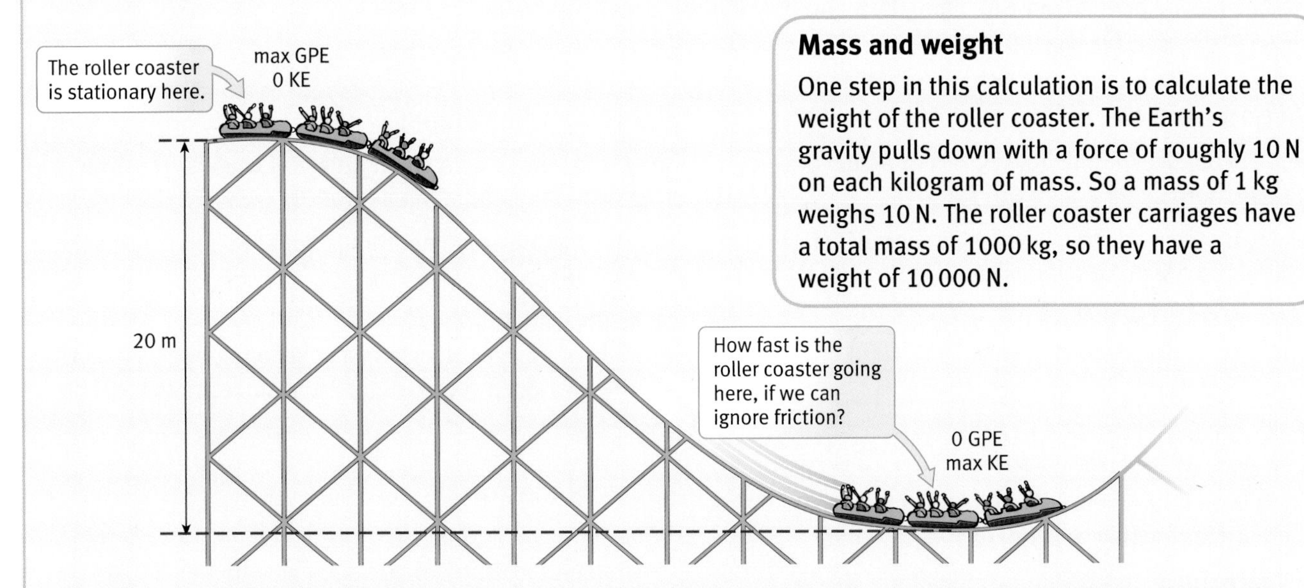

Mass and weight

One step in this calculation is to calculate the weight of the roller coaster. The Earth's gravity pulls down with a force of roughly 10 N on each kilogram of mass. So a mass of 1 kg weighs 10 N. The roller coaster carriages have a total mass of 1000 kg, so they have a weight of 10 000 N.

The simplest way to solve this problem is to use energy ideas. As there is no friction, the gravitational potential energy that the roller coaster loses as it goes down the slope is equal to the kinetic energy it gains:

1 loss of gravitational potential energy = weight × vertical height change
= 10 000 N × 20 m
= 200 000 J

2 loss of gravitational potential energy = gain in kinetic energy

3 So gain in kinetic energy = 200 000 J

4 But gain in kinetic energy = $\frac{1}{2}$ × mass × (velocity)^2

So $\frac{1}{2}$ × mass × (velocity)^2 = 200 000 J

Multiply both sides by 2: mass × (velocity)^2 = 400 000 J

or 1000 kg × (velocity)^2 = 400 000 J

Divide both sides by 1000 kg: (velocity)^2 = $\frac{400\,000 \text{ J}}{1000 \text{ kg}}$

= 400 (m/s)^2

Take the square root of both sides: velocity = 20 m/s

Questions

5 A ten-pin bowling ball has a mass of 4 kg. It is moving at 8 m/s. How much kinetic energy does it have?

6 Which of the following has more kinetic energy?

a a car of mass 500 kg travelling at 20 m/s

b a car of mass 1000 kg travelling at 10 m/s

⑦ Repeat the calculation above for a roller coaster that is only half as heavy (weight 5000 N). What do you notice about its speed at the bottom? How would you explain this?

P4 Explaining motion

Summary

This module has all been about explaining how and why things move as they do. A few laws of motion can account for all the kinds of motion you see around you.

Interactions and forces

- When two objects interact (by contact or action at a distance), both experience a force.
- These two forces are equal in size but opposite in direction. Each acts on a different object.
- Vehicles (and people) move by pushing back on something. This interaction causes a forward force to act on them.

Friction and normal reaction

- Friction is an interaction between two objects that are sliding (or tending to slide) past each other.
- The friction force matches the applied force that is making the objects slide – up to a limit.
- When an object sits on a surface, it distorts it slightly. The 'springiness' of the surface then causes a reaction force on the object – matching its downward push on the surface.

Describing motion

- The average speed of a moving object is

$$\frac{\text{distance}}{\text{time taken}}$$

- The instantaneous speed of a moving object is its speed at a particular instant. To estimate this, you measure its average speed over a very short distance (or time).
- Velocity means speed in a particular direction.
- Distance–time, speed–time, and velocity–time graphs are useful for summarizing and analysing the motion of an object.

Forces and motion

- When a force acts on an object, it causes a change in its momentum. Momentum is 'mass × velocity'.
- The change of momentum is equal to 'force × time for which it acts'.
- Many vehicle safety features work by making the time of an event (such as a collision) longer, so that the average force is less, for the same change of momentum.

Laws of motion

- If the resultant force acting on an object is zero, the momentum of the object does not change. If it is stationary, it does not move. If it is moving, it will keep moving at a constant speed in a straight line.
- If the resultant force acting on an object is not zero, this will cause a change in its momentum, in the direction of the resultant force.

Work and energy

- Work is done when a force makes an object move. The amount of work is 'force × distance'.
- When something does work, its energy decreases by that amount. If work is done on it, its energy increases by that amount.
- Doing work on an object can increase its gravitational potential energy (by lifting it up) or its kinetic energy (by making it move faster).
- Change in gravitational potential energy is 'weight × vertical height difference'. Change in kinetic energy is '½ × mass × velocity^2'
- When an object drops to a lower level, it loses gravitational potential energy. If friction can be ignored, it gains the same amount of kinetic energy.
  This can be used to work out its speed at the bottom. H

Questions

1 Think about the following situations:

   i Amjad on his skateboard, throwing a heavy ball to his friend (main objects to consider: Amjad + skateboard; the ball).

   ii A furniture remover trying to pull a piano across the floor – but it will not move (main objects to consider: the furniture remover; the piano; the floor).

   iii A hanging basket of flowers outside a café (main objects to consider: the basket; the chain it is hanging from).

   For each of them:

   a Sketch a diagram (looking at it from the side).

   b Then sketch separate diagrams of the main objects in the situation (these are listed for each).

   c On these separate diagrams, draw arrows to show the forces acting on that object. Use the length of the arrow to show how big each force is.

   d Write a label beside each arrow to show what the force is.

2 Imagine what it would be like to wake up one morning and discover that friction had disappeared. Write a short story about what it would be like to live in a friction-free world.

3 A tin of beans on a kitchen shelf is not falling, even though gravity is still acting on it. The shelf exerts an upward force, which balances the force of gravity. Explain in a short paragraph how it is possible for a shelf to exert a force. Draw a sketch diagram if it helps your explanation.

4 **a** The winner of a 50 m swimming event completes the distance in 80 s. What is his average speed?

   b How far could Leonie cycle in 10 minutes if her average speed is 8 m/s?

   c The average speed of a bus in city traffic is 5 m/s. How much time should the timetable allow for the bus to cover a 6 km route?

5 What is the momentum of:

   a a hockey ball of mass 0.4 kg moving at 5 m/s?

   b a jogger of mass 55 kg, running at 4 m/s?

   c a van of mass 10 000 kg, travelling at 15 m/s?

   d a car ferry of mass 20 000 000 kg, moving at 0.5 m/s?

6 Which of the following would cause the biggest change of momentum? And which would cause the smallest?

   a a force of 35 N acting for 4 s

   b a force of 3 N acting for 50 s

   c a force of 1500 N acting for 0.1 s

   d a force of 8 N acting for 20 s

   Explain how you worked this out.

7 Your uncle, who last studied science many years ago at school, thinks that you obviously need a force to keep something moving. As he says, 'If you stop pushing something, it stops moving'. Write him a note, explaining why he is wrong. Include plenty of examples, to convince him of your argument.

8 A weightlifter raises a bar of mass 50 kg until it is above his head – a total height gain of 2.2 m. How much gravitational potential energy has it gained? How much more work must he do to hold it there for 5 s?

9 A packet of mass 2 kg falls from the upstairs window of a flat, 45 m above the ground.

   a How much gravitational potential energy has the packet lost just before it hits the ground?

   b If we ignore the effect of air resistance, how much kinetic energy does the packet have just before it hits the ground?

   c With what speed does it hit the ground?

   d Would the speed be different for a similar packet of mass 5 kg?

   e If the packet had slid down a smooth chute instead of falling, how would its speed at the bottom compare?

Why study electric circuits?

Imagine life without electricity – rooms lit by candles or oil lamps, no electric cookers or kettles, no radio, television, computers, or mobile phones, no cars or aeroplanes. Electricity has transformed our lives, but you need to know enough to use it safely. More fundamentally, electric charge is one of the basic properties of matter – so anyone who wants to understand the natural world around them needs to have some understanding of electricity.

The science

The particles of which atoms are made carry an electric charge. An electric current is a flow of charges. A useful model of an electric circuit is to imagine the wires full of charges, being made to move around together by the battery. The size of current depends on the battery voltage and the resistance of the circuit. A voltage can also be produced by moving a magnet near a coil. This is used to generate electricity on a large scale.

Physics in action

The scientific understanding of electricity was developed over quite a short period, from about 1800 to 1840. Nowadays scientists use electricity, or instruments that depend on electricity, in almost every aspect of their work. One important focus of research in the 21st century is on the development of new ways of generating electricity, using renewable energy sources such as sunlight, wind, and waves.

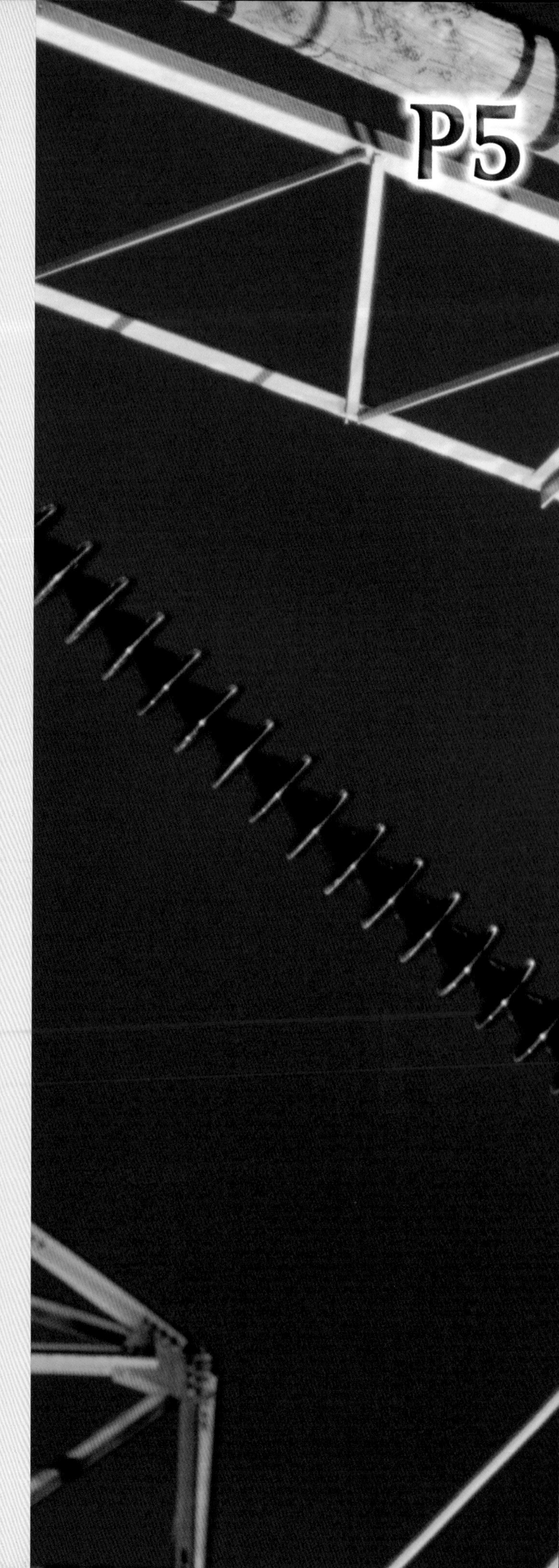

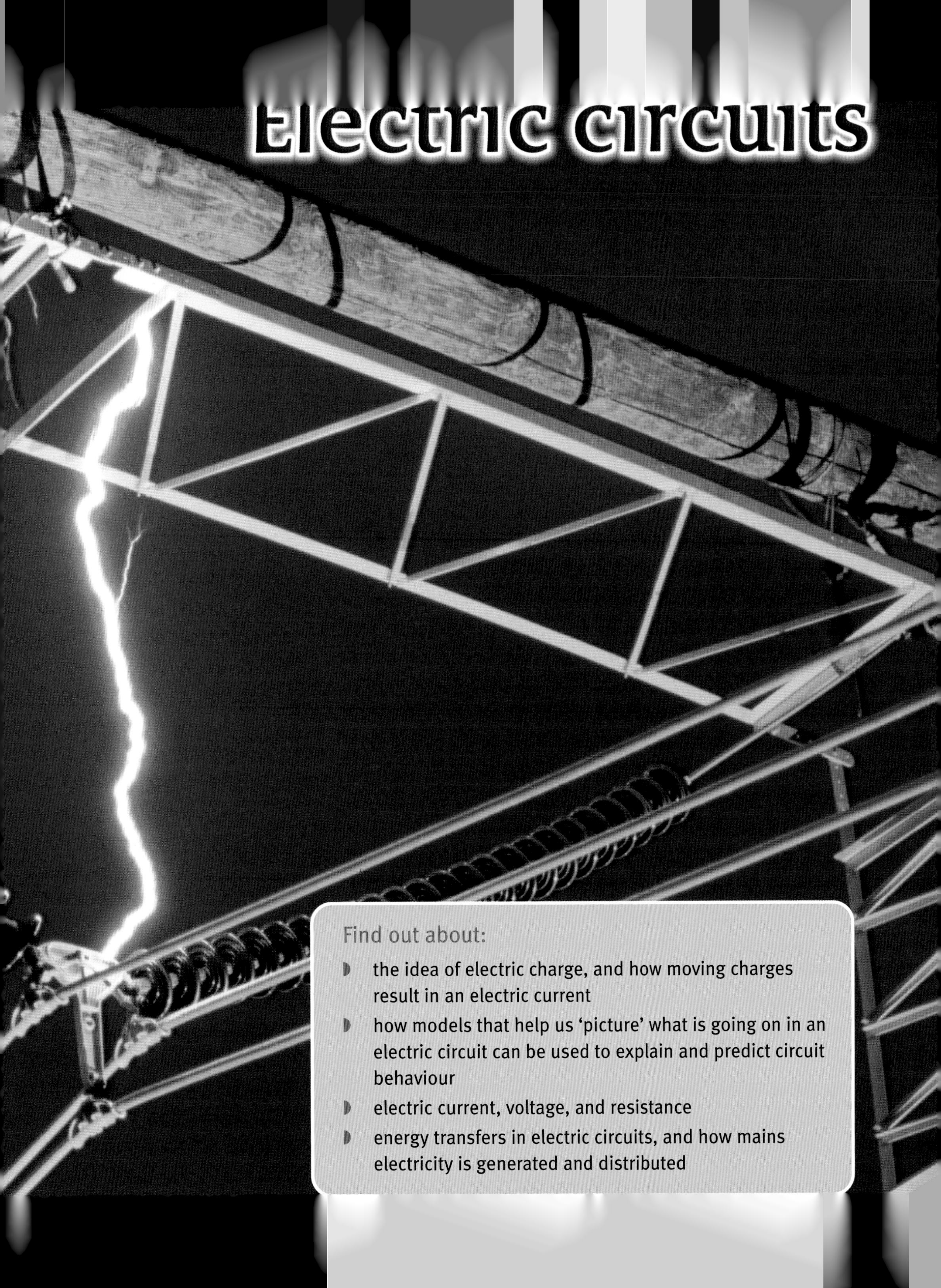

Electric circuits

Find out about:

- the idea of electric charge, and how moving charges result in an electric current
- how models that help us 'picture' what is going on in an electric circuit can be used to explain and predict circuit behaviour
- electric current, voltage, and resistance
- energy transfers in electric circuits, and how mains electricity is generated and distributed

Find out about:
- electric charge and how it can be moved when two objects rub together
- the effects of like and unlike charges on each other

A Static electricity

When you get out of a car, you sometimes get a small electric shock when you touch the metal door – and you might hear a little 'crack' as a spark jumps between your hand and the car door. You sometimes hear the same sort of little crackles when you take off a jumper. In a dark room, you may be able to see the sparks. They are caused by **static electricity**. Electricity is part of our everyday world.

Small sparks caused by static electricity are harmless, but on a larger scale they can be much more dangerous. Lightning is an electrical spark between a thundercloud and the ground.

Charging by rubbing

Electrical effects can be produced by rubbing two materials together. If you rub a balloon against your jumper, the balloon will stick to a wall. If you rub a plastic comb on your sleeve, the comb will pick up small pieces of tissue paper – they are attracted to the comb. In both cases, the effect wears off after a short time.

When you rub a piece of plastic, it is somehow changed: it can then affect objects nearby. The more it is rubbed, the stronger the effect. It seems that something is being stored on the plastic. If a lot is stored, it may escape by jumping to a nearby object, in the form of a spark. We say that the plastic has been charged.

Charging by rubbing also explains the examples discussed at the top of the page. When you get out of a car, you slide across the seat, rubbing your clothes against it. As you pull off a jumper, it rubs against the shirt or blouse you are wearing beneath it.

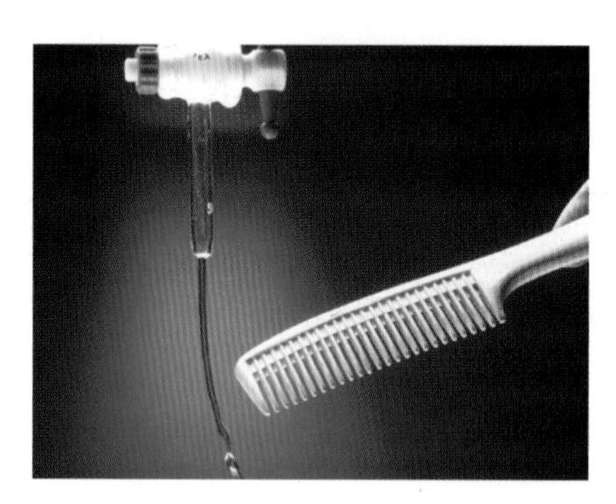

A charged comb attracts a stream of water.

Two types of charge

If you rub two identical plastic rods and then hold them close together, the rods push each other apart – they **repel**. The forces they exert on each other are very small, so you can only see the effect if one of the rods can move easily.

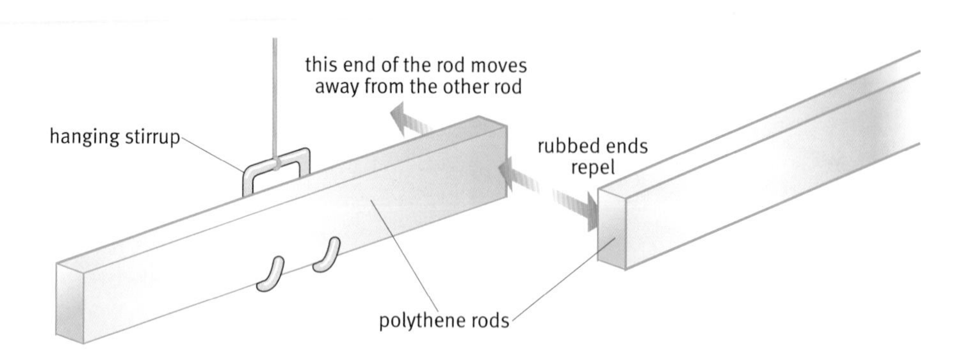

Two rubbed polythene rods repel each other. The hanging rod moves away from the other one.

Why is it called 'charge'?

In the late 1700s, experimenters working on elecricity thought that the effect you produced when you rubbed something was a bit like preparing a gun. In those days, you had to prime a gun by pushing down an explosive mixture into the barrel. This was called the 'charge'. When you had done this, the gun was 'charged'. It could be 'discharged' by firing.

If you try this with two rods of different plastics, however, you can find some pairs that **attract** each other. Scientists' explanation of this is that there are two types of **electric charge**. If two rods have the same type of charge, they repel each other. But if they have charges of different types, they attract. The early electrical experimenters called the two types of charge **positive** and **negative**. These names are just labels. They could have called them red and blue, or A and B.

Where does charge come from?

Scientists believe that charge is not *made* but is *moved around* when two things are rubbed together. If you rub a plastic rod with a cloth, both the rod and the cloth become charged. (To see this, you need to wear a polythene glove on the hand holding the cloth, otherwise the charge will escape through your body.) Each object gets a different charge: if the rod has a positive charge, the cloth has a negative one. Rubbing does not make charge. It separates charges that were there all along.

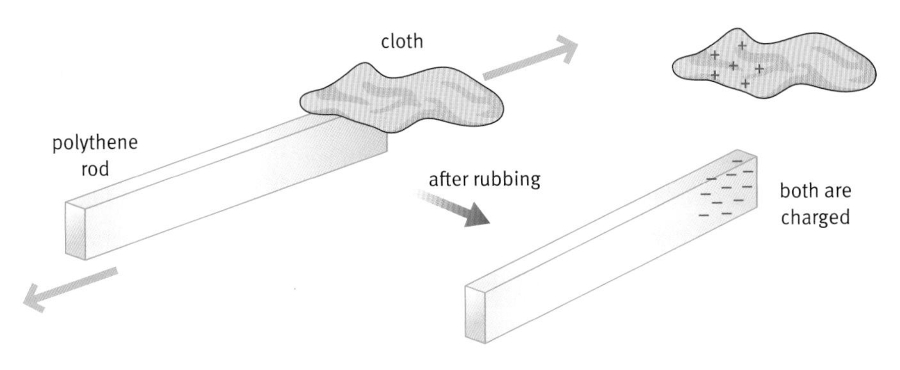

After it has been rubbed, the rod has a negative charge and the cloth has a positive charge. A possible explanation is that some electrons have been transferred from the cloth to the rod.

Attracting light objects

An object with a positive charge attracts another object with a negative charge. But why does a charged rod also attract light objects, such as little pieces of paper? The reason is that there are charges in the paper itself, all the time. Normally these are mixed up together, with equal amounts of each. So a piece of paper is uncharged. If a negatively charged rod comes near, it repels negative charges in the paper to the end farthest away. This leaves a surplus of positive charges at the near end. The attraction between these positive charges and the rod is stronger than the repulsion between the negative charges (at the far end) and the rod. So the little piece of paper is attracted to the rod.

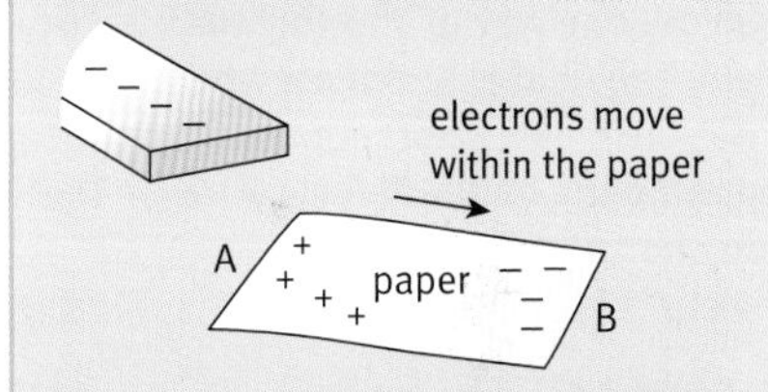

A charged rod separates the charges in a piece of paper nearby – and the paper is attracted to the rod.

Questions

1 Imagine that you have two plastic rods which you know get a positive and a negative charge when you rub them with a cloth. Now you are given a third plastic rod. Explain how you could test whether it gets a positive or a negative charge when rubbed.

2 Some picture frames are made with plastic rather than glass. If you clean the plastic with a duster, it may get dusty again very quickly. Use the ideas on these pages to explain why this happens.

What is electric charge?

Charge is a basic property of matter, which cannot be explained in terms of anything simpler. All matter is made of atoms, which in turn are made out of protons (positive charge), neutrons (no charge), and **electrons** (negative charge). In most materials there are equal numbers of positive and negative charges, so the whole thing is neutral. When you charge something, you move some electrons to it or from it.

Chemists think of the atom as a tiny nucleus, with a positive charge, surrounded by a cloud of electrons, which have negative charge. As the electrons are on the outside, they can be 'rubbed off', on to another object.

Although they cannot explain charge, scientists have developed useful ideas for predicting its effects. An example is the idea of an **electric field**. Around every charge there is an electric field. In this region of space, the effects of the charge can be felt. Another charge entering the field will experience a force.

electric field lines

The lines around this positively charged ball are one way of showing the electric field. The field is strongest where the lines are close. Another positive charge entering the field will feel a force in the direction of the arrows.

Moving charge = current

When a van de Graaff generator is running, charge collects on its dome. As the charge builds up, the electric field around the dome gets stronger. The charge may 'jump' to another nearby object, in the form of a spark. If you hold a mains-testing screwdriver close to the dome and touch its metal end-cap, the indicator lamp inside the screwdriver lights up. So there is an **electric current** through the lamp, making it light. Charge on the dome is escaping across the air gap, through the indicator lamp, and through you to the Earth. This (and other similar observations) suggests that an electric current is a flow of charge.

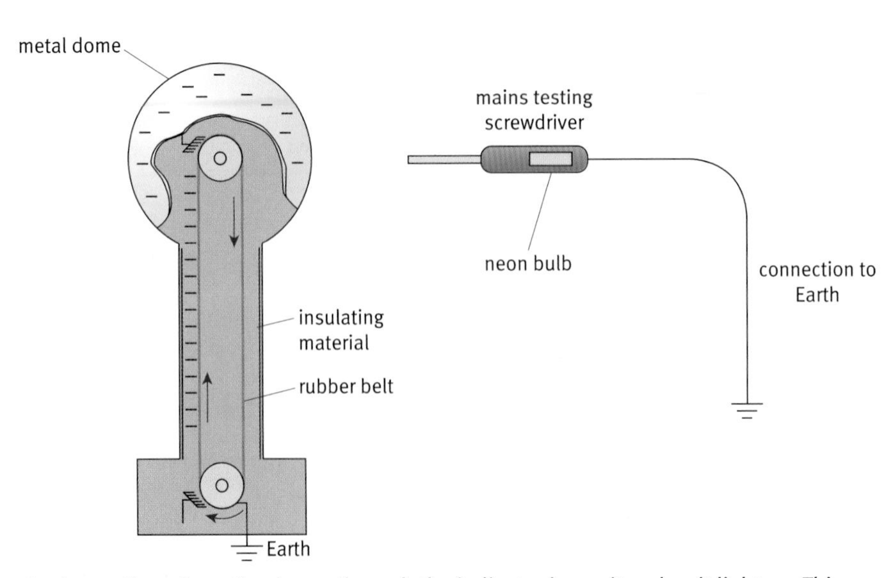

As charge flows from the dome, through the indicator lamp, it makes it light up. This suggests that an electric current is a flow of charge.

Key words

static electricity
electric charge
repel, attract
positive, negative
electron
electric field
electric current

B Simple circuits

Find out about:

- how simple electric circuits work
- models that help explain and predict the behaviour of electric circuits

A closed loop

The diagram below shows a simple **electric circuit**. If you make a circuit like this, you can quickly show that:

- If you make a break *anywhere* in the circuit, *everything* stops.

When you open the switch in this electric circuit, both bulbs go off.

When you unscrew one bulb, you make a break in the circuit and all the lights go out.

This suggests that something has to go all the way round an electric circuit to make it work. This 'something' is electric charges. If it was enough for the charges simply to go from the battery to the lamp, then one of the lamps would be lit, even with the switch open. But this does not happen. There has to be a complete loop – from one terminal of the battery, through the lamps and switch, and back to the other battery terminal.

You will also notice that:

- Both lamps come on *immediately* when the circuit is completed. And they go off *immediately* if you make a break in the circuit.

Perhaps this is because the circuit is small. So imagine making a 'big circuit' with much longer wires than usual. When you turn on the switch, it is still impossible to see any delay before the lamp lights. So the size of the circuit makes no difference. Remember that an electric current is moving charges. So perhaps charges move very quickly through the wires, from the battery to the lamp, as soon as the circuit is switched on. However, there is a better explanation. Imagine that there are charges in all the components of the circuit (wires, lamp filaments, batteries) *all the time*. Closing the switch just allows these charges to move. They all move together, so the effect is immediate, even though the charges themselves do not move very fast.

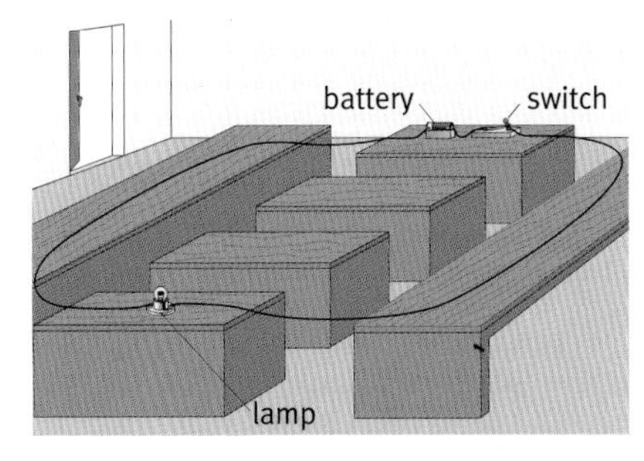

A 'big circuit' with longer wires than usual. When we switch on, the bulb lights immediately. There is no delay.

Some ways of thinking about charges in an electric circuit

Peas in a pipe

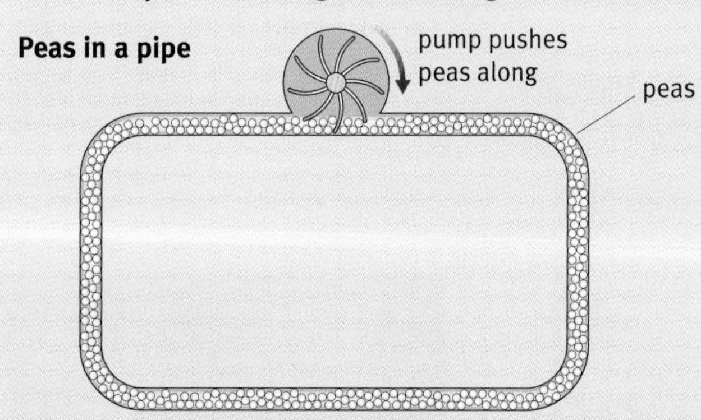

Imagine a plastic pipe full of dried peas, formed into a closed loop, with a pump that can push the peas along. Because the peas are close together, when one moves, they all move. The effect is immediate. If a barrier is inserted anywhere, it will stop them moving everywhere – instantly.

Chain on a bike

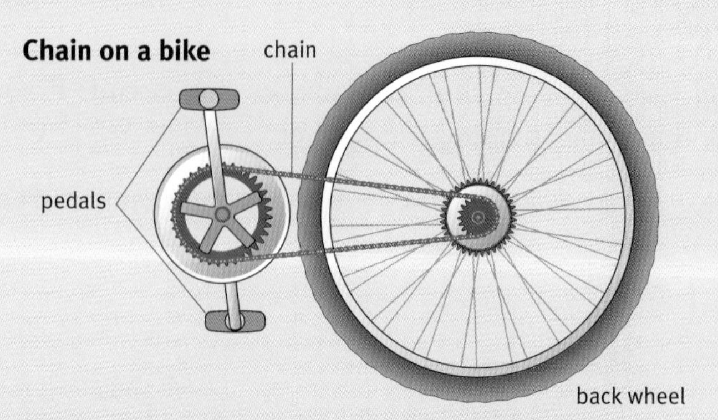

When you turn the pedals of a bike, this makes the chain move. When one part of the chain moves, it all moves. So it immediately makes the back wheel turn. There is no delay. The effect of your pedalling is transmitted immediately to the back wheel.

Moving rope role-play

Here is another way to think about an electric circuit. Imagine a group of about six to eight people standing in a circle, about 1 metre apart. Each person holds out both hands, palms upwards and fingers slightly bent. A continuous loop of rope passes from person to person, right round the circle.

One person now acts as the 'battery'. She makes the rope slowly move round, passing through everyone's hands as it goes. Notice that the effect is instantaneous. As soon as the rope moves anywhere, it moves everywhere. If everyone keeps this going for a short time, they will notice two things happening. First, the 'battery' begins to tire – because she is transferring stored energy from her muscles to make the rope move. Second, the hands of the people around the circle begin to get slightly hot – because of the friction force between their hands and the rope. The battery is doing work against this friction force, and this is causing some heating.

This is a very similar to an electric circuit. The battery makes current move round the circuit – and this does work on all the other components round the circuit.

Questions

1. Imagine making the circuit at the top of page 125, but with the switch between the two batteries instead of between the two lamps. Explain what would now happen when the switch is opened and closed.
2. Look at the 'peas in a pipe' and 'chain on a bike' diagrams. For each of these, say what corresponds to:
   - **a** the battery in the circuit
   - **b** the electric current
3. In the electric circuit role-play, how could you illustrate the effect of putting a second battery into a circuit?
   - **a** pointing in the same direction as the first;
   - **b** pointing in the opposite direction to the first?

   Does the role-play situation correctly predict what happens in a real circuit?

An electric circuit model

A model is a way of thinking about how something works, under the surface. The diagram on the right summarizes the scientific model of an electric curcuit. The key ideas are:

- Charges are present throughout the circuit all the time.
- When the circuit is a closed loop, the battery makes the charges move.
- All of the charges move round together.

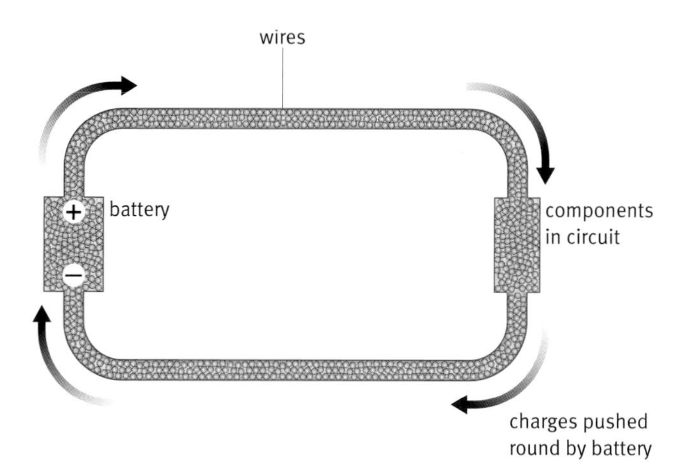

You can think of an electric circuit as a flow of charges, which are present in all materials (and free to move in conductors), moving round a closed conducting loop, pushed by the battery.

The battery makes the charges move in the following way. Chemical reactions inside the battery have the effect of separating electric charges, so that positive charge collects on one terminal of the battery and negative charge on the other. If the battery is connected into a circuit, the charges on the battery terminals set up an electric field in the wires of the circuit. This makes free charges in the wire drift slowly along. However, even though the charges move slowly, they all begin to move at once, as soon as the battery is connected. So the effect of their motion is immediate. Notice too that the flow of charge is continuous, all round the circuit. Charge also flows through the battery itself.

Key words
electric circuit

Conventional current, electron flow

In the model above, the charges in the circuit are shown moving away from the positive terminal of the battery, through the wires and other components, and back to the negative terminal of the battery. This assumes that the moving charges are positive. In fact there is no simple way of telling whether the moving charges are positive or negative, or which way they are moving. Several decades after this model was first proposed and had become generally accepted, scientists came to realize that the moving charges in metals were electrons, which have negative charge. In all metals, the atoms have some electrons that are only loosely attached to their 'parent' atom and are relatively free to wander through the metal. It is these that move, in the weak electric field that the battery sets up in the wire.

To explain and predict how electric circuits behave, it makes no difference whether you think in terms of a flow of electrons in one direction or positive charges in the other. Although scientists now believe it is electrons that flow in metals, in this course use the model of conventional current going the other way, as most physicists and engineers do.

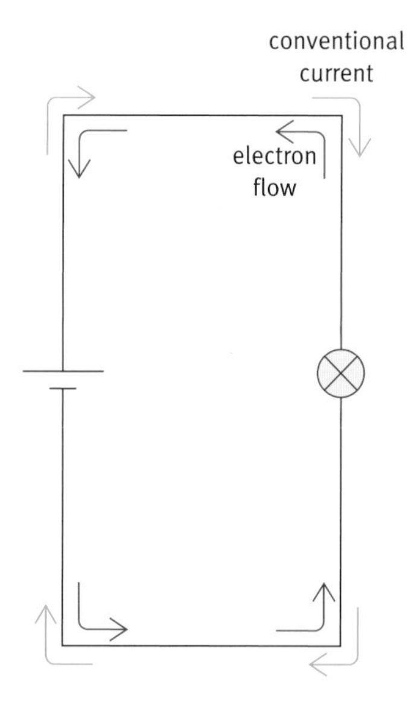

An electric current is a flow of electrons through the wires of the circuit. You can think of it equally well as a 'conventional current' of positive charges going the other way.

Find out about

- how to measure electric current
- the size of the electric current at different places in a circuit

Typical currents

Currents smaller than 1 amp are often measured in milliamps (mA) or microamps (μA).

1 A = 1000 mA = 1 000 000 μA

Typical size of the current in some common applications:

in an electric kettle: 8 A

in a torch bulb: 0.3 A or 300 mA

in a radio: 0.1 A or 100 mA

in a calculator: 0.005 A or 5 mA

in a digital watch: 0.000 05 A or 50 μA

C Electric current

An electric current is a flow of charge. You cannot see a current, but you can observe its effects. The current through a torch bulb makes the fine wire of the filament heat up and glow. The bigger the current through a bulb, the brighter it glows (unless, the current gets too big and the bulb 'blows'). So the brightness of a bulb shows the size of the current through it.

A better way to measure the size of an electric current is to use an **ammeter**. The reading (in amperes, or amps (A) for short) indicates the amount of charge going through the ammeter every second.

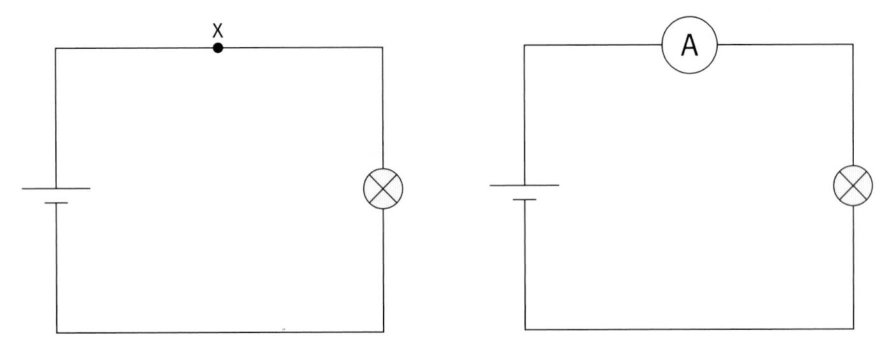

To measure the current at point X, you have to make a gap in the circuit at X and insert the ammeter in the gap, so that the current flows through it.

Current around a circuit

If you use an ammeter to measure the size of the electric current at different points around a circuit, you get a very important result.

- The current is the same everywhere in a simple (single-loop) electric circuit.

This may seem surprising. Surely the bulbs must use up current to light. But this is not the case. Current is the movement of charges in the wire, all moving round together like dried peas in a tube, or a moving belt or chain. So the current at every point round the circuit must be the same.

Of course, *something* is being used up. It is the energy stored in the battery. This is getting less all the time. The battery is doing work to push the current through the filaments of the light bulbs, and this heats them up. The light then carries energy away from the glowing filament. So the circuit is transferring energy from the battery, to the bulb filaments, and then on to the surroundings (as light). The current enables this energy transfer to happen. But the current itself is not used up.

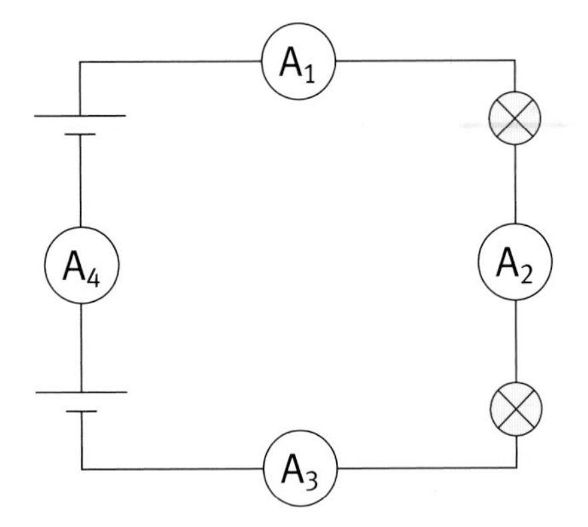

The current is the same size at all these points – even between the batteries. Current is not used up to make the bulbs light.

Branching circuits

Often, you want to run more than one thing from the same battery. One way to do this is to put them all in a single loop, one after the other. Components connected like this are said to be **in series**. The moving charges then have to pass through each of them in turn.

Another way is to connect components **in parallel**. In the circuit on the right, the two bulbs are connected in parallel. This has the advantage that each bulb now works independently of the other. If one burns out, the other will stay lit. This makes it easy to spot a broken one and replace it.

In this portable MP3 player, the battery has to run the motor that turns the hard drive, the head that reads the disk, and the circuits that decode and amplify the signals.

Currents in the branches

Look at the circuit below, which has a motor and a buzzer connected in parallel. A student, Nicola, was asked to measure the current at points a, b, c, and d. Her results are shown in the table below on the right.

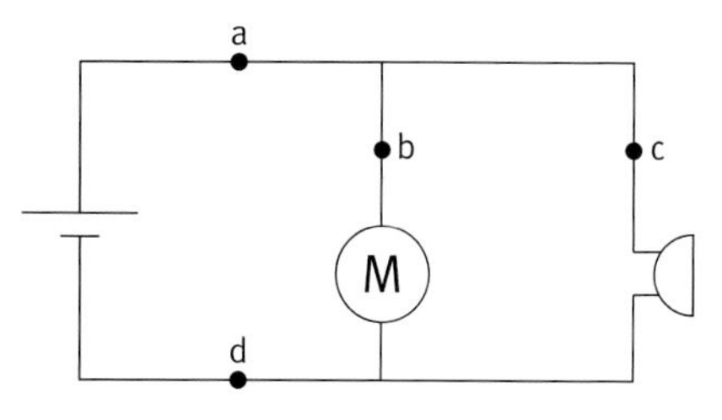

Point in circuit	Current (mA)
a	230
b	150
c	80
d	230

Measuring currents in a circuit with two parallel branches.

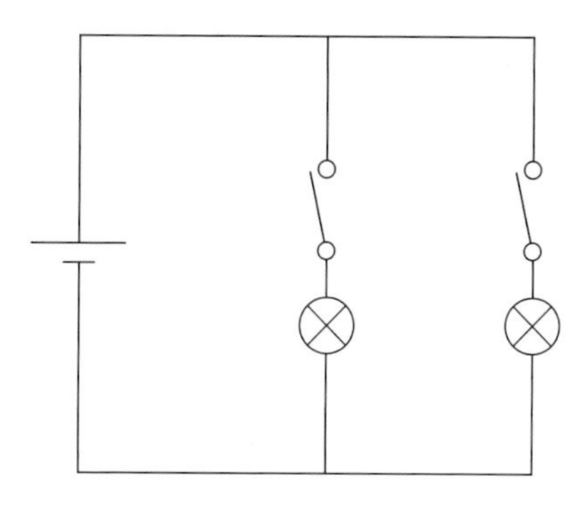

An advantage of connecting components in parallel is that each can be switched on and off independently.

Nicola noticed that the current is the same size at points a and d: 230 mA. When she added the currents at b and c, the result was also 230 mA. This makes good sense if you think about the model of charges moving round. At the junctions, the current splits, with some charges flowing through one branch and the rest flowing through the other branch. Current is the amount of charge passing a point every second. So the amounts in the two branches must add up to equal the total amount in the single wire before or after the branching point.

Key words

ammeter
in series
in parallel

Questions

1 Look at the circuit at the top of page 128. Draw a circuit diagram showing how you would connect an ammeter to measure the current in the wire between the bulb and the negative terminal of the battery.

2 When we want to run several things from the same battery, it is much more common to connect them in parallel than in series. Write down three advantages of parallel connections.

3 Look at the circuit above on the right. If you wanted to switch both bulbs on and off together, where would you put the switch? Draw two diagrams showing two possible positions of the switch that would do this.

4 In the circuit above with the motor and buzzer, what size is the electric current:
a in the wire just below the motor?
b in the wire just below the buzzer?
c through the battery itself?

Find out about:

- how the battery voltage and the circuit resistance together control the size of the current
- what causes resistance
- the links between battery voltage, resistance, and current

All the batteries on the front row are marked 1.5 V – but are very different sizes. The three at the back are marked 4.5 V, 6 V, and 9 V

D Controlling the current

The scientific model of an electric circuit imagines a flow of charge round a closed conducting loop, pushed by a battery. The next step is to ask how the size of the current can be controlled. As the current in a circuit is caused by the battery, this is a good place to begin.

Battery voltage

Batteries come in different shapes and sizes. They usually have a **voltage** measured in volts (V), marked clearly on them, for example, 1.5 V, 4.5 V, 9 V. To understand what voltage means and what this number tells you, look at the following diagrams, which show the same bulb connected first to a 4.5 V battery and then to a 1.5 V battery.

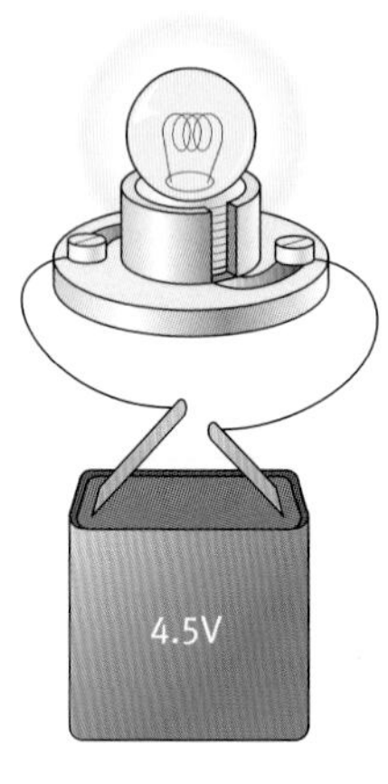

With a 4.5 V battery, this bulb is brightly lit.

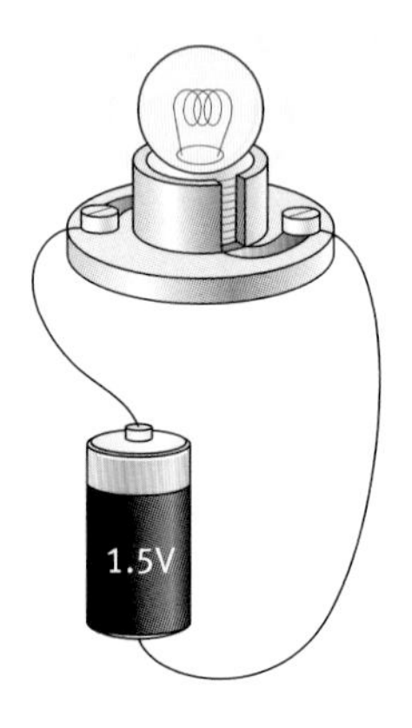

With a 1.5 V battery, the same bulb is lit, but very dimly.

The bigger the current through a light bulb, the brighter it will be (up to the point where it 'blows'). So the current through the bulb above is bigger with the 4.5 V battery. You can think of the voltage of a battery as a measure of the 'push' it exerts on the charges in the circuit, or the amount of work it does pushing charges round the circuit. The battery sets up an electric field in the wires of the circuit, and this makes the free charges move round. The bigger the voltage, the bigger the 'push' – and the bigger the current as a result.

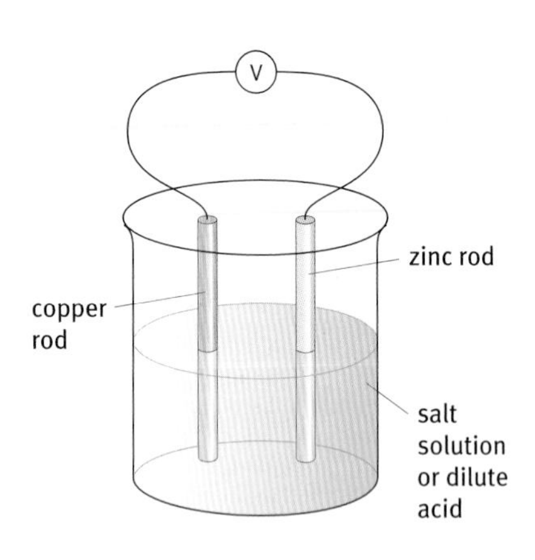

A simple battery. The voltage depends on the metals and the solution you choose.

The battery voltage depends on the choice of chemicals inside it. To make a simple battery, all you need are two pieces of different metal and a beaker of salt solution or acid. The voltage quickly drops, however. The chemicals used in real batteries are chosen to provide a steady voltage for several hours of use.

Resistance

The size of the current in a circuit depends on the battery voltage, but this is not the only thing that matters. The components in the circuit provide a **resistance** to the flow of charge. The battery 'pushes' against this resistance. You can see the effect of this if you compare two circuits with different resistors. Resistors are components designed to control the flow of charge.

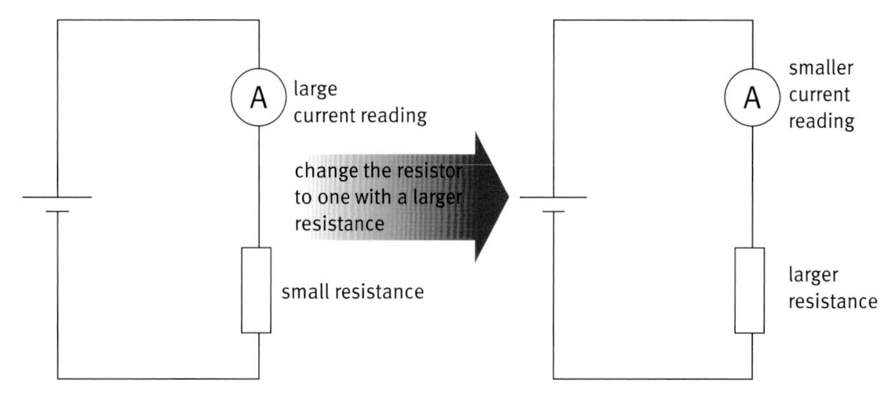

The size of the current is smaller if the resistance is larger

Changing resistance changes the size of the current. The bigger the resistance, the smaller the current.

What causes resistance?

Everything has resistance, not just special components called resistors. The resistance of connecting wires is very small, but not zero. Other kinds of metal wire have larger resistance. The filament of a light bulb has a lot of resistance. This which is why it gets so hot when there is a current through it. A heating element, like that in an electric kettle, is just a resistor.

All metals get hot when charge flows through them. In metals, the moving charges are free electrons. As they move round, they collide with the fixed array, or lattice, of atoms in the wire. These collisions make the atoms vibrate a little more, so the temperature of the wire rises. In some metals, the fixed atoms provide only small targets for the electrons, which can get past them relatively easily. In other metals, the fixed atoms present a much bigger obstacle – and so the resistance is bigger.

flow of electrons
wire
atom centres (in a fixed lattice)
moving electrons

Why the temperature of a wire rises when a current flows through it

Key relationships in an electric circuit: a summary

The size of the electric current (I) in a circuit depends on the battery voltage (V) and the resistance (R) of the circuit.

- If you make V bigger, the current (I) increases.
- If you make R bigger, the current (I) decreases.

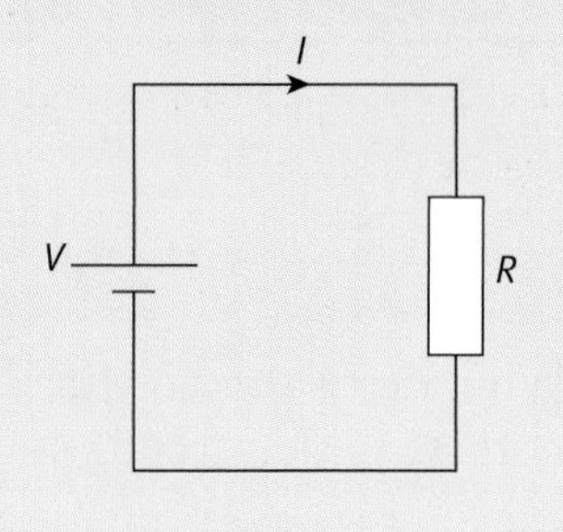

Questions

1 Look back at the moving rope role-play described on page 126. How would you role-play an increase in resistance? What effect would this have? (You should be able to think of at least two.) Explain how the role-play helps to predict the behaviour of a real circuit.

2 Suggest two different ways in which you could change a simple electric circuit to make the electric current bigger.

Ohm's law

Ohm's law says that the current through a conductor is proportional to the voltage across it – provided its temperature is constant.
It applies only to some types of conductor (such as metals).
An electric current itself causes heating, which complicates matters. For example, the current through a light bulb is not proportional to the battery voltage. The *I*–*V* graph is curved. The reason is that the current through the bulb filament heats it up – and its resistance increases with temperature.

Measuring resistance

You can explore the relationship between battery voltage and current in more detail. Keiko did this by measuring the current through a coil of wire with different batteries.

1. Keiko connected a coil of resistance wire to a 1.5 V battery and an ammeter. She noted the current.

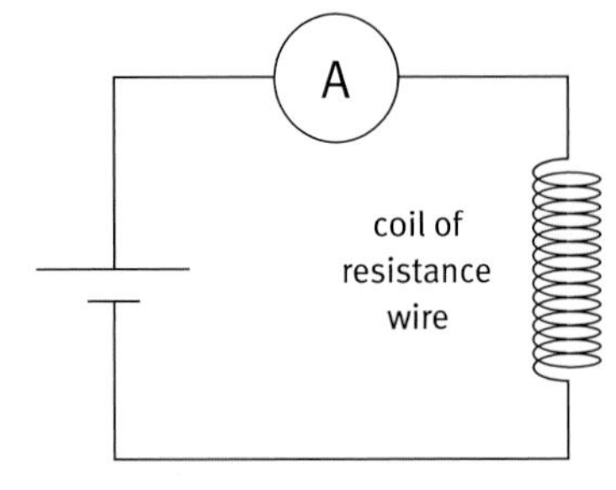

2. She then added a second battery in series. She noted the current again.

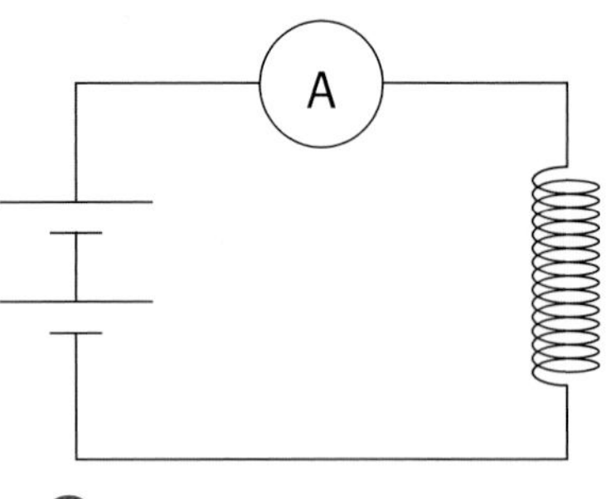

3. Keiko repeated this with 3, 4, 5, and 6 batteries, to get a set of results:

Number of 1.5 V batteries	Battery voltage (V)	Current (mA)
1	1.5	75
2	3.0	150
3	4.5	225
4	6.0	300
5	7.5	375
6	9.0	450

4. Finally, she drew a graph of current against battery voltage:

The straight-line graph means that the current in the circuit is proportional to the battery voltage. This result is known as **Ohm's law**. The number you get if you divide voltage by current is the same every time. The bigger the number, the larger the resistance. This is how to measure resistance:

$$\text{resistance of a conductor} = \frac{\text{voltage across the conductor}}{\text{current through the conductor}}$$

$$R = \frac{V}{I}$$

The units of resistance are called ohms (Ω).

Rearranging this equation gives $I = V/R$. You can use this to calculate the current in a circuit, if you know the battery voltage and the resistance of the circuit.

Questions

3 In Keiko's investigation:

a How many 1.5 V batteries would she need to use to make a current of 600 mA flow through her coil?

b What is the resistance of her coil, in ohms?

4 In the circuit below, a 9 V battery is connected to a 45 Ω resistor.

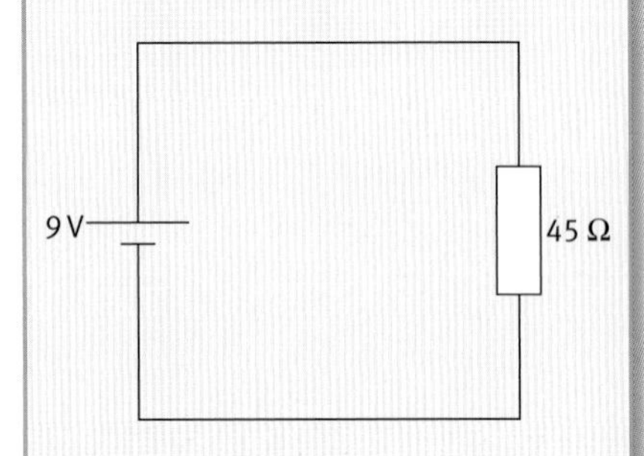

What size is the electric current in the circuit?

Variable resistors

Resistors are used in electric circuits is to control the size of the current. Sometimes we want to be able to vary the current easily, for example to change the volume on a radio or CD player. A variable resistor is used. This is a resistance whose size can be steadily changed by turning a dial or moving a slider.

The circuit diagram on the right shows the symbol for a variable resistor. As you alter its resistance, the brightness of the light bulb changes, and the readings on *both* ammeters increase and decrease together. The variable resistor controls the size of the current everywhere round the circuit loop.

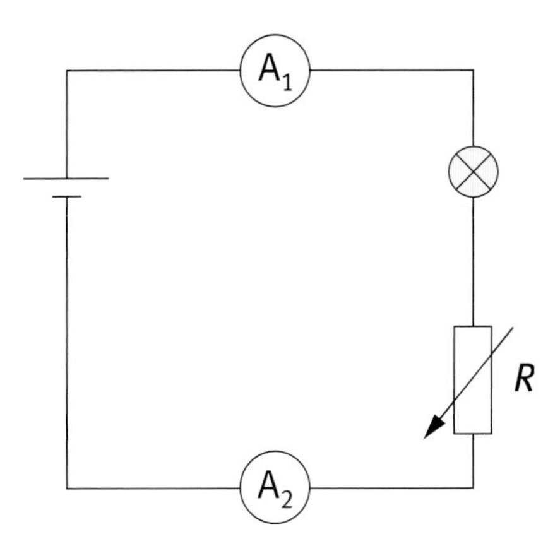

Using a variable resistor to control the current in a series circuit

Each of these sliders adjusts the value of a variable resistor.

Key words
voltage
resistance
Ohm's law

Some useful sensing devices are really variable resistors. For example, a light-dependent resistor (LDR) is a semiconductor device whose resistance is larger in the dark but gets smaller as the light falling on it gets brighter. An LDR can be used to measure the brightness of light or to switch another device on and off when the brightness of the light changes. For example, it could be used to switch an outdoor light on in the evening and off again in the morning.

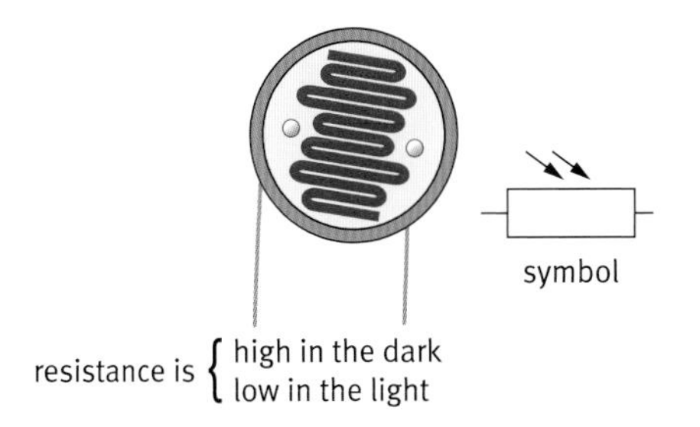

A light-dependent resistor (LDR)

A thermistor is another device made from semiconductor material. Its resistance changes rapidly with temperature. The commonest type has a lower resistance when it is hotter. Thermistors can be used to make thermometers (to measure temperature) or to switch another device on or off as temperature changes. For example, a thermistor could be used to switch an immersion heater on when the temperature of water in a tank falls below a certain value and off again when the water is back at the required temperature.

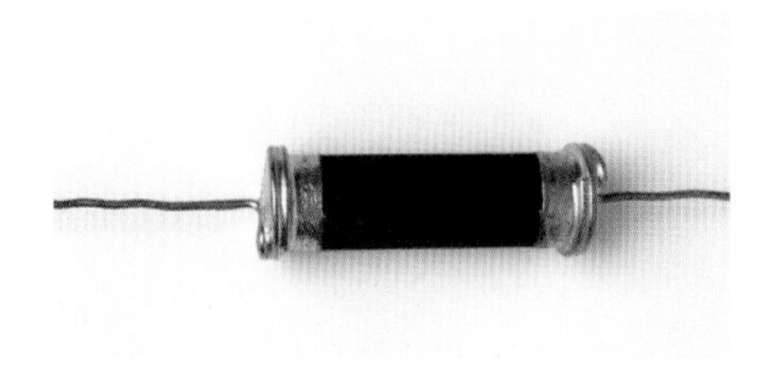
One type of thermistor

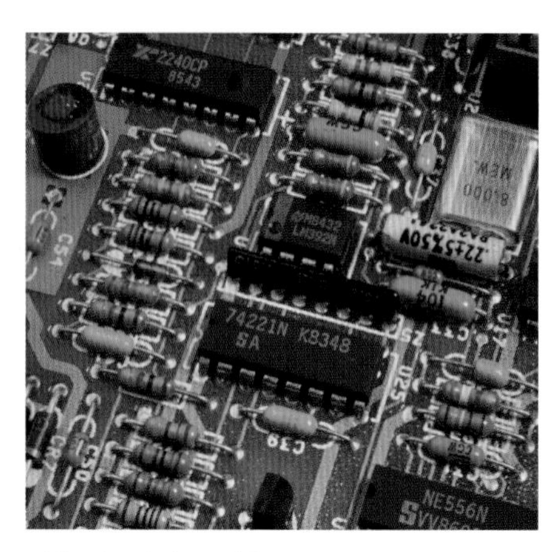

This circuit board from a computer contains a complex circuit, with many components. The small cylindrical ones are resistors.

Combinations of resistors

Most electric circuits are more complicated than the ones discussed so far in this chapter. Circuits, usually contain many components, connected in different ways. There are just two basic ways of connecting circuit components: in series or in parallel.

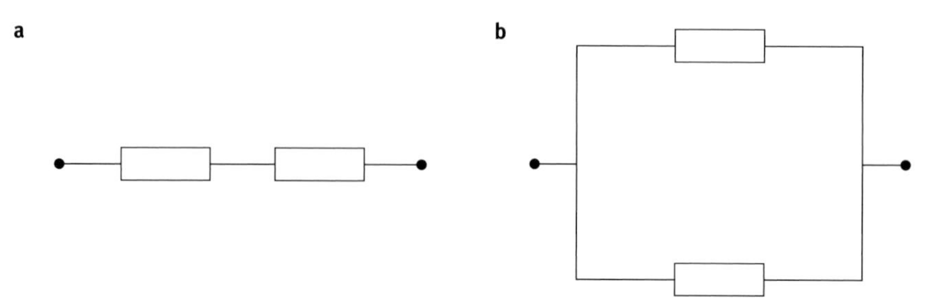

Two resistors connected **a** in series, **b** in parallel

Two resistors in series have a larger resistance than one on its own. The battery has to push the current through both of them. But connecting two resistors in parallel makes a smaller total resistance. There are now two paths that the moving charges can follow. Adding a second resistor in parallel does not affect the original path but adds a second equivalent one. It is now easier for the battery to push charges round, so the resistance is less.

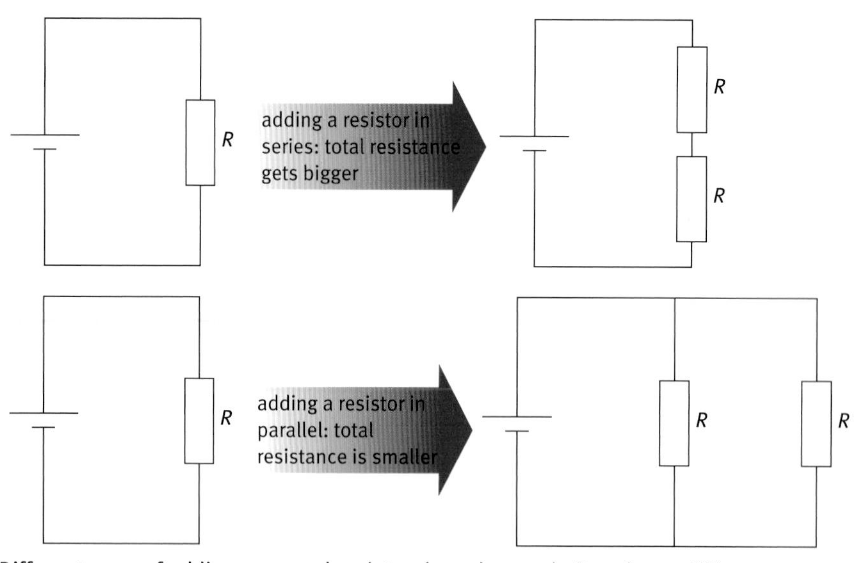

Different ways of adding a second resistor: how do you do it makes a difference

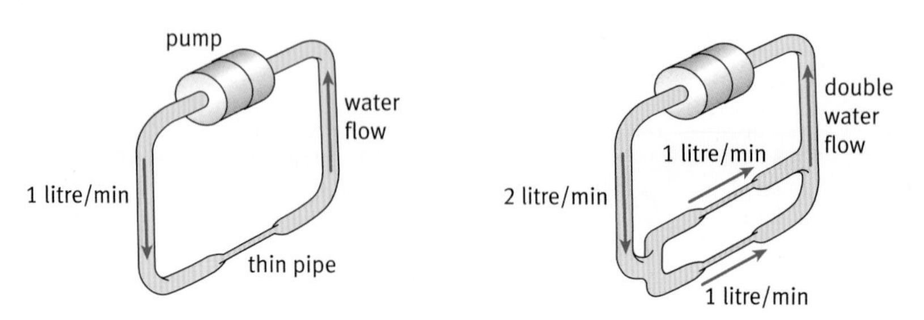

A water flow model shows how the total resistance gets less when a second parallel path is added.

Questions

5 All the resistors in the three diagrams below are identical. Put the groups of resistors in order, from the one with the largest total resistance to the one with the smallest total resistance.

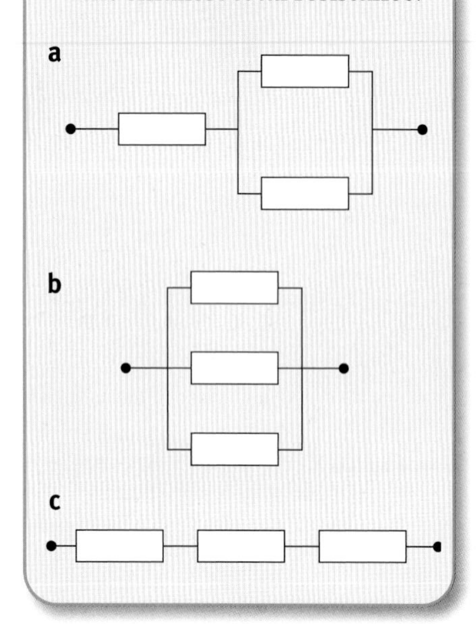

E Potential difference

Find out about:

- how voltmeters measure the potential difference between two points in a circuit
- how height provides a useful model for thinking about electrical potential
- how current splits between parallel branches

You can think of the voltage of a battery as a measure of the 'push' it exerts on the charges in a circuit. But a voltmeter also shows a reading if you connect it across any resistor (or bulb) in a working circuit. Resistors and bulbs do not 'push'. So the **voltmeter** reading must be indicating something else.

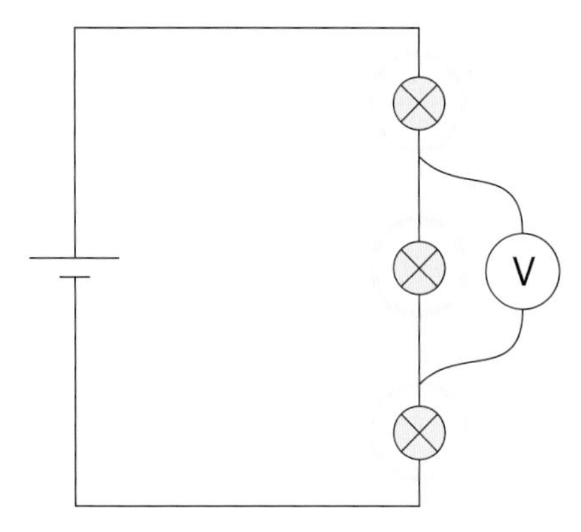

There is a reading on this voltmeter. This cannot be a measure of the strength of a 'push' – so what is it telling us?

A useful picture is to think of the battery as a pump, lifting water up to a higher level. The water then drops back to its original level as it flows back to the inlet of the pump. The diagram below shows how this would work for a series circuit with three resistors (or three lamps). The pump increases the potential energy of the water. The water then loses this energy in three steps. The total amount of energy lost has to be equal to the amount of energy gained.

In the electric circuit, the battery does work on the electric charges, to lift them up to a higher 'energy level'. They then transfer energy in three stages as they drop back to their starting level. A voltmeter measures the difference in 'level' between the two points it is connected to. This is called the potential difference between these points. **Potential difference (p.d.)** is measured in volts (V).

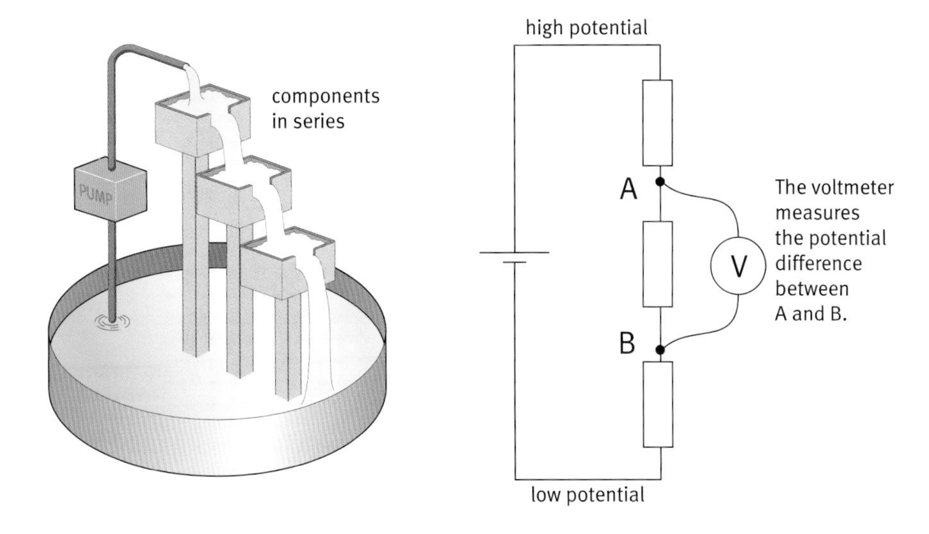

The voltage of a battery is the potential difference between its terminals. If you put a battery with a larger voltage into the circuit above, this would mean a bigger potential difference across its terminals. The potential difference across each lamp (or resistor) would also now be bigger. Going back to the water pump model, this is like changing to a stronger pump that lifts the water up to a higher level. The three downhill steps then also have to be bigger, so that the water ends up back at its starting level.

There is a potential difference of 12 V across the terminals of a car battery.

The same idea works for a parallel circuit. In this case, the water divides into three streams. Each loses all its energy in a single step.

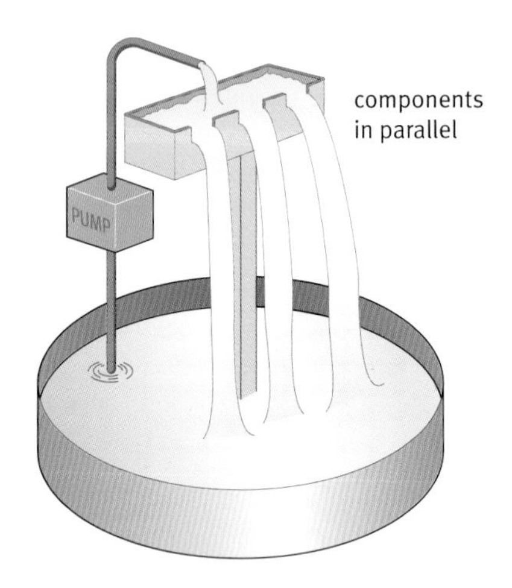

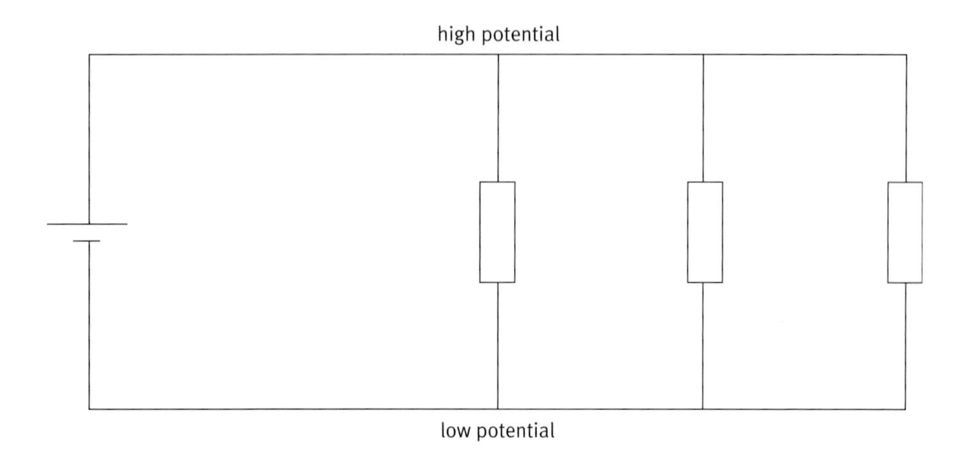

Voltmeter readings across circuit components

This water pump model helps to explain and to predict voltmeter readings across resistors in different circuits. If several resistors are connected in parallel to a battery, the potential difference across each is the same. It is equal to the battery voltage.

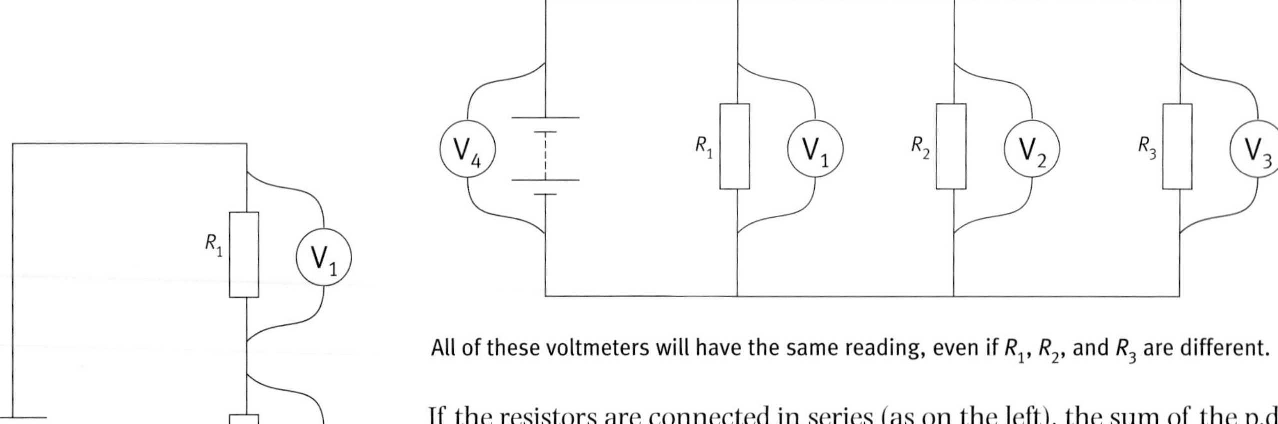

All of these voltmeters will have the same reading, even if R_1, R_2, and R_3 are different.

If the resistors are connected in series (as on the left), the sum of the p.d.s across them is equal to the battery voltage. This is exactly what you would expect from the 'waterfall' picture on page 135.

In the series circuit, the p.d. across each resistor depends on its resistance. The biggest voltmeter reading is across the resistor with biggest resistance. Again this makes sense. More work has to be done to push charge through a big resistance than a smaller one.

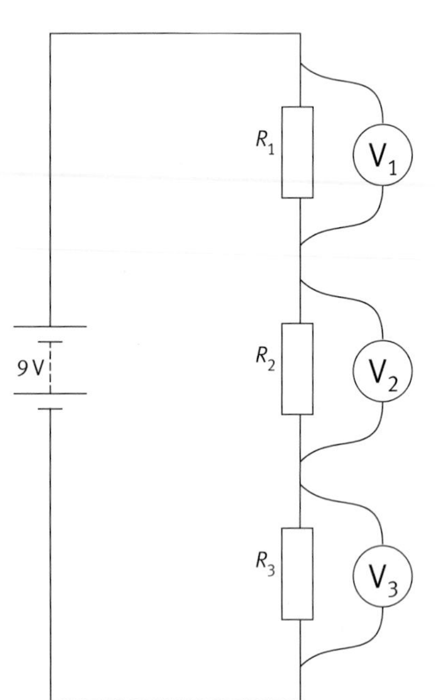

The voltages are in proportion to the resistances. And their sum is equal to the battery voltage.

Currents in parallel branches

The potential difference across resistors R_1, R_2, and R_3 in the parallel circuit the opposite page is exactly the same for each. It is equal to the p.d. across the battery itself. But the currents through the resistors are not necessarily the same. This will depend on their resistances. The current through the biggest resistor will be the smallest. There are two ways to think of this:

1 Imagine water flowing through a large pipe. The pipe then splits in two, before joining up again later. If the two parallel pipes have different diameters, more water will flow every second through the pipe with the larger diameter. The wider pipe has less resistance than the narrower pipe to the flow of water So the current through it is larger.

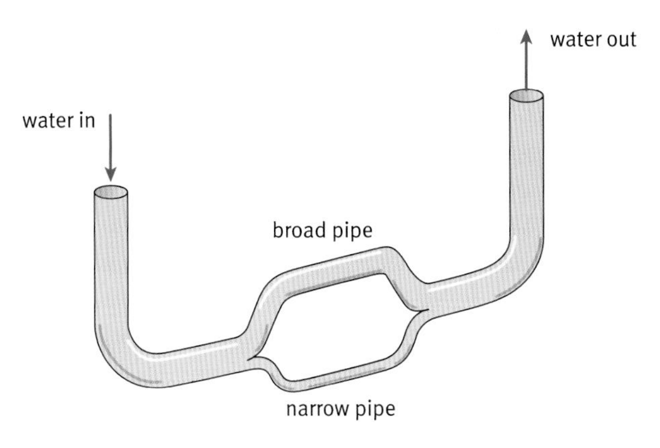

More water flows each second through the larger pipe. It has less resistance to the water flow.

2 Think of two resistors connected in parallel to a battery as making two separate simple loop circuits that share the same battery. The current in each loop is independent of the other. The smaller the resistance in a loop, the bigger the current. Some wires in the circuit are part of both loops, so here the current will be biggest. The current here will be the sum of the currents in the loops.

Key words
voltmeter
potential difference (p.d.)

Questions

1 Imagine removing the red resistor from the circuit below leaving a gap. What would happen to:
a the current through the purple resistor?
b the current from (and back to) the battery?
Explain your reasoning each time.

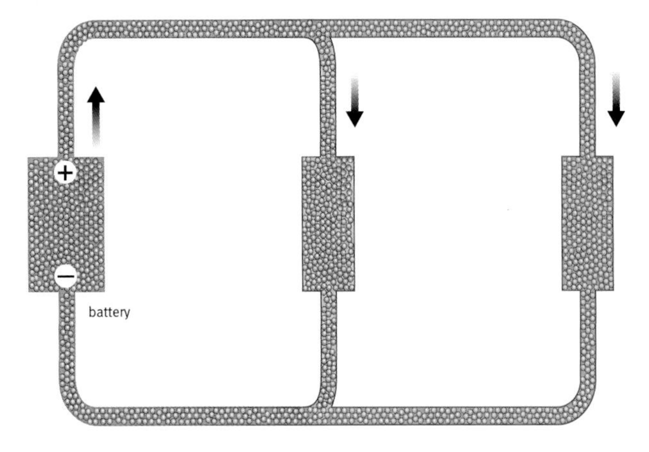

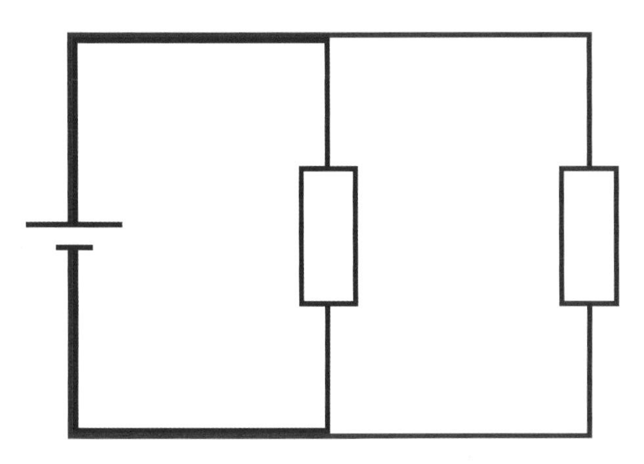

A parallel circuit like this behaves like two separate simple loop circuits.

Find out about:

- how the power produced in a circuit component depends on both current and voltage

F Electrical power

An electric circuit is primarily a device for doing work of some kind. It transfers energy initially stored in the battery to somewhere else. A key feature of any electric circuit is the rate at which work is done on the components in the circuit – that is, the rate at which energy is transferred from the battery to the other components. This is called the **power** of the circuit.

Measuring the power of an electric circuit

Imagine starting with a simple battery and bulb circuit and trying to double, and treble, the power. You could do this in two ways.

- One is to add a second bulb, and then a third, in parallel with the first. In the circuits down the left-hand side of the diagram below, the p.d. is the same, but the current supplied by the battery doubles and trebles. The power is proportional to the current.
- Another is to add a second bulb, and then a third, in series with the first. Now you need to add a second battery, and then a third, to keep the brightness of the bulbs the same each time. In these circuits (across the diagram below), the current is the same, but the p.d. of the battery doubles and trebles. The power is proportional to the current.

This is summarized in the box in the bottom right-hand corner.

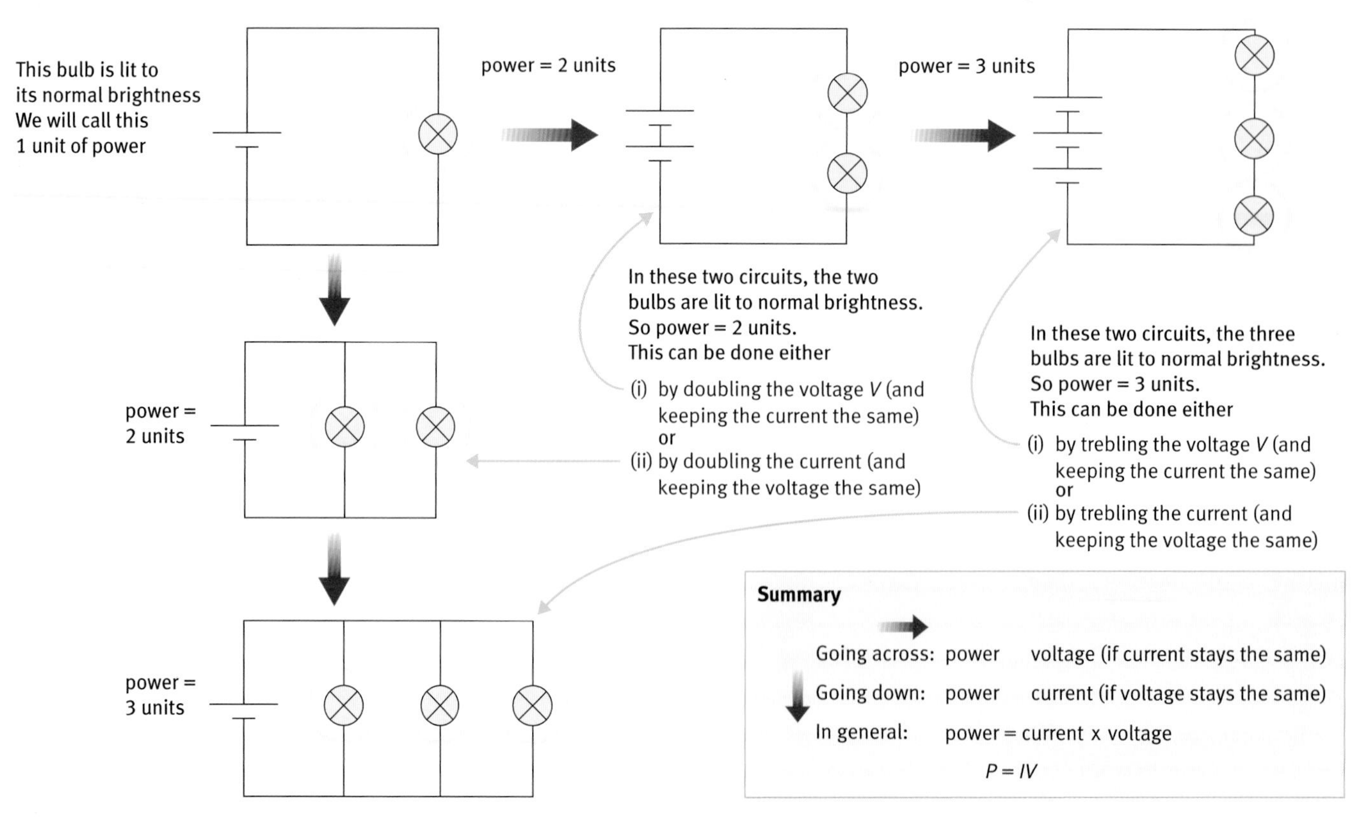

In general, the power dissipated in an electric circuit depends on both the current and the voltage:

power	=	current	×	voltage
P	=	I		V
(watt, W)		(ampere, A)		(volt, V)

The unit of power is the watt (W). One watt is equal to one joule per second. So if you know the power, it is easy to calculate how much work is done (or how much energy is transferred) in a given period of time:

work done (or energy transferred)	=	power	×	time
(joule, J)		(watt, W)		(second, s)

To see how this equation for power makes sense, look back at the explanation of resistance and heating on page 131. If the battery voltage is increased, the electric field in the wires gets bigger. So the free charges (the electrons) move twice as fast. When a charge collides with an atom in the wire, twice as much energy is transferred in the collision. These collisions also happen more often, simply because the charges are moving along faster. So when the voltage is increased, collisions between electrons and the lattice of atoms are *both* harder *and* more frequent.

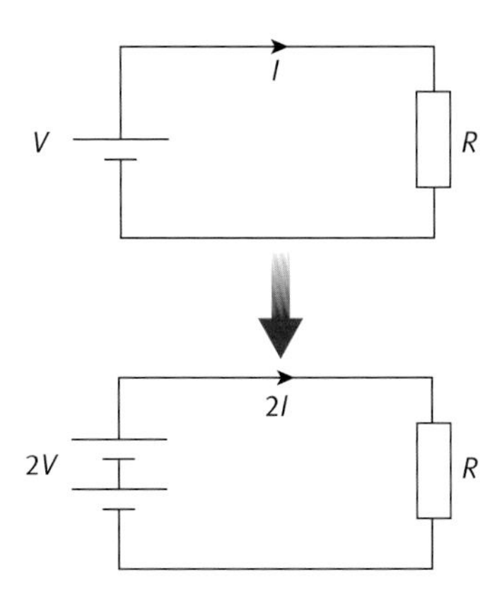

Doubling the battery voltage makes the current double. So the power ($P = IV$) is four times as big.

The power of the electric motor in this tube train is much greater than the power of the strip light above the platform. Both the voltage and the current are bigger.

Questions

1 In these two circuits, resistor R_1 has a large resistance and resistor R_2 has a small resistance. If each circuit is switched on for a while, which resistor will get hotter? Explain your answers.

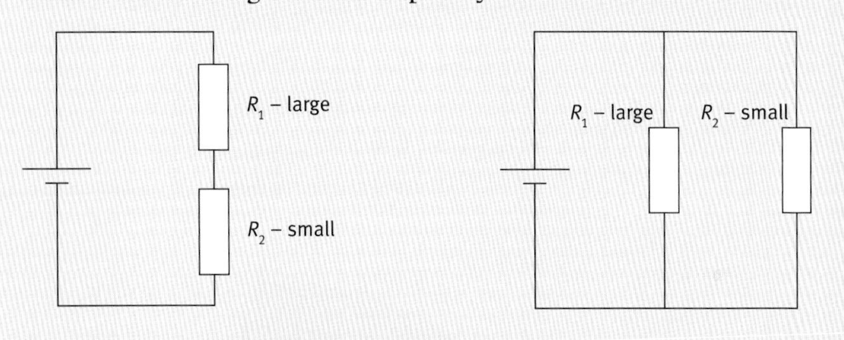

2 In circuit A, a battery is connected to a resistor with a small resistance. In circuit B, the resistor has a large resistance. The two batteries are identical. Which will go 'flat' first? Explain your answer.

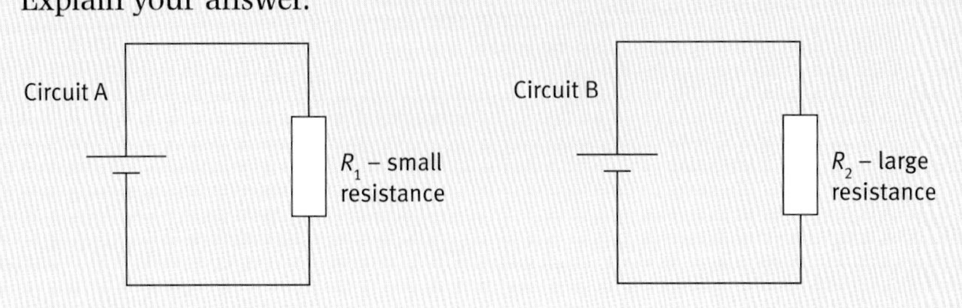

Key words

power

Find out about:

- how to calculate the energy transferred by a domestic appliance
- what is meant by 'efficiency'
- how to calculate the correct fuse value for an appliance

G Domestic appliances

Electricity bills are based on the amount of 'electrical energy' used. This is the amount of work done by the electricity supply on all the appliances we run. The work done on an appliance depends on

- its power rating
- the time it is on for

You could calculate this in joules, but the result would usually be a very large number. So, for domestic appliances, it is more convenient to use the **kilowatt-hour** as the unit of energy:

energy transferred when device is on	=	power rating	×	time
(joule, J)		(watt, W)		(second, S)
(kilowatt-hour, kW h)		(kilowatt, kW)		(hour, h)

Every electrical appliance has a power rating (in W) marked on it. This tells you how much work is done by the electricity supply every second it is on.

On an electricity bill, '1 unit' means 1 kilowatt-hour. The electricity meter in your house measures the number of kilowatt-hours of electrical energy you buy.

Efficiency

In all electrical appliances, some of the energy transferred does not end up where it is wanted, or in the form it is wanted. For example, in a filament light bulb, less than 10% of the work done on the filament is carried away as light. The rest goes into heating the bulb and its surroundings.
The **efficiency** of an electrical appliance is defined as follows:

$$\%\text{ efficiency} = \frac{\text{energy transferred to the place (or in the manner) we want}}{\text{work done on the appliance by the electricity supply}}$$

The compact fluorescent (energy-saving) bulb on the left is more efficient than the filament lamp on the right. The electricity supply has to do less work to produce the same amount of light output per second.

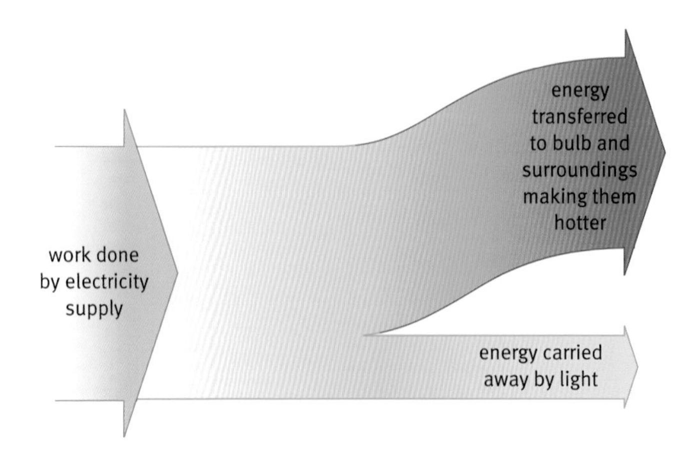

The efficiency of a light bulb is about 10%.

Working out the current

If you know the power rating of a mains appliance, you can easily work out the current through it when it is switched on, using the equation.

$$\underset{(\text{watt, W})}{\text{power}} = \underset{(\text{ampere, A})}{\text{current}} \times \underset{(\text{volt, V})}{\text{voltage}}$$

For all UK mains appliances, the operating voltage is 230 V. So

$$\text{power (in W)} = \text{current (in A)} \times 230\text{ V}$$

Divide both sides by 230 V to find the current:

$$\text{current (in A)} = \frac{\text{power (in W)}}{230\text{ V}}$$

This is important to know when you have to choose the right fuse for a plug. A fuse is a short piece of wire made of a metal that melts at a low temperature. An electric current through it makes it heat up. Its length and thickness are chosen so that it melts if the current goes above the value marked on the fuse. So it is, the 'weakest link' in the circuit. It will melt first if the current for any reason gets bigger than it should be. Fuses for mains plugs normally come in two values: 3 A and 13 A. The electricity companies recommend using a 3 A fuse for any appliance with a power rating below 690 W, and a 13 A fuse for those with a higher power rating.

Key words
kilowatt-hour
efficiency

Questions

1 What is the power of a mains appliance that needs a current of 3 A to make it run? Use your answer to explain the advice of the electricity companies on choosing fuse values.

2 Look carefully at each of the tasks in the table on the right which use electricity, and put them in the order you think they would come – from the cheapest to the most expensive. Then calculate the number of kilowatt-hours which each involves and see if your estimate of the cost was correct.

Task	Appliance used	Power rating (W)	Time for which it is on
watch television for the evening	television	300	5 hours
dry your hair	hairdryer	700	5 minutes
make a pot of tea	electric kettle	2000	4 minutes
write a homework assignment	computer	250	2 hours
keep a front door light on overnight	light bulb	100	10 hours
listen to a football match on the radio	radio	10	2 hours
heat your bedroom while you do your homework	electric fan heater	1500	2 hours
wash a load of dirty clothes	washing machine	1850	$1\frac{1}{2}$ hours

3 A 20 W energy-saving light bulb costs £2.99. It gives the same amount of light as a 100 W filament light bulb, costing 45p. The lifetime of the energy-saving bulb is 5000 hours. The filament bulb has a lifetime of 1000 hours. If 1 unit (1 kilowatt-hour) of electrical energy costs you 5p, how much do you pay for 5000 hours of lighting with each bulb? Which is the better buy?

Find out about

- how a magnet moving near a coil can generate an electric current
- the factors that determine the size of this current
- how this is used to generate electricity on the large scale

H An electricity supply

Electricity reaches last village in Britain

Nowadays most of us in Britain take a mains electricity supply for granted. But in fact it was only in May 2003 that Cym Brefi in mid-Wales became the last village in Britain to get a mains electricity supply.

Of course, not having a mains electricity supply does not mean you cannot use electrical appliances. Many can be run from batteries. But this works only for relatively low-power devices. For others, you might use a diesel-powered **generator**.

This is how the inhabitants of Cym Brefi ran their washing machines and vacuum cleaners before they got mains electricity. But generators are noisy, and each 'unit of electricity' is much more expensive than from the mains. So they can only be run for a short time.

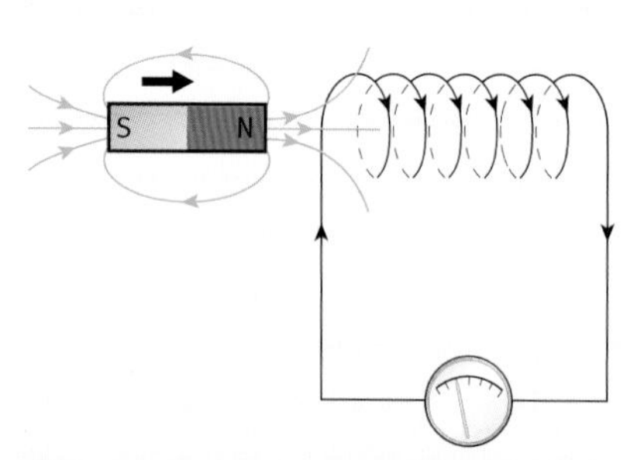

1 While the bar magnet is moving into the coil, there is a small reading on the sensitive ammeter.

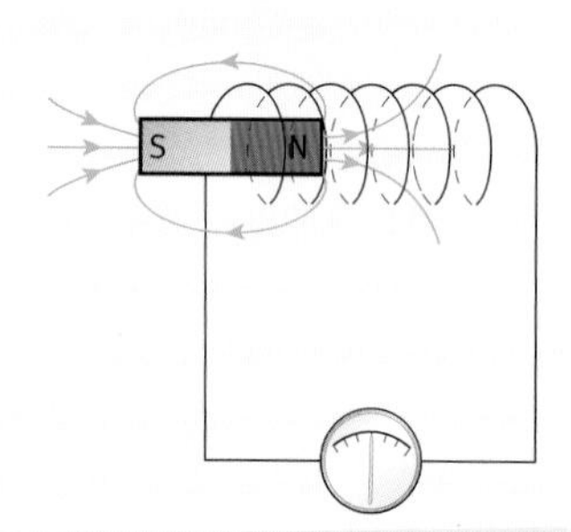

2 There is no current while the magnet is stationary inside the coil.

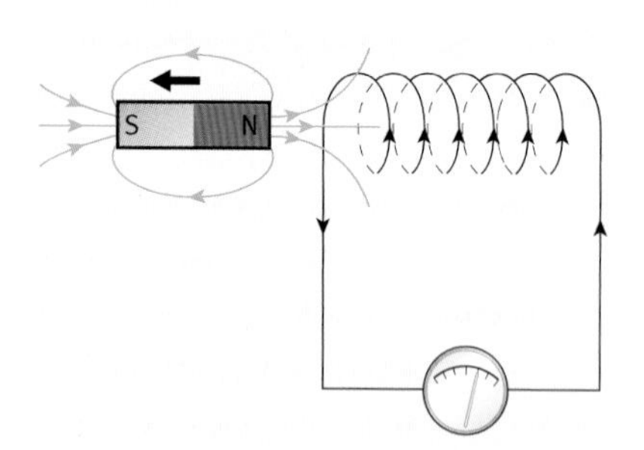

3 While the magnet is being removed from the coil, there is again a small current, but now in the opposite direction.

Moving a magnet into, or out of, a coil generates a current.

Generating electricity

Generators work on the principle of **electromagnetic induction**. This phenomenon, which does so much to make our lives comfortable and convenient, was discovered in the 1830s by Michael Faraday.

About 10 years earlier, a Danish physicist, Hans Christian Oersted, had shown that an electric current in a wire produces a magnetic field in the region around it. Faraday wondered if he could do this in reverse: use a magnetic field to produce an electric current. After several years of experimentation, using homemade coils of wire and magnets, he succeeded.

One way to generate a current is to move a magnet into, or out of, a coil. The movement of the magnet causes an induced voltage across the ends of the coil. 'Induced' means that it is caused by something else – in this case, the movement of the magnet. The coil, for a brief time, is like a small battery. If the coil is part of a closed circuit, this induced voltage makes a current flow.

The size of the current can be increased by

- moving the magnet in and out more quickly
- using a stronger magnet
- using a coil with more turns

Key words
generator
electromagnetic induction
alternating current (a.c)
direct current (d.c)

Now imagine what would happen if the magnet were rotated near the end of the coil. The magnetic field around the coil would be constantly reversing direction. This changing magnetic field induces a current in the coil, first in one direction, then the other. This is called an **alternating current** (**a.c.**). For many applications, this works just as well as a **direct current** (**d.c.**). A direct current is one which is always in the same direction.

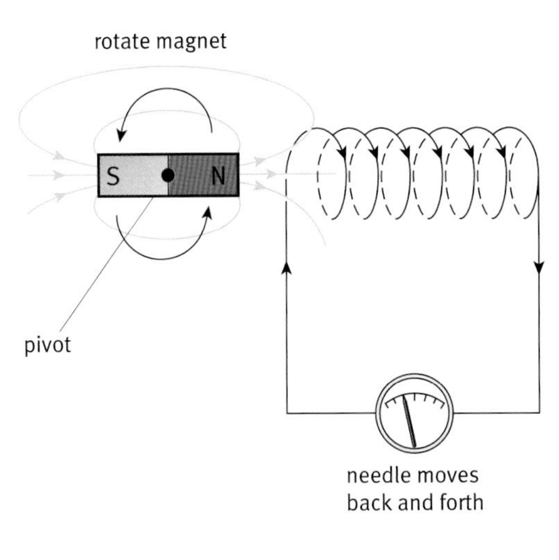

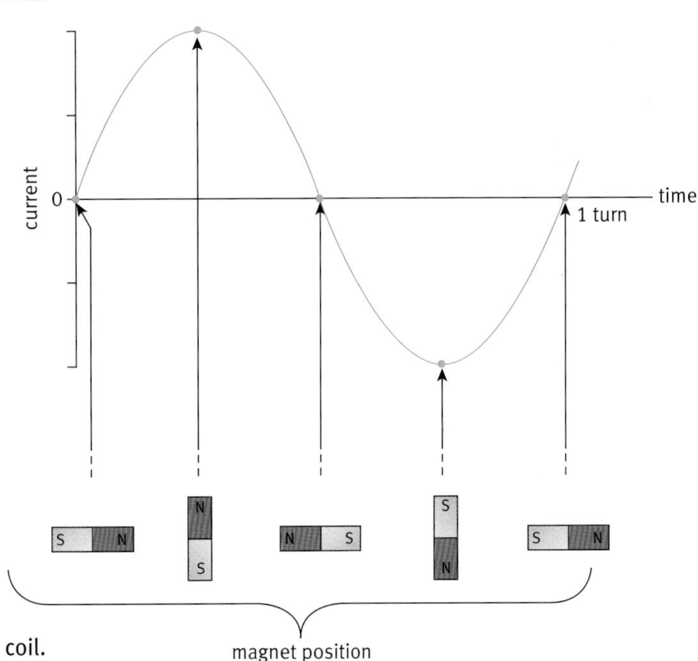

If the magnet is rotated, it will induce an alternating current in the coil.

The size of the alternating voltage (and current) produced by a generator of this sort can be increased by:

- using a stronger rotating magnet or electromagnet
- rotating the magnet or electromagnet faster (though this also affects the frequency of the a.c. produced)
- using a fixed coil with more turns
- putting an iron core inside the fixed coil (this makes the magnetic field a lot bigger – as much as 1000 times)

In a real generator, an electromagnet is rotated inside a fixed coil. As it spins, a.c. is generated in the coil. In power stations in the UK, the rate of turning is set at 50 cycles per second. The generator is turned by a turbine, which is driven by steam. The steam is produced by burning gas, oil, or coal, or by the heating effect of a nuclear reaction.

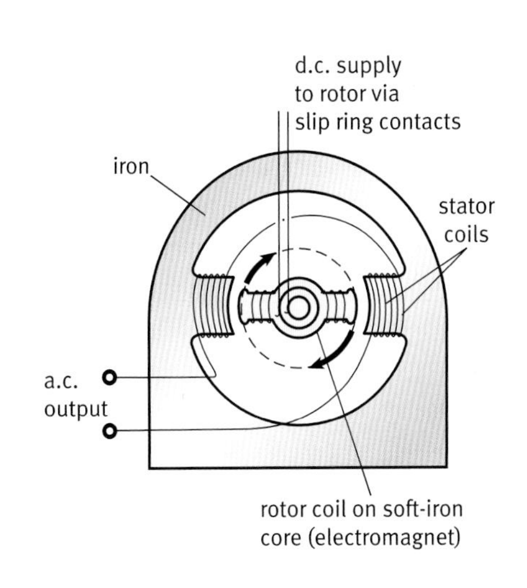

A simplified diagram of an a.c. generator

1 Distributing electricity

Transformers

An electric current can be generated by moving a magnet into (or out of) a coil of wire (see page 142). The moving magnet could be replaced by an electromagnet. If a coil is wound round an iron core, it becomes quite a strong magnet when a current flows through it.

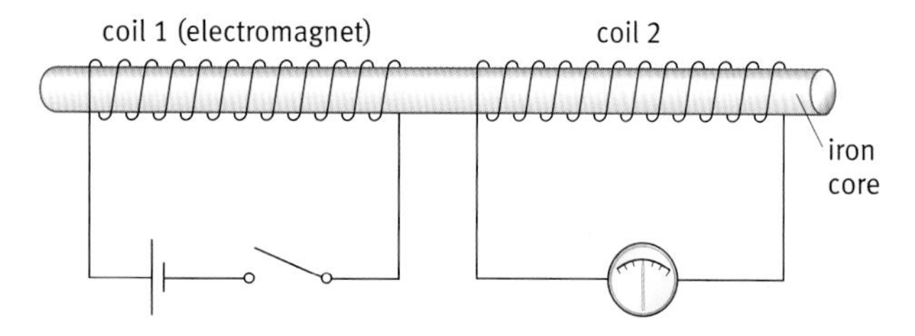

When the current in the electromagnet (coil 1) is switched on, this has the same effect as plunging a bar magnet into coil 2. So a current is generated in coil 2, whilst the current in coil 1 is changing. This arrangement of two coils on the same iron core is called a **transformer**. Changing the current in the primary coil induces a voltage across the secondary coil. If the current in the primary is a.c., it is changing all the time. So an alternating voltage is induced across the secondary coil.

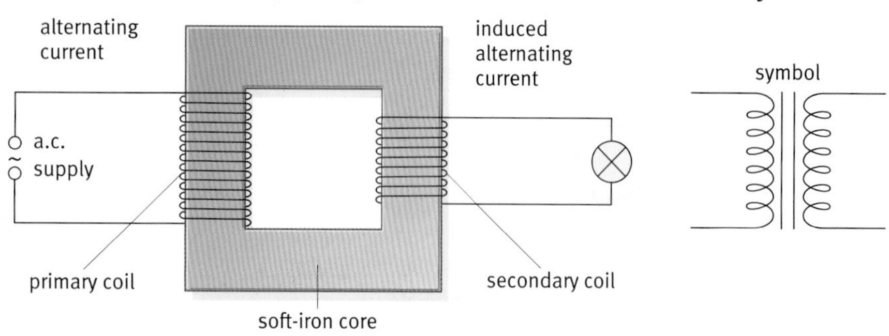

The transformer. When the current in the primary coil is changing, a voltage is induced across the secondary coil. This makes a current flow round the right-hand circuit. Notice that there is no direct electrical connection between the coils of a transformer. The only connection is through the magnetic field.

The behaviour of a transformer depends on the number of turns of wire on the two coils.

H

The equation that links the two is:

$$\frac{\text{voltage across secondary coil } (V_S)}{\text{voltage across primary coil } (V_P)} = \frac{\text{number of turns on secondary coil } (N_S)}{\text{number of turns on primary coil } (N_P)}$$

If there are more turns on the secondary coil, then the induced voltage across this coil is bigger than the applied voltage across the primary coil. However, this is not something for nothing! The current in the secondary will be correspondingly less, so that the power available from the secondary is no greater than the power supplied to the primary (remember: power = IV).

Find out about:

- how transformers are used to alter the voltage of a supply
- the main components of the National Grid

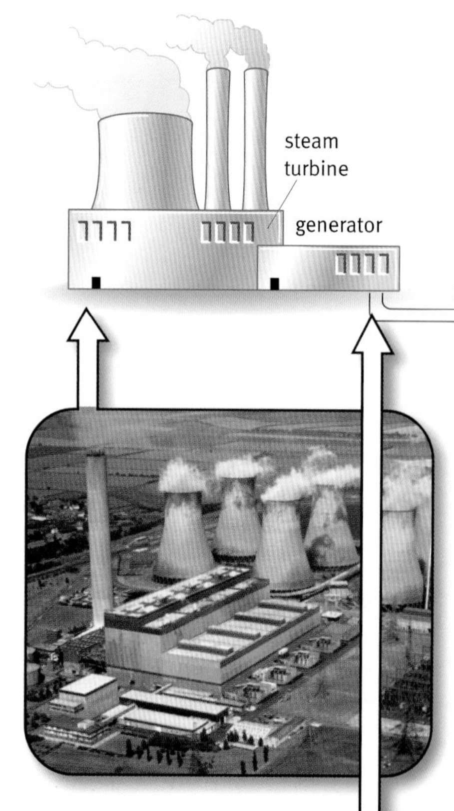

In a power station, a primary fuel is used to generate electricity. The commonest fuel is gas, but coal, nuclear fuel, water in high dams, and wind are also used.

The heart of a power station is a turbine. The primary fuel is used to drive this, and as it turns it makes the coil of a generator rotate.

The National Grid

Transformers play an important role in the National Grid system for distributing electricity. All the electricity power stations in Britain are connected into the National Grid, which is used to distribute electricity to all the places where we want to use it. The Grid connects every power socket in your home back to the power station. It does this by means of a long chain of wires and magnetic fields in transformers.

Key words
transformer

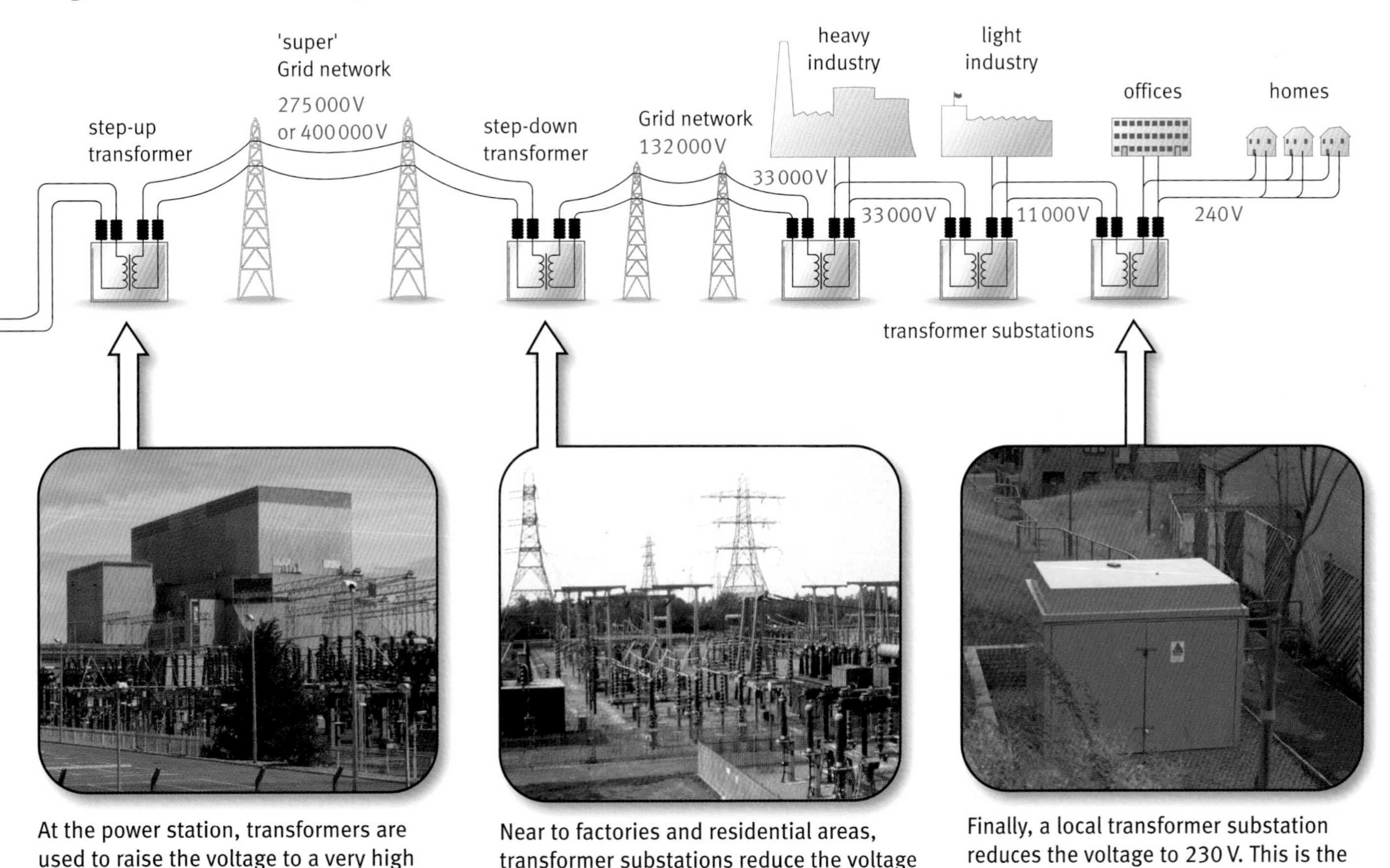

At the power station, transformers are used to raise the voltage to a very high value (sometimes as much as 400 000 V). This means that the current in the pylon lines is small. So relatively little energy is wasted heating the pylon cables themselves.

Near to factories and residential areas, transformer substations reduce the voltage to a lower level, around 33 000 V.

Finally, a local transformer substation reduces the voltage to 230 V. This is the voltage at which electricity is supplied to homes in the UK. There is likely to be one of these transformer substations close to where you live.

Questions

1 A transformer has 100 turns on its primary coil, and 25 turns on its secondary coil. A 12 V a.c. supply is connected to the primary coil. What will be the voltage across the secondary coil?

2 If you had a 6 V a.c. supply and wanted to use it to operate a 12 V bulb, explain how you could make a simple transformer to enable you to do this.

3 In the National Grid, transformers are used to 'step-up' the voltage from 25 000 V to 400 000 V. What gets smaller as a result (and stops us getting something for nothing)?

Summary

In this module you have met some of the basic ideas that scientists use to explain electrical phenomena: charge, current, voltage, resistance, and potential difference. The module has introduced you to some models that are useful for explaining and predicting the behaviour of electric circuits.

Electric charge

- Electric charge is a fundamental property of matter.
- Charge cannot be created or destroyed. But positive and negative charges can be separated, and moved from one object to another, for example by rubbing.

Electric current

- A working electric circuit always consists of a closed loop (or loops) of conducting material, between the positive and negative terminals of a battery.
- An electric current is a flow of charges, which are already present in the materials of the circuit. The battery makes the charges move round the circuit.
- Current is not used up as it goes round – but it does work on the components it passes through, and so transfers energy from the battery to the other components.

Current, voltage, and resistance

- The voltage of a battery is a measure of the strength of its 'push' on the charges. The bigger the voltage, the bigger the current.
- The components in a circuit resist the flow of charge. The bigger the resistance, the smaller the current.
- Together, the battery voltage and the circuit resistance determine the current in the circuit.

Resistors in series and parallel

- The total resistance of resistors in series is their sum.
- The total resistance of resistors in parallel is less than that of any single resistor – as the group provides more loops for charges to flow round. H

Potential difference

- It is useful to think of a battery as raising charges to a higher level (giving them more potential energy). They then lose this potential energy as they go round the circuit.
- A voltmeter measures the potential difference (p.d.) between the two points it is connected to.
- Resistors in parallel have the same p.d. across each of them.
- The p.d. across resistors in series is proportional to their resistance.

Electrical power

- The power (energy per second) transferred by an electric circuit is equal to 'current × voltage'.

Electromagnetic induction

- A potential difference (p.d.) is induced across the ends of a wire or coil placed in a changing magnetic field.
- If this wire or coil is part of a circuit, there is an induced electric current in the circuit.
- This phenomenon is called electromagnetic induction. It is the basis of the electrical generator, and the transformer – both of which are key components of the mains electricity supply system (the National Grid).

Questions

1 Look at the three electric circuit models on page 126. Copy and complete the following table:

Model	What corresponds to the battery?	What corresponds to electric current?	What corresponds to the resistors or lamps in the circuit?
'peas in a pipe'			
'chain on a bike'			
'moving rope'			

2 In a simple single-loop electric circuit, the current is the same everywhere. It is not used up. How does each of the models above help to account for this?

3 Imagine a simple electric circuit consisting of a battery and a bulb. For each of the following statements, say if it is true or false (and explain why):

a Before the battery is connected, there are no electric charges in the wire. When the circuit is switched on, electric charges flow out of the battery into the wire.

b Collisions between the moving charges and fixed atoms in bulb filament make it heat up and light.

c Electric charges are used up in the bulb to make it light.

4 In shops, you can buy batteries labelled 1.5 V, 4.5 V, 6 V, or 9 V. But you cannot buy batteries labelled 1.5 A, 4.5 A, 6 A or 9 A. Explain why not.

5 You are given four 4 Ω resistors. Draw diagrams to show how you could connect all four together to make a resistance of:

a 16 Ω

b 1 Ω

c 10 Ω

d 4 Ω

Note that there is more than one possible way to do parts c and d.

6 A family's electric shower has an electrical power of 6 kW. In a typical week, it is used for twelve showers, each lasting ten minutes.

a Use the equation energy (joules) = power (watts) × time (seconds) to calculate the energy (in joules) transferred by the shower in a typical week.

b Use the equation energy (kilowatt hours) = power (kilowatts) × time (hours) to calculate the energy (in kW h) transferred by the shower in a typical week.

c The electricity company charges 8p per kW h. Calculate how much the twelve showers cost.

7 By my bed I have a 60 W spotlight, which I use for reading. It is only about 10% efficient. My tortoise has an identical 60 W spotlight which he basks under. It is about 90% efficient. Explain why both of these statements can be true.

8 Copy and complete these sentences:

When a magnet is moved into a coil of wire, a voltage is ____________ in the coil. The voltage is produced only when the magnet is ____________ . This is used in an a.c. generator, which has an ____________ rotating near a fixed coil (see diagram on page 171). To increase the size of the induced voltage, you could use a ____________ electromagnet, have more ____________ on the fixed coil, turn the rotor coil ____________ , or put a core of ____________ inside it.

The current in the external circuit constantly changes direction, so it is called ____________ current (_____). This is differerent from the current from a battery, which always goes in one direction and is called ____________ current (_____).

Why study waves?

A wave transfers energy through a medium, without the medium itself having to move as a whole. Scientists are interested in waves because many types of radiation behave in ways that are similar to waves we can see. Light and sound are two examples. Thinking of these as waves helps us explain and predict how they behave in different situations.

The science

When you disturb a 'springy' medium, the movement of one bit of the material makes the neighbouring bit move, after a slight delay. This causes a pulse to travel along. A wave is a continuous series of pulses moving through a medium. Waves are reflected when they hit a barrier, refracted when they cross from one medium into another, and are diffracted at corners and edges of obstacles. Two waves in the same region interfere. Light also has these properties. Light is one small part of the electromagnetic spectrum, a 'family' of waves whose properties depend on their frequency (or wavelength).

Physics in action

Physicists' understanding of waves has enabled them to develop new instruments, such as radio telescopes and infrared telescopes, for exploring the galaxy and beyond. Waves are also the foundation of modern communications, using radio and microwaves for television and radio broadcasts and for telephones, and optical fibres instead of wires to carry signals.

The wave model of radiation

Find out about:

- how vibrations can travel through a medium, and how this causes waves
- the speed at which a wave travels, its frequency, and its wavelength
- reflection, refraction, diffraction, and interference
- why scientists think of light and sound as waves
- the different radiations of the electromagnetic spectrum, how they differ, and how this affects their properties
- how waves of various kinds are used to carry information

Find out about:
- how waves travel through a medium
- the two main types of wave (transverse and longitudinal)

A What is a wave?

A wave is a disturbance moving through a material. Imagine a long elastic rope lying on a flat table, with one end tied to a fixed point. If you move the other end from side to side, a pulse travels along the rope. The diagram on the left shows the shape of the rope at different instants.

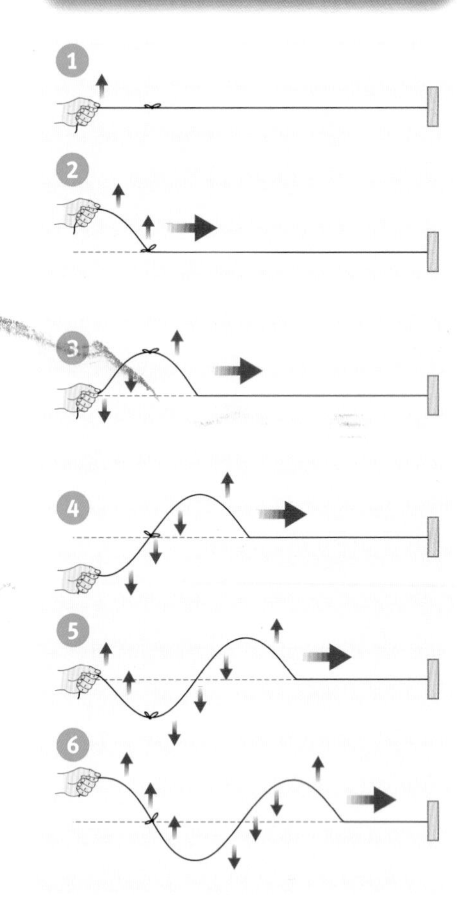

'Snap shots' of a pulse travelling along an elastic rope. The small arrows show how each bit of rope is moving. The large arrows show how the pulse moves.

What causes a wave?

You may have seen what happens if you topple a line of dominoes. As each one falls, it knocks the next one over – and so on, right down the line.

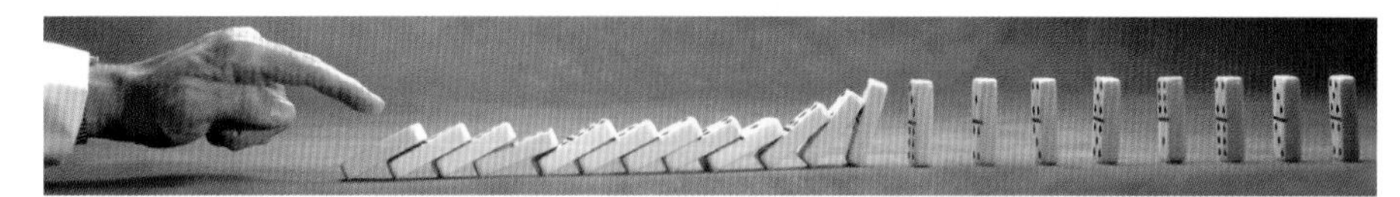
Toppling a line of dominos. The disturbance moves down the line, with a short delay from one domino to the next.

This is not a wave, however, just a single pulse. But now imagine that the dominoes are hinged at the bottom end, and connected to each other at the top by springs.

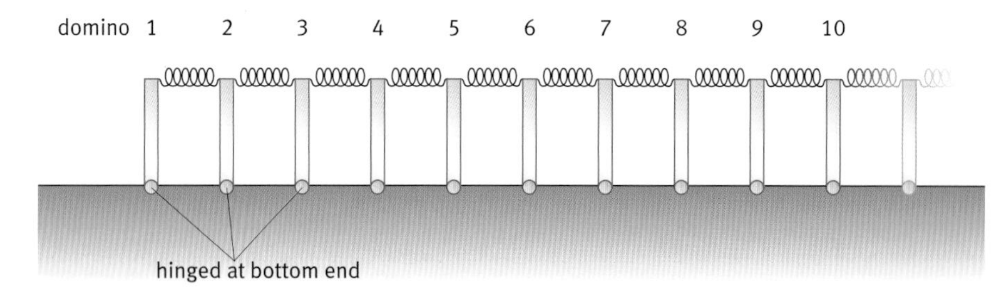

A line of dominoes hinged at the bottom and connected by springs at the top.

Think about what would happen if you moved the top of domino 1 to the right. As it moves, it compresses the first spring, which then pushes domino 2 to the right. This then compresses the second spring, pushing domino 3 to the right – and so on, along the line. The movement of domino 1 is repeated along the row, with a small time delay as it passes from one to the next. Now imagine moving the top of domino 1 slowly backwards and forwards. This sends a stream of pulses along the line – a continuous wave.

The source of a wave is always something that **vibrates**. The material that the wave travels through is called the **medium**. The medium has to be 'springy' and come back to its original position after being disturbed. In the example above, the line of dominoes and springs is the medium. For water waves, the medium is water. Gravity pulls the water back into position if it is disturbed.

Two meanings of 'medium'

Sometimes the word 'medium' is used to mean in the middle, between two extremes – like an average value. When we are talking about waves, the medium means the material the wave is travelling through. The plural of 'medium' is 'media'.

Transverse and longitudinal

Two kinds of wave can travel through a medium. In one, the particles of the medium vibrate at right angles to the direction in which the wave moves. This is called a **transverse wave**. Water waves and waves on a rope are examples. But these can also be a wave where the particles of the medium vibrate in the same direction as the wave moves. Imagine a Slinky spring stretched across a table and fixed at one end. You can make a pulse by pushing the free end inwards, along the line of the spring, and back again. This compression pulse then travels along the spring. If you keep moving the free end in and out, this produces a continuous wave of pulses. This is called a **longitudinal wave**.

Key words
- vibrates
- medium
- transverse wave
- longitudinal wave

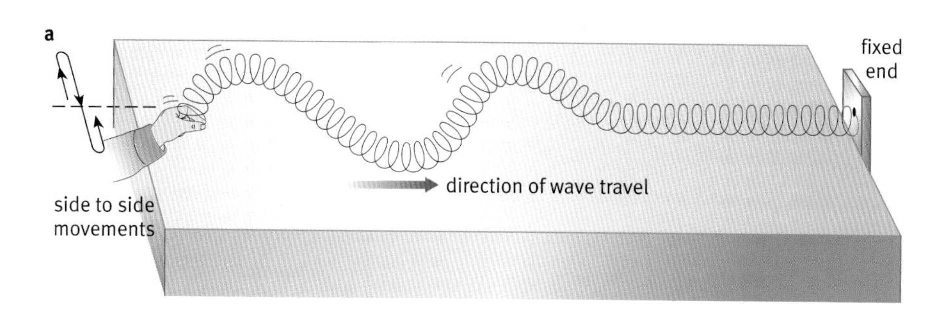

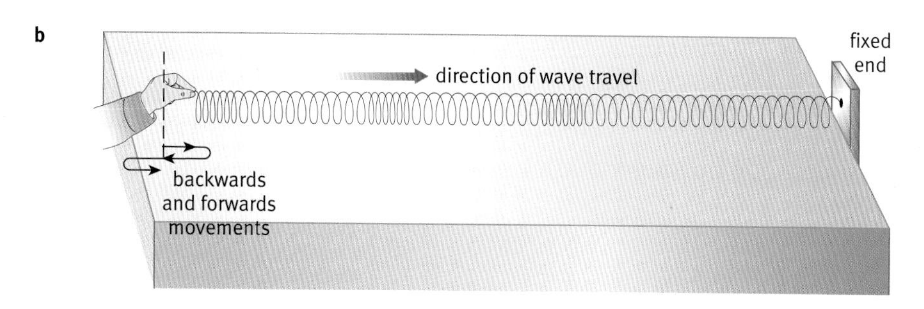

Two types of wave on a Slinky spring: **a** transverse; **b** longitudinal.

Sound is a longitudinal wave. A sound source vibrates, causing little compression pulses in the air nearby. Compressed air is 'springy'. As it recoils, it pushes on the neighbouring bit of air, which compresses it – and so on. A continuous wave of pulses travels through the air. Sound waves can travel through any gas, and also through liquids (you can hear underwater) and solids.

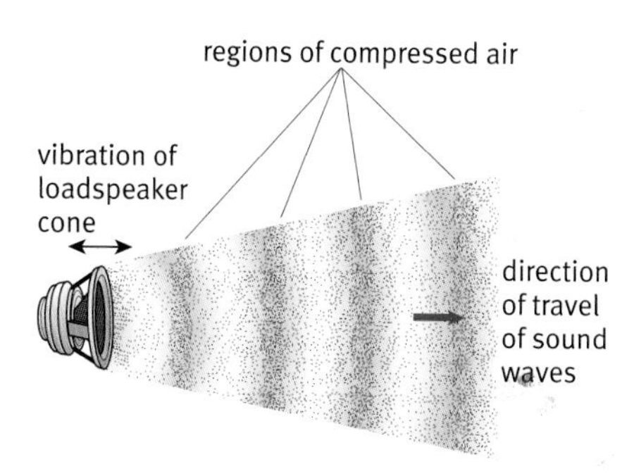

The vibrations of the loudspeaker cone send a stream of compression pulses through the air. This is a sound wave.

Questions

1 Look at the knot on the rope in the diagrams in the left margin.

a Describe how it moves as the wave passes along the rope.

b Sketch a graph to show how its displacement (its distance from the centre line) varies with time as the wave passes.

Find out about:

- how to describe a wave clearly
- how wave speed, frequency, and wavelength are connected

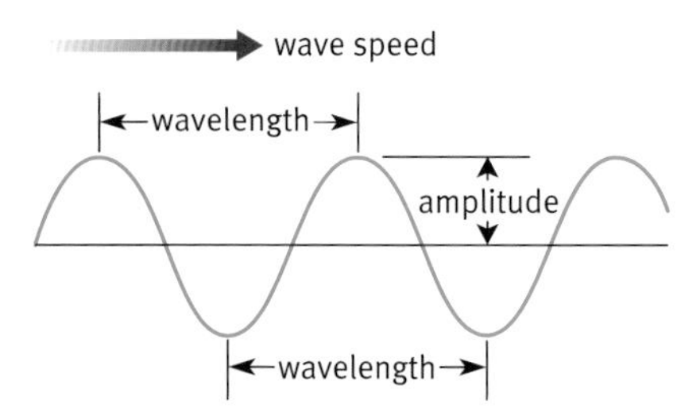

Important information when describing a wave

B Describing waves

To describe a wave clearly, there are several things you need to know:

1. Its **amplitude**. For water waves or waves on a rope or spring, this is the maximum distance that each point in the medium moves from its normal position as the wave passes. It is measured in metres (m).
2. Its **frequency**. This is the number of waves that pass any point in the medium every second. So it is the same as the number of vibrations per second of the source. Frequency is measured in hertz (Hz). 1 Hz means 1 wave per second.
3. Its **wave speed**. This is the speed at which each wave crest moves through the medium. It is measured in metres per second (m/s).
4. Its **wavelength**. This is the length of a complete wave. You might measure it from one wave crest to the next, but any point on the wave will do just as well. The distance is the same from any point on one wave to the corresponding point on the next. It is measured in metres (m).

It is important to realize that frequency and wave speed are two completely different things. The frequency depends on the source – how many times it vibrates every second. The wave speed depends on the medium the wave is travelling through. Once the wave has left the source, the source can no longer affect its speed through the medium.

Questions

1 Look at the diagrams of transverse and longitudinal waves on a Slinky on page 151. What would you change if you wanted to increase:

a the frequency of the transverse wave?

b the wave speed of the transverse wave?

c the amplitude of the longitudinal wave?

2 Estimate the wavelength of the sound waves in the diagram on page 151. Assume that the diagram is drawn to scale (life size).

Sound waves

For sound waves, the bigger the amplitude, the louder the sound.
The greater the frequency, the higher the pitch of the sound. So, for example, the musical note C one octave above middle C has twice the frequency of middle C. The wave speed of sound (often just called the speed of sound) depends on the medium only. The bar graph below shows the speed of sound in some common media.

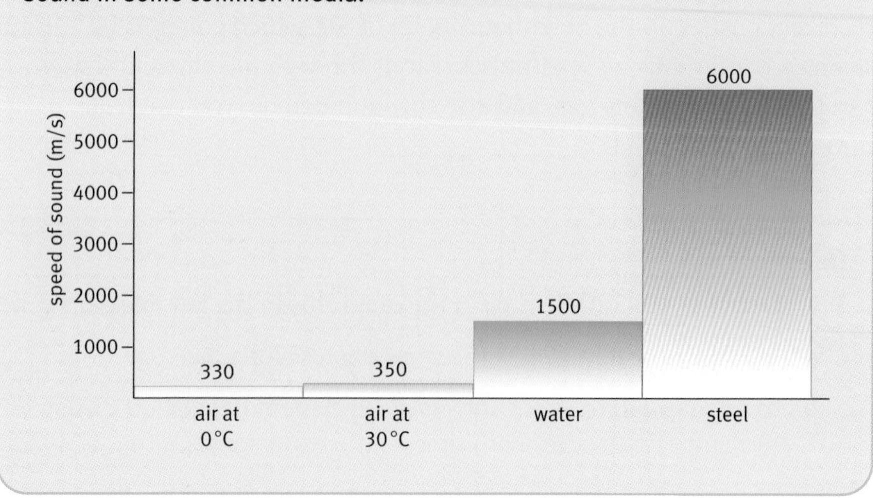

The wave equation

The frequency of a wave, its wave speed, and its wavelength are not independent of each other. There is a link between them, which applies to all waves of every kind. Imagine a source that vibrates five times per second. So it produces waves with a frequency of 5 Hz. If these have a wavelength of 2 metres in the medium they are travelling through, then every wave moves forward by 10 metres (5 × 2 m) in one second. The wave speed is 10 metres per second. In general,

	wave speed	=	frequency	×	wavelength
	(metre per second, m/s)		(hertz, Hz)		(metre, m)
or	(millimetre per second, mm/s)		(hertz, Hz)		(millimetre, mm)

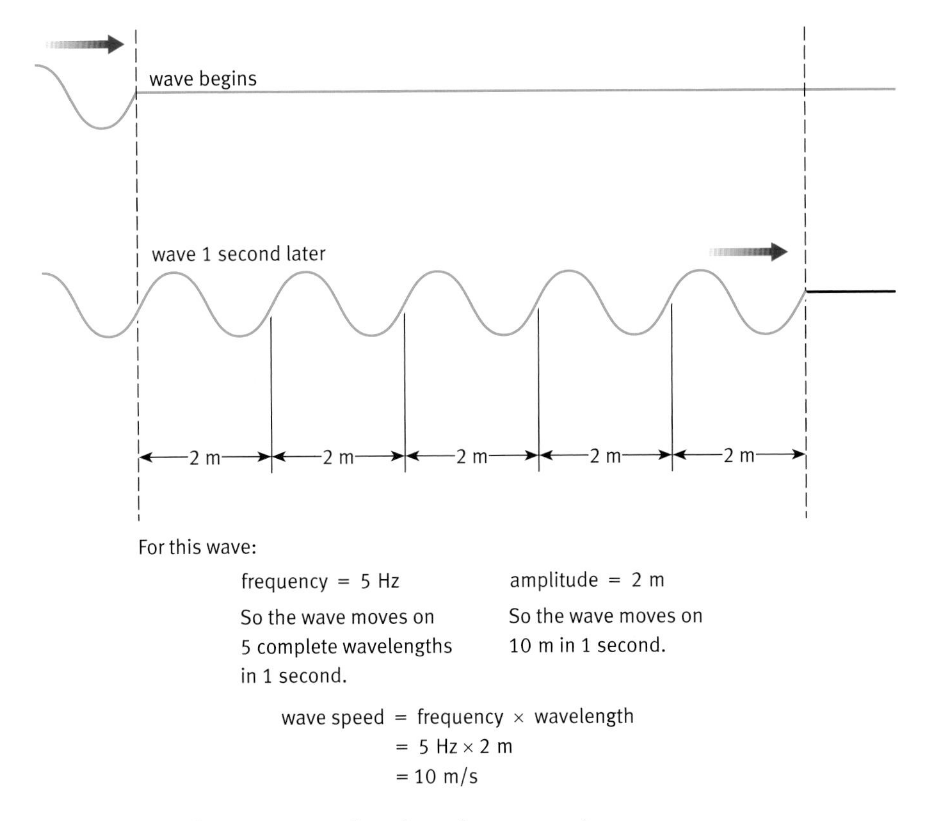

The link between frequency, wavelength, and wave speed.

Is wave speed the same for all frequencies?

The frequency of a wave and its wave speed are two quite separate things. In fact the speed of sound waves in air is almost exactly the same for all audible frequencies. And the same is true of the speed of light in air. However, light waves of different frequencies travel at slightly different speeds in media like glass or water. This has some important consequences (as you will see later on page 159).

Key words

amplitude
frequency
wave speed
wavelength

Questions

3 If you increase the frequency of waves through a medium, what will happen to their wavelength? Explain your reasoning.

4 When a stone is dropped into a pond, waves travel outwards at a speed of 500 mm/s. The wavelength of these waves is 100 mm. What is their frequency?

⑤ The speed of sound in air is 330 m/s. The note middle C has a frequency of 256 Hz. What is the wavelength of the sound wave when a musician plays this note? (You might find a calculator useful for this as the numbers do not work out neatly.)

⑥ If the speed of sound waves were very different for different frequencies, what effect would this have on the sounds we hear?

Find out about:
- four properties shared by all waves
- how waves behave at barriers, boundaries, edges, or obstacles

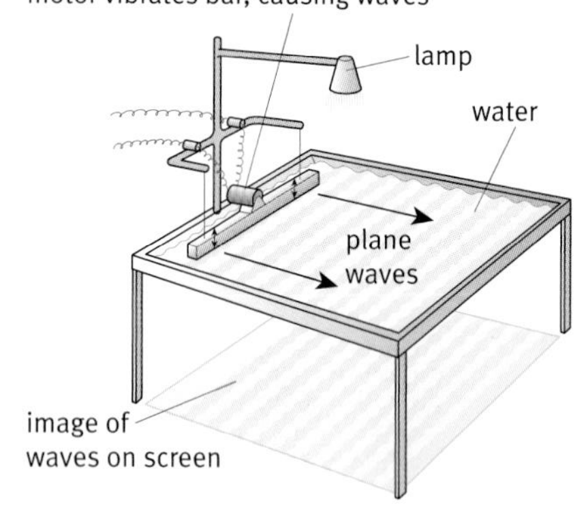

A ripple tank producing a steady stream of plane waves.

C Wave properties

Studying waves

Waves on a spring or a rope can travel in only one direction. Waves can be studied more fully by using a ripple tank. Waves are created on the surface of water and a lamp projects an image of the waves on to a screen.

Waves in water of constant depth are equally spaced. This shows that waves do not slow down as they travel. The wave speed stays the same. As the wave travels, it loses energy (because of friction). Its amplitude gets less but not its speed.

Using a ripple tank, you can observe four key properties of all waves.

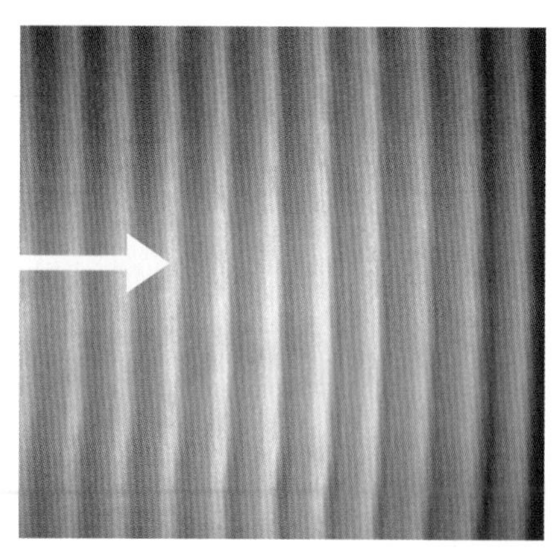

Plane waves travelling at constant speed

Reflection

Water waves are reflected by a straight barrier placed in their path. If you draw a line at right angles to the barrier, the reflected waves make the same angle with this line as the incoming waves, but on the other side.

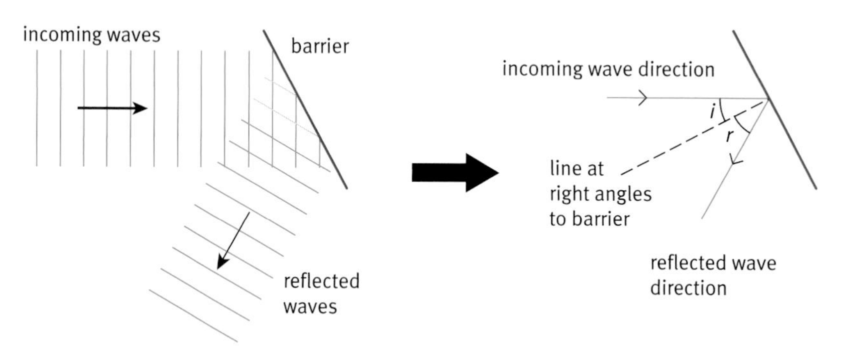

Reflection of water waves at a plane barrier. The angle of reflection (*r*) is equal to the angle of incidence (*i*).

Refraction

If waves cross a boundary from a deeper to a shallower region, they are closer together in the shallow region. The wavelength is smaller. This effect is called **refraction.** It happens because water waves travel slower in shallower water. The frequency (f) is the same in both regions, so the slower wave speed (v) means that the wavelength (λ) must be less (as $v = f\lambda$).

If the waves are travelling at an angle to the boundary between the two regions, their direction also changes. You can work out which way they will bend by thinking about which side of the wave gets slowed down first.

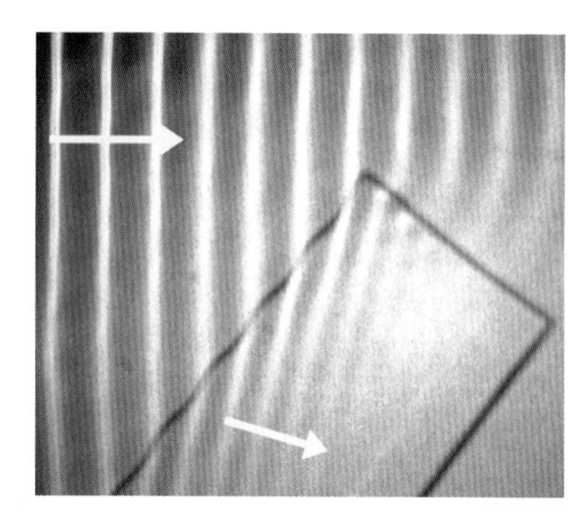

Refraction of water waves at a boundary between deep and shallow regions.

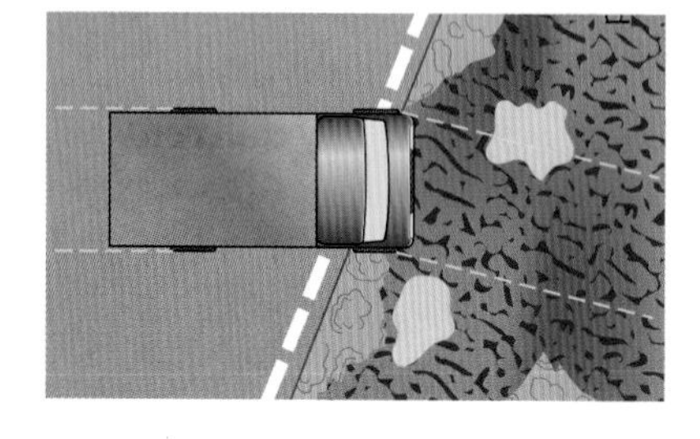

A model to help explain refraction. The truck goes more slowly on the muddy field. One wheel crosses the boundary first – so one side slows down before the other. This makes the truck change direction.

Diffraction

When water waves hit a barrier, they bend a little at the edge, and travel into the 'shadow' region behind the barrier. This effect is called **diffraction**. The longer the wavelength of a wave, the more it diffracts. At a gap between two barriers, waves bend a little at both edges. If the width of the gap is similar to their wavelength, the waves beyond the gap are almost perfect semicircles. If the gap is really tiny, much less than the wavelength of the waves, the waves do not go through at all.

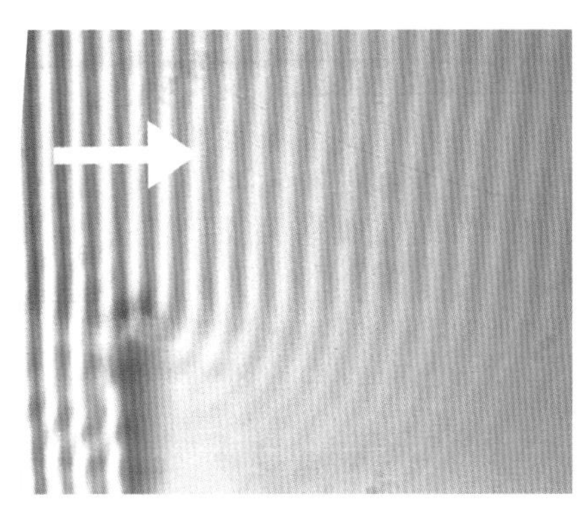

Waves diffracting at the edge of a barrier. Notice how some waves get round the corner.

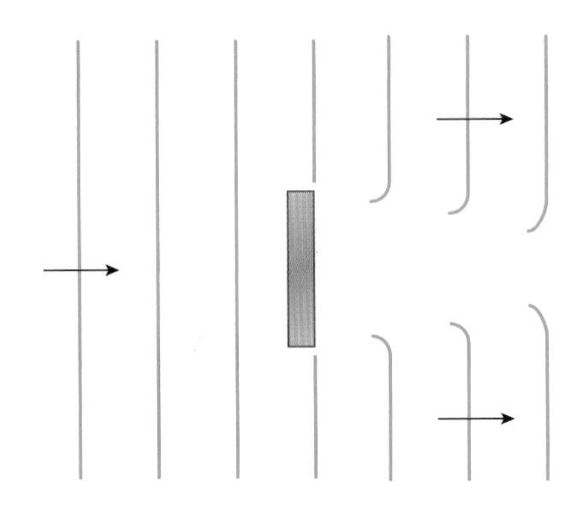

Diffraction occurs when waves meet an obstacle. The edges of the waves bend round the obstacle, into the shadow region behind.

Plane waves arrive at this harbour mouth. The waves inside the harbour are semicircular, because of diffraction at the narrow gap between the two piers.

Questions

1 Draw diagrams like the one on the left to show what happens as plane waves pass through a gap in a barrier that is:

a bigger than their wavelength

b about the same as their wavelength

Interference

When two waves meet, their effects add. If both waves have the same frequency, this causes an **interference** pattern. Where a crest of one wave meets a crest of the other, they add to make a bigger wavecrest. Where a crest of one meets a dip (or trough) of the other wave, the two cancel each other out.

The photograph on the right shows the surface of a ripple tank when two circular waves meet. Along some lines there is a large disturbance of the water surface. These are the points where the waves are 'in step' when they meet. Along other lines in between these, there is almost no disturbance of the water surface. Here the waves are 'out of step' when they meet, so they cancel each other out. The result is calm water.

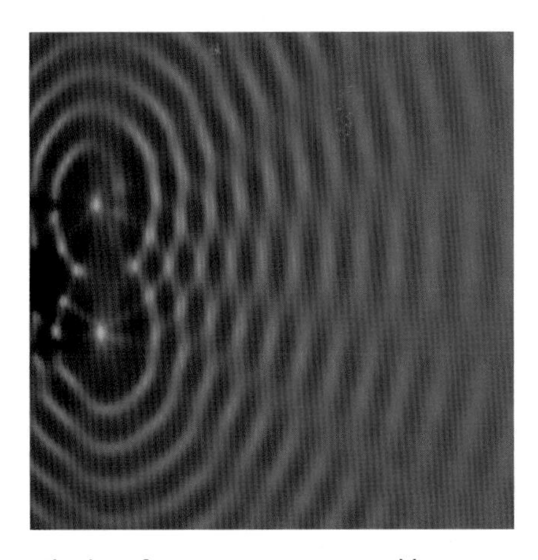

The interference pattern caused by circular waves spreading out from two dippers. Notice the lines of disturbed and calm water.

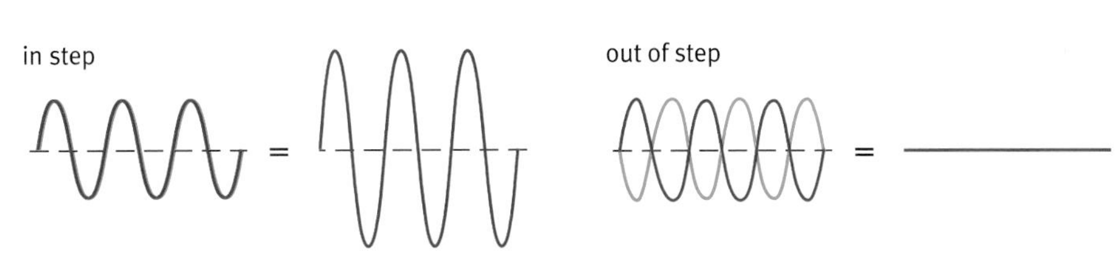

Two waves that are in step add to make a bigger disturbance. Two waves that are exactly out of step cancel each other out.

Key words

reflection
refraction
diffraction
interference

Find out about:

- a useful model for thinking about light and other radiations
- the evidence that light behaves like a wave

D Radiation and waves

Light is a type of **radiation**. Like all types of radiation, it travels out in all directions from a source until it hits another object. Here it might be **reflected** (bounce off), **transmitted** (go through), or **absorbed** (transfer energy to it), or a combination of all three. When radiation is absorbed, it causes changes in the absorber, which allow us to detect it.

There are many sources of light – electric light bulbs, candles, the Sun. Detectors include the retina of the eye, photographic film, light-dependent resistors (LDRs), and charge-coupled devices (CCDs) in digital cameras.

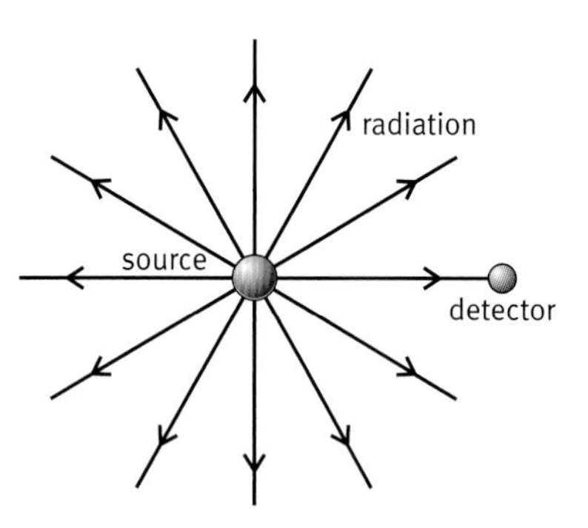

Radiation spreads out from a source and can affect another object (the detector) some distance away.

Light waves

People have always wondered what light really is. For a long time, scientists were unsure how best to think about light.

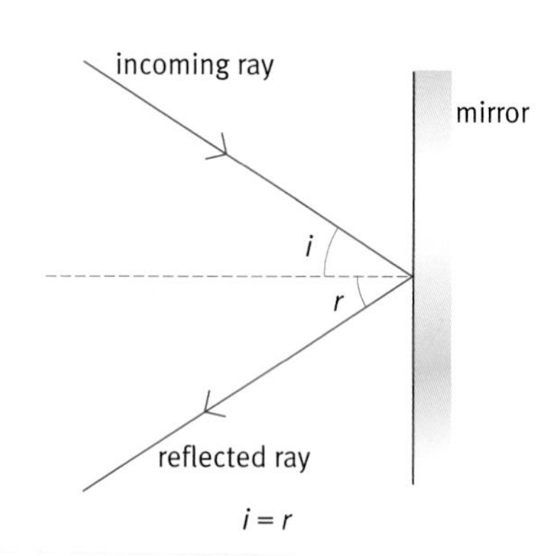

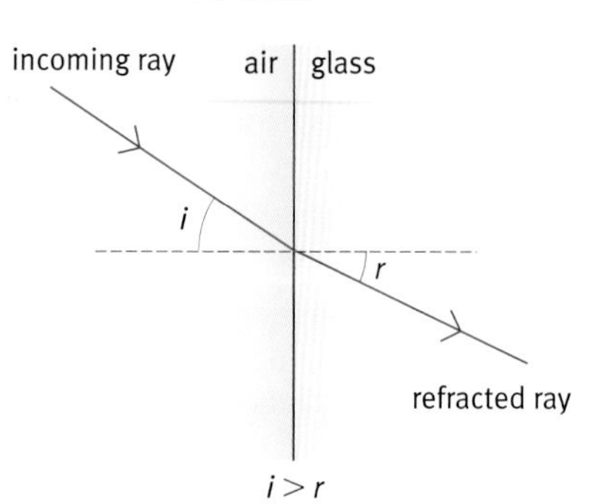

Light is reflected and refracted. This is consistent with light being a wave but does not really provide strong evidence, because a stream of particles would also do the same.

Two of the wave properties – reflection and refraction – do not really provide conclusive evidence either way. A wave or a stream of particles would also be reflected by a barrier. And both would be refracted at a boundary where their speed changed.

The evidence that convinced scientists it was useful to think of light as a wave came from an experiment carried out by Thomas Young in 1801. Young used a narrow slit to produce a fine beam of light from a bright lamp.

He shone this on a slide with a double slit (two parallel clear lines on a black slide). On a screen about 1 metre away Young saw a pattern of bright and dark vertical lines. To understand this, look back at the photograph on page 155 of interference in a ripple tank. With water waves, there are lines of disturbance and lines of calm water. If the same happens with light, you would expect lines of brightness and lines of dimness, spreading out behind the double slit. On the screen, you should see a series of bright and dark patches.

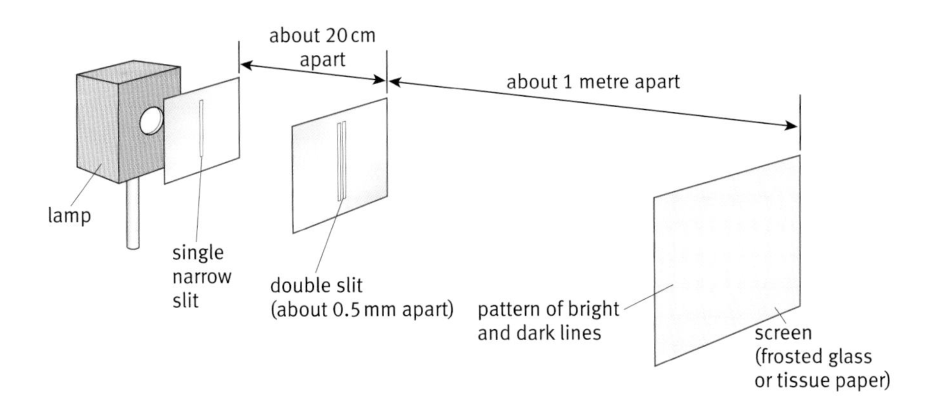

Young's experiment. The double slits are two sources of light of the same frequency. Light from these interferes to produce the pattern on the screen.

This is strong evidence that light behaves as a wave. If light were a stream of tiny particles, it could not produce an interference pattern in this way. Two streams of particles would not cancel each other if they met!

Young's experiment also involves diffraction. This the other property that is characteristic of waves. Light is diffracted by the two narrow slits. It spreads out into the region behind the slits, so that waves from the two sources overlap and interfere.

A wave model of radiation

If you look at water waves, you can see the medium (water) moving up and down. With sound waves, you cannot see the compression pulses in the medium, but you can detect them fairly easily. A candle flame in front of a loudspeaker flickers, showing the movements in the air around it. But with light, it is not possible to see or detect anything 'waving' or vibrating.

It is useful to think of light (and other radiations discussed later in this chapter) as a wave because it behaves in the same way as waves we can observe directly. These provide a useful model – a way of imagining what light is like. The wave model helps us explain our observations and predict what will happen in new situations.

What is the medium?

All the waves discussed earlier in this chapter need a medium to move through. But light can travel through a vacuum. It seems impossible to have a wave in a region where there is nothing at all. But in fact space (a vacuum) is not completely empty. Although it contains no matter, there can be fields in a vacuum. (Think, for example, of the gravitational field between the Sun and the Earth, which must act through empty space.) Scientists think that light is an **electromagnetic wave**. It consists of constantly varying electric and magnetic fields, which are able to move together through space.

Key words

radiation
reflected
transmitted
absorbed
electromagnetic wave

Questions

1 If you carefully covered one of the slits (of the double slit) in Young's experiment, what would you expect to see on the screen? Explain your answer.

2 What provides good evidence that sound is a wave?

Find out about:

- what happens to light crossing a boundary
- how optical fibres work
- the link between colour and frequency

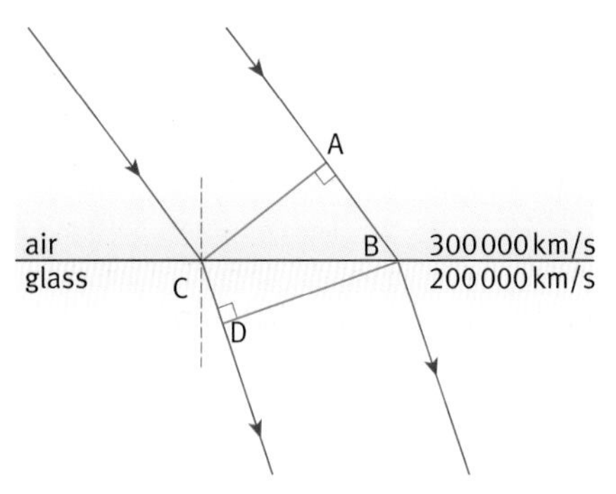

Light is refracted because its speed is different in the two media. Light travels the distance AB in air in the same time as it travels the distance CD in glass.

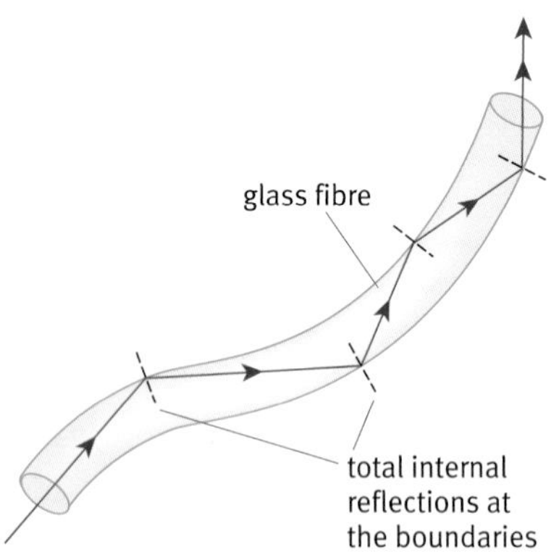

Light travels along an optical fibre because of total internal reflection.

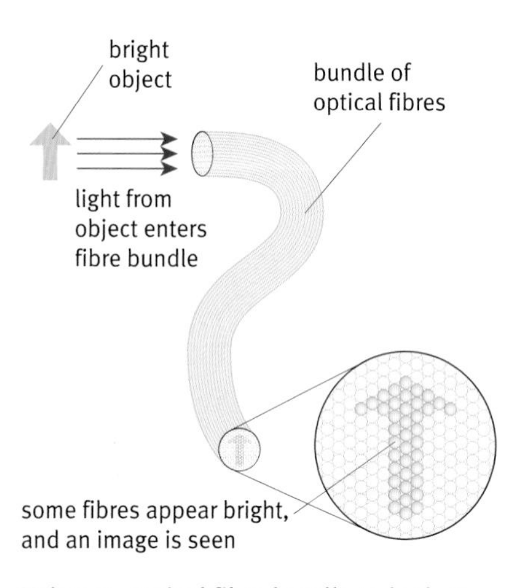

Using an optical fibre bundle to look at an inaccessible object.

E Bending light beams

Refraction and change of speed

A beam of light is refracted when it passes from one medium into another. This happens because the speed of light is different in different media. If the beam hits the boundary at an angle, one side is slowed down (or speeded up) before the other. This makes the beam change direction.

Total internal reflection

When a light beam crosses into a medium where it moves faster, it is bent towards the boundary between the media. By changing the direction of the ray in the first medium, you can make the emerging beam get closer to the boundary. At some point, it emerges right along the boundary. If you make the light hit the boundary at an even shallower angle, the boundary behaves as though it were a mirror. The light is reflected back into the first medium. This is called **total internal reflection (TIR)**.

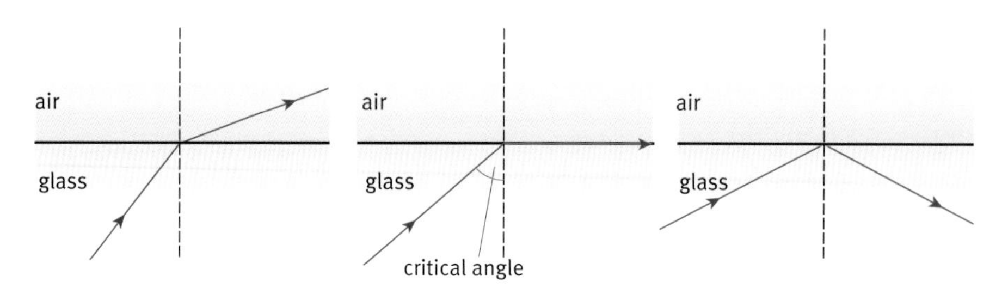

Light striking a boundary from glass to air at different angles.

Optical fibres

Optical fibres are at the forefront of the communications revolution. Thousands of telephone conversations are carried along a glass or clear plastic fibre, no thicker than a hair! An optical fibre makes use of total internal reflection. Light entering one end of a fibre is reflected repeatedly from the sides until it comes out the other end. Once inside the fibre, the light cannot escape as it always hits the surface at an angle that is bigger than the critical angle. It even follows bends in the fibre. The reflection is 'total', meaning that very little energy is lost through the sides. So the light can travel a long way down a fibre without getting much weaker.

Optical fibres enable doctors to look inside the body without using surgery. A bundle of optical fibres is used. Light is shone down some of the fibres to illuminate the object they want to inspect. A camera then takes a picture, using the light reflected from the object, coming back up the other fibres. Light from the object stays within the fibre it enters. So a clear image is visible at the other end.

Dispersion

A change of direction is not the only thing that happens to a light beam when it crosses a boundary between two transparent media. If a beam of white light passes through a triangular block (a prism), the emerging ray is coloured. This is called a **spectrum**. Newton carried out a famous series of experiments with prisms. He concluded that white light is really a mixture of the colours of the spectrum. This splitting of white light into colours is called dispersion.

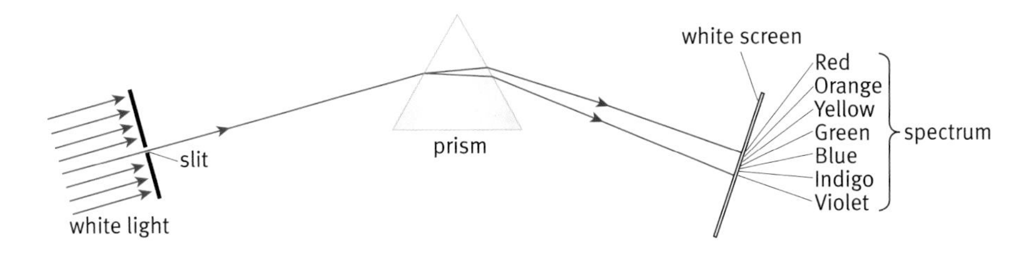

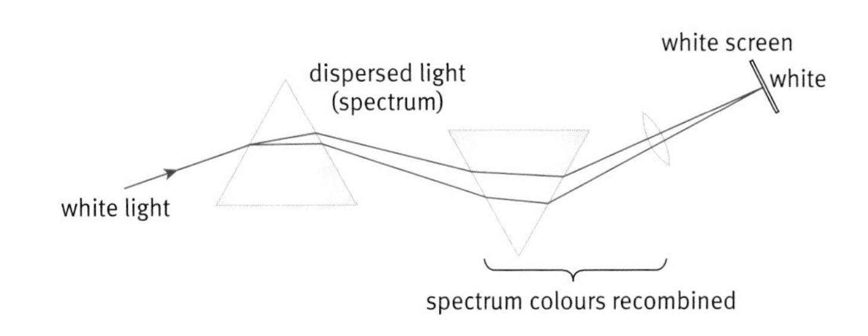

A second prism can recombine the coloured light rays to make white light again. So the colours are not caused by the glass of the prism.

Newton did not know what caused the different colours. Following Young's evidence that light is a wave (page 157), scientists were able to deduce that the colour of light depends on its frequency. (It therefore also depends on its wavelength, as the two are linked.) The visible band (waves that the human eye is able to detect) extends from red through to violet. Red has the lowest frequency (longest wavelength) and violet has the highest frequency (shortest wavelength).

Dispersion happens because light of different frequencies travels through glass (and other transparent media) at different speeds. The differences are small but they are enough to split the light up. In glass, red light travels faster than violet light. So it is not bent as much when it enters or leaves.

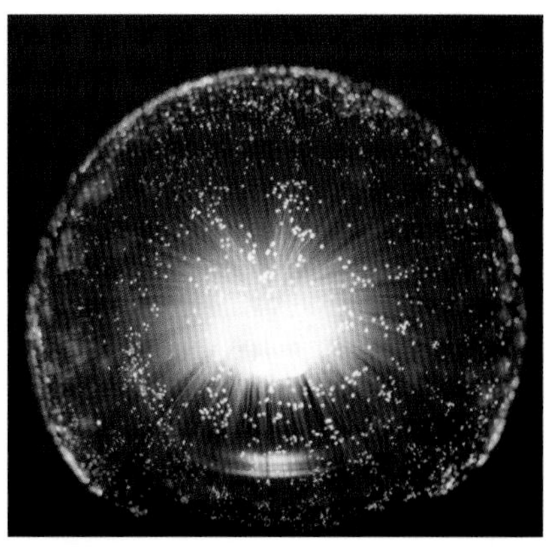

An optical fibre lamp. Light from a bulb inside the base travels through each fibre, so that each shows a tiny pinpoint of light at the end.

Key words

total internal reflection (TIR)
optical fibres
spectrum

Questions

1 These diagrams show rays of light travelling from one medium into another. Put the materials in order, from the one in which light travels slowest to the one in which it travels fastest.

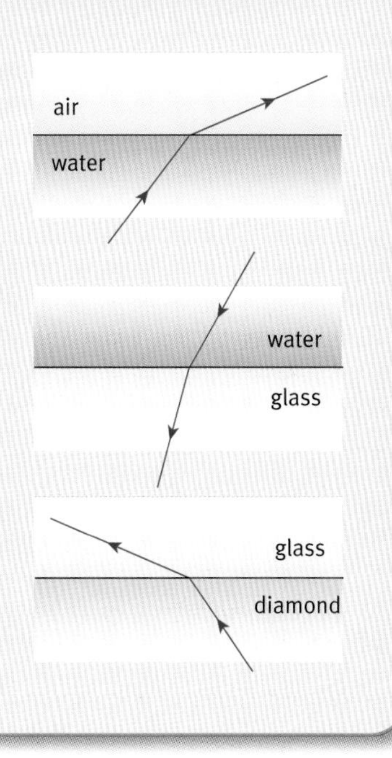

Find out about:

- the family of waves known as the electromagnetic spectrum
- the properties shared by all electromagnetic waves

The Sun emits electromagnetic waves, including infrared, light, and ultraviolet. All take the same time to reach the Earth because all travel through space at the same speed (see Question 4).

Key words

speed of light
photons
intensity
electromagnetic spectrum

F Electromagnetic waves

The human eye can detect light with frequencies between those of red and violet light. But there are other waves with frequencies bigger and smaller than this, which our eyes cannot detect. Visible light is just one member of a much larger family of electromagnetic waves – most of which cannot be detected by the human eye but require special equipment. These have several features in common:

- They can travel through empty space (a vacuum).
- They travel through space at a speed of 300 000 kilometres per second. This is usually called the **speed of light**, although it is the speed of all electromagnetic waves.
- They are transverse waves. However, it is tiny electric and magnetic fields that vibrate, rather than any material.
- They transfer energy. A source loses energy when it emits (sends out) electromagnetic waves. A material gains energy when it absorbs them.

Electromagnetic waves are very different from sound waves. Sound waves are longitudinal, travel much more slowly, and need a material (solid, liquid, or gas) to travel through.

Photons

Around 1905, Albert Einstein and some other physicists carried out investigations which led them to think that electromagnetic radiation is always emitted and absorbed in 'packets', or **photons**.

- The higher the frequency of an electromagnetic wave, the more energy each photon has.
- The total amount of energy which a beam of radiation transfers each second to the absorber (its **intensity**) depends on:
  - the energy of each photon in the beam
  - the number of photons arriving every second

Questions

1 *radio waves* *sound waves* *light* *gamma rays* *ultraviolet*

a Which of the above types of radiation is the odd one out, and why?

b State three ways in which the other radiations are similar.

2 Name a type of electromagnetic radiation which

a is visible to the eye

b is emitted by hot objects

c is used for radar

d is emitted by radioactive materials

3 For each type of electromagnetic wave in the diagram on page 161, name one source and one detector.

4 The Sun is 150 000 000 km from Earth. Estimate how long it takes for the Sun's light to reach us.

The electromagnetic spectrum

Below, you can see the full range of electromagnetic waves. It is called the **electromagnetic spectrum**. At the bottom are the lowest-frequency radio waves. These have wavelengths of several kilometres. At the top are the highest-frequency gamma rays. These have wavelengths of less than a billionth of a millimetre.

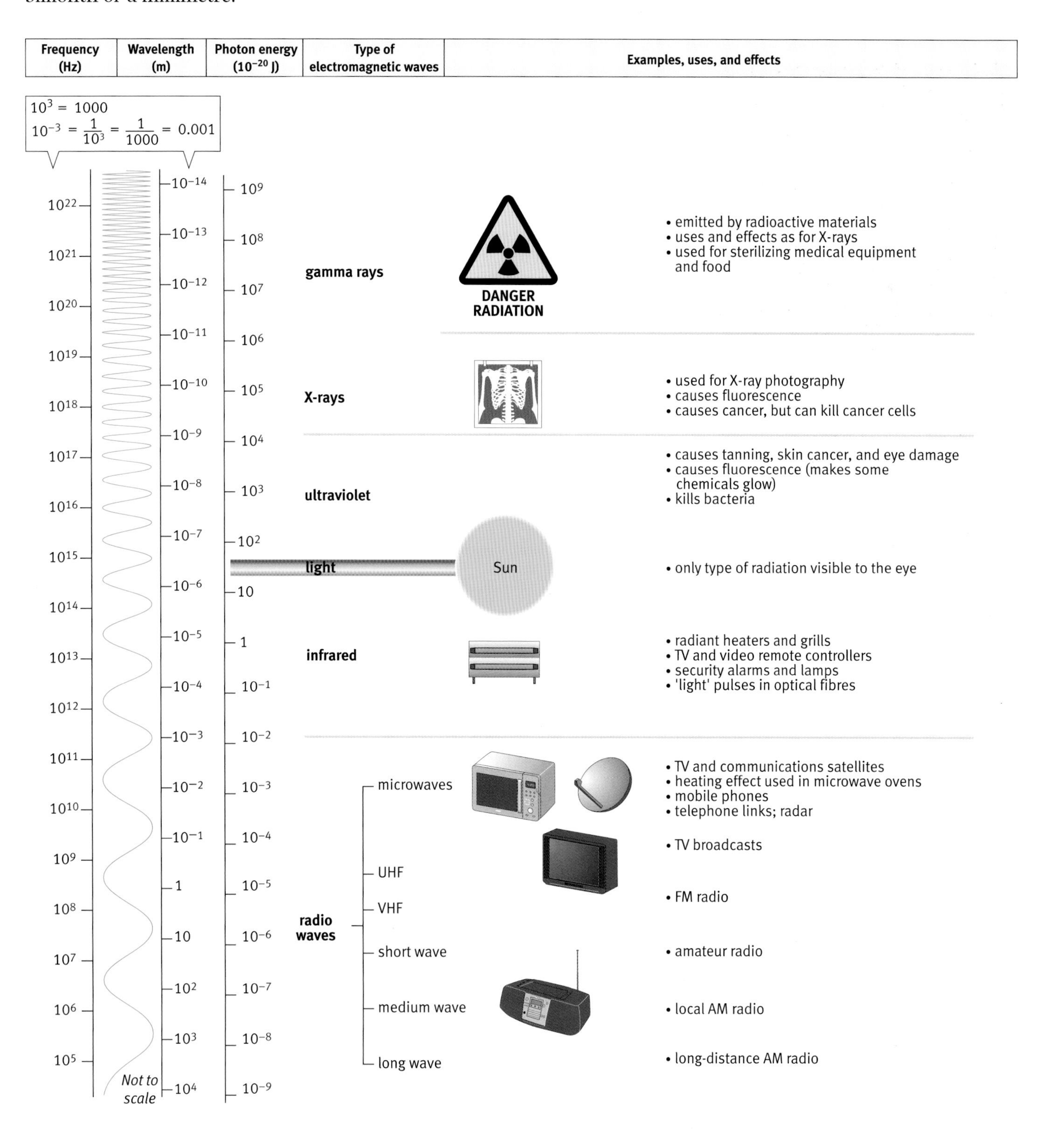

Find out about:
- electromagnetic waves with higher frequency than visible light
- ionizing radiations

G Above the visible

The human eye is sensitive to only a narrow range of frequencies within the electromagnetic spectrum. This is the range we call light. Together, the eye and brain sense the different frequencies as different colours. Violet light has the highest frequency and red the lowest. There are other kinds of electromagnetic radiation beyond the visible range.

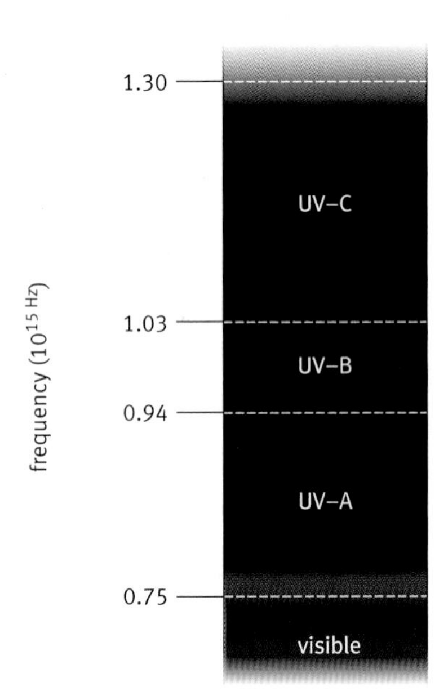

Ultraviolet is classified into three ranges according to its effect on skin. UV-C is the most dangerous but much of it is absorbed by ozone in the Earth's upper atmosphere. UV-A is not harmful in normal doses.

Ultraviolet

Very hot objects, such as the Sun, emit some of their radiation beyond the violet end of the visible spectrum. This is **ultraviolet** radiation.

Most of the Sun's ultraviolet radiation does not reach the Earth's surface. High in the atmosphere, a band of gases called the ozone layer absorbs most of it. Only the 'near' ultraviolet, with frequencies closer to visible violet light, gets through. This is fortunate, as ultraviolet is harmful to living cells. If too much is absorbed by the skin, it can cause skin cancer. If you have black or dark skin, the ultraviolet is absorbed near the surface. But with fair skin, the ultraviolet can go deeper. Skin develops a tan to try to protect itself against ultraviolet. Ultraviolet can also damage the retina in the eye. On the other hand, ultraviolet is essential for the chemical reaction by which the body produces vitamin D. Vitamin D is needed by the immune system to resist diseases, including cancers.

Some materials fluoresce when they absorb ultraviolet. That is, they emit visible light and glow. Fluorescent paints look bright because they absorb the ultraviolet in sunlight and emit visible light. In fluorescent lamps the ultraviolet is produced by passing an electric current through the gas (mercury vapour) in the tube. The inside of the tube is coated with a white powder that emits light when it absorbs ultraviolet.

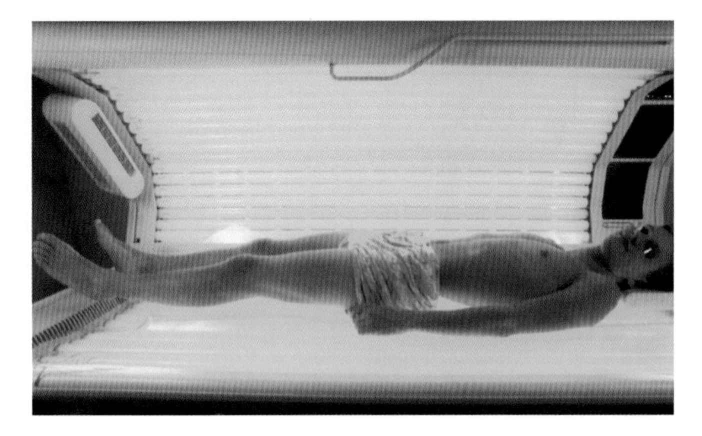

Sunbeds tan you because they emit UV-A.

Mars has a thin atmosphere and almost no ozone layer. Ultraviolet levels on its surface are much higher than on Earth. So simple life, such as bacteria, would find it very difficult to exist there.

Ionizing radiations

Ultraviolet and electromagnetic waves with still higher frequencies are ionizing. Each photon has enough energy to strip an electron from an atom in its path (remember: electrons are tiny charged particles in atoms). **Ionizing radiation** can alter the materials it strikes and, in the case of living cells, may damage or destroy them.

X-rays

X-rays have higher frequencies than ultraviolet. In an X-ray tube, X-rays are emitted when a beam of fast-moving electrons hits a metal target.

High-frequency X-rays are extremely penetrating. They can even pass through dense metals like lead. Lower-frequency X-rays are less penetrating. For example, they can pass through flesh but are absorbed by bone, so bones will show up on an X-ray photograph. X-rays are dangerous because they damage living cells deep in the body and can cause cancer. For this reason, exposure times are kept very short. However, concentrated beams of X-rays can be used to treat cancer by destroying abnormal cells.

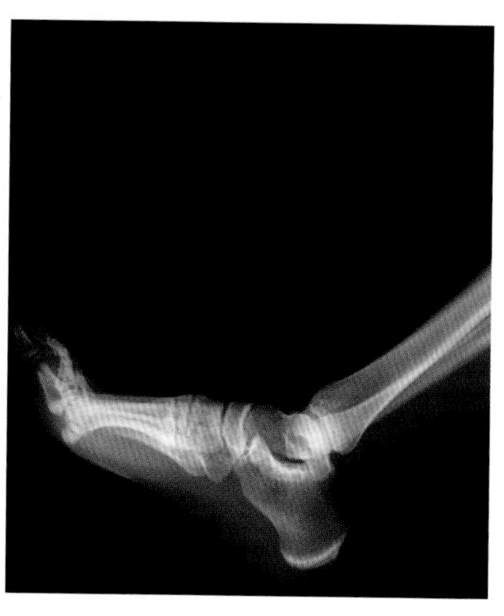

X-ray photograph. The X-rays pass through the flesh but not the bone. This is why these materials are seen as different shades in the picture.

At airports, X-ray machines are used for security checks on baggage. The wavelength is chosen so that the X-rays are absorbed by metal objects but pass through the softer, less-dense materials. So objects such as knives and guns can be seen on a screen.

Gamma rays

Gamma rays come from some radioactive materials. Most have shorter wavelengths than X-rays. However, the two types overlap (in frequency, and wavelength) in the electromagnetic spectrum, and there is no difference in behaviour between X-rays and gamma rays of the same wavelength. Like X-rays, gamma rays can be used in the treatment of cancer, and for taking X-ray-type photographs. As they can kill cells, they are used to kill harmful bacteria, for example to sterilize food and medical equipment.

Key words

ultraviolet
ionizing radiation
X-rays
gamma rays

Questions

1 The Sun emits many ultraviolet frequencies. Here are three of them:
0.8×10^{15} Hz 1.0×10^{15} Hz 1.2×10^{15} Hz

Which of these:

a is the most dangerous?

b is normally absorbed in the Earth's upper atmosphere?

c is safe in normal doses?

2 Someone claims to have invented a machine that produces X-rays so penetrating that they can pass right through any known material. Why would these X-rays be of no use for medical photos or security checks at airports?

3 a State two ways in which gamma rays are similar to X-rays.

b A key difference is in how two kinds of waves are produced. Explain what this difference is.

Find out about:
- electromagnetic waves with lower frequency than visible light
- how the wave model helps explain their behaviour

H Below the visible

Infrared

Beyond the red end of the spectrum there is radiation that the human eye cannot detect. This is **infrared** radiation.

When a radiant heater or grill is switched on, you can detect the infrared radiation coming from it. You notice the heating effect in your skin when it absorbs the radiation. In fact, all objects emit some infrared because of the motion of their atoms or molecules. Most emit a wide range of frequencies.

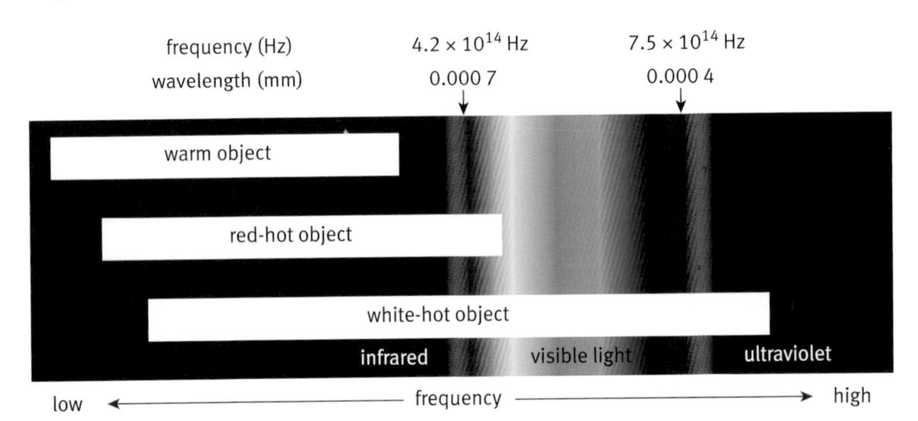

All objects emit some infrared. As the temperature rises, the wavelengths get shorter and enter the visible part of the spectrum.

As something heats up, it radiates more and more infrared, and at higher frequencies. At about 700 °C, the highest frequencies can be detectedby the eye, so the object glows 'red hot'. Above about 1000 °C, the emitted radiation includes the whole of the visible spectrum, so the object is 'white hot'.

A greenhouse 'traps' energy by letting high-frequency infrared in but not letting low-frequency infrared out. This is sometimes called 'the greenhouse effect'.

Most solids absorb infrared: this makes their temperature rise. However, high-frequency infrared can pass through glass and clear plastics, although longer wavelengths are absorbed or reflected. The inside of a greenhouse heats up in the sunshine because the glass lets the Sun's light and high-frequency infrared through but reflects back the lower frequencies coming from the warmed materials inside. In other words, the energy is 'trapped'.

Using infrared

Grills, toasters, and radiant heaters all use the heating effect of infrared. At night, warm things go on emitting infrared even though no light comes from them. Night-vision goggles and cameras detect this radiation and use it to produce a visible image. You can see an example on the left.

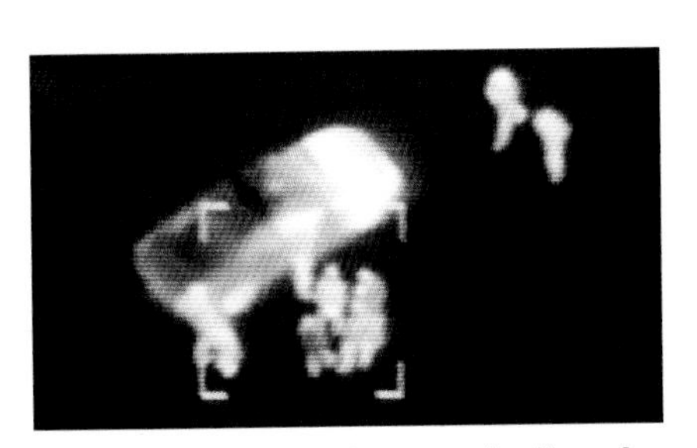

Although it is dark, the warm bodies of the people standing by the car and walking towards it are emitting infrared. This is detected by the camera and turned into a visual image.

Most security lamps are switched on by motion sensors that detect the changing pattern of infrared caused by the warm body of an approaching person. Pulses of infrared can also transmit information. For example, a remote control uses them to send coded instructions to a TV. Infrared pulses can also be carried by optical fibres.

Microwaves

Like all electromagnetic waves, **microwaves** have a heating effect when they are absorbed. This principle is used in microwave ovens, where microwaves penetrate deep into food and heat up the water in it.

In a microwave oven, the part that emits the radiation is called a magnetron. It is designed to emit microwaves with a frequency of 2450 MHz. This is lower than the frequency at which water absorbs best, but there is a good reason for this. If the absorption were much better, the microwaves would just heat the surface of the food, and not penetrate deep inside.

Microwaves are not ionizing radiation, however. A microwave photon does not have enough energy to knock an electron out of an atom.

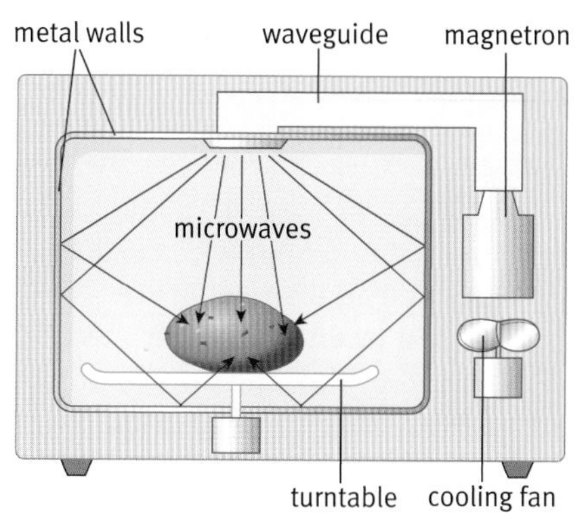

Microwaves produced by the magnetron pass along the waveguide and into the cooking chamber. The metal walls of the chamber reflect the microwaves inwards. The waves are also reflected by the metal grid behind the glass door because their wavelength is much larger than the holes in it. The food is rotated on a turntable to even out the effects of any 'hot spots'.

Microwave communications

Microwaves have the highest frequencies (and shortest wavelengths) of all radio waves. They are used by mobile phones and satellite TV. Satellites can relay (pass on) microwaves from one part of the Earth to another because microwaves are not reflected or absorbed by the ionosphere. This is a layer of ions (electrically charged atoms) in the upper atmosphere. It reflects lower-frequency radio waves (see page 166). Microwaves can also be beamed across country between dish aerials on tall towers. But they do not diffract readily round obstacles because of their relatively short wavelength. So many towers are needed to give good coverage.

Mobile phones rely on a network of transmitting and receiving aerials on masts all over the country. The areas between masts are called cells.

Some clever electronics in your phone keeps switching the carrier frequency (see page 168) as you move from one cell to another. By doing this, it is possible for thousands of people to use mobile phones at the same time without overhearing other people's conversations or needing to keep their own special frequency.

Aerials for sending and receiving radio signals to mobile phones. They are high up to provide good lines of site to other aerials

Questions

1 *light*
long-wavelength infrared
short-wavelength infrared

Which of the above:

a can be detected by the human eye?

b can pass through glass?

c is (or are) emitted by red hot objects?

2 Explain why a metal bar starts to glow red when it is heated up.

3 Infrared has a heating effect on most solids. What does this tell you about whether this radiation is reflected, transmitted, or absorbed?

4 A TV remote control emits an infrared beam that carries coded information to the TV set. Explain how you might use this to find out which everyday materials absorb infrared and which transmit it (allow it to pass through).

5 List three uses of microwaves.

Radio waves

Radio waves are produced by making an electric current oscillate (making charges move to and fro) in a transmitting aerial. All radio waves have a lower frequency (and longer wavelength) than infrared. But this still covers a large range of wavelengths, so radio waves are further sub-divided into the groups shown on page 161. To carry information, radio waves must be varied in some way (see page 168). The variations, called signals, can carry sound, text, and TV pictures.

Radio waves normally travel in straight lines. As the Earth is curved, you would expect that no radio station would transmit farther than 30 or 40 miles. This is true for ground-based TV transmissions. But radio stations using the short and medium wave bands can travel much further. The reason is that the ionosphere reflects short and medium radio waves. The waves bounce between the ground and the ionosphere and so travel further from the transmitter.

Radar

Radar also uses microwaves. The name is shorthand for 'Radio Detection And Ranging'. Pulses sent out by a dish aerial are reflected by an object (such as a plane or a ship). The time it takes for reflected pulses to come back is measured. From this the distance to the object can be estimated.

An air traffic control radar

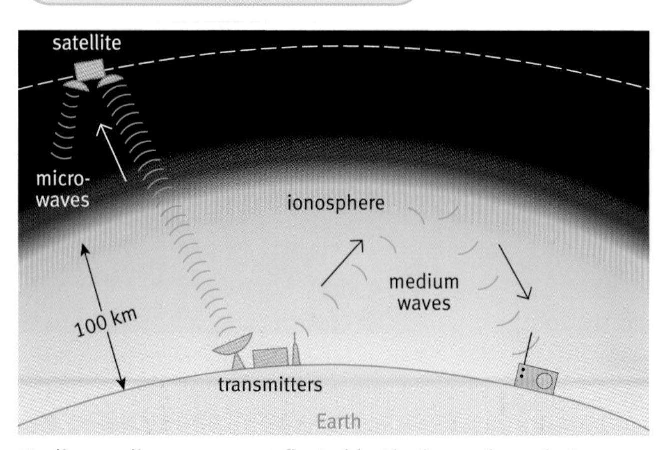

Medium radio waves are reflected by the ionosphere, but microwaves can pass through.

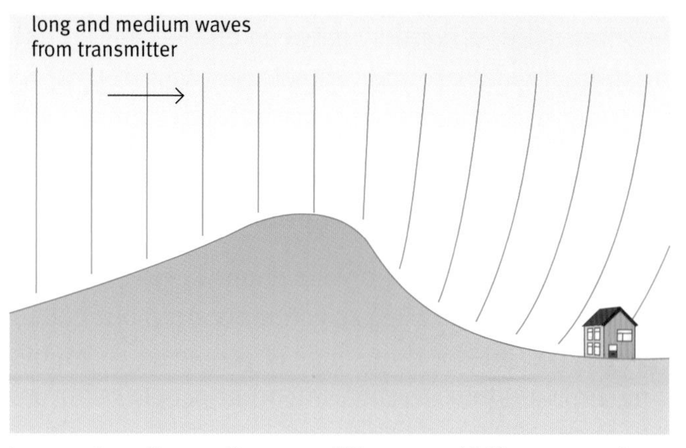

Long and medium radio waves diffract round hills, so a receiver down in a valley can still pick them up.

Diffraction also plays an important part in determining how different kinds of radio wave behave. Remember (page 155) that longer-wavelength waves diffract more readily than shorter-wavelength ones. Long and medium waves diffract (bend) around hills and other obstacles. VHF (very high frequency) is used for stereo radio and UHF (ultra-high frequency) for TV broadcasts. Because their wavelength is much shorter, these waves do not diffract so much round hills. So for good reception, there normally needs to be a straight path to your aerial. This means that more transmitting aerials are needed to give good reception everywhere.

Transmit

In science, the word 'transmit' has two different meanings. When we say that glass 'transmits' light that shines on it, we mean that the light passes through the glass and (most of it) is not absorbed. But when we say that an aerial 'transmits' radio waves, we mean that it sends them out.

Radio interference

If you drive along a motorway with the car radio on, you may sometimes notice that the strength of the radio signal changes. This is caused by interference between radio signals from two transmitters. As you drive along, you pass through places where the two waves add to give a stronger wave – and other places where they cancel each other out (partly) and the signal is weaker.

People who live near airports also sometimes find the radio signal varying as a plane flies overhead. This is due to interference between the radio wave reflected off the metal body of the plane and the one reaching them directly.

Bluetooth is a new idea using microwaves. It enables information to be sent between mobile phones, computers, and other devices, without any wires.

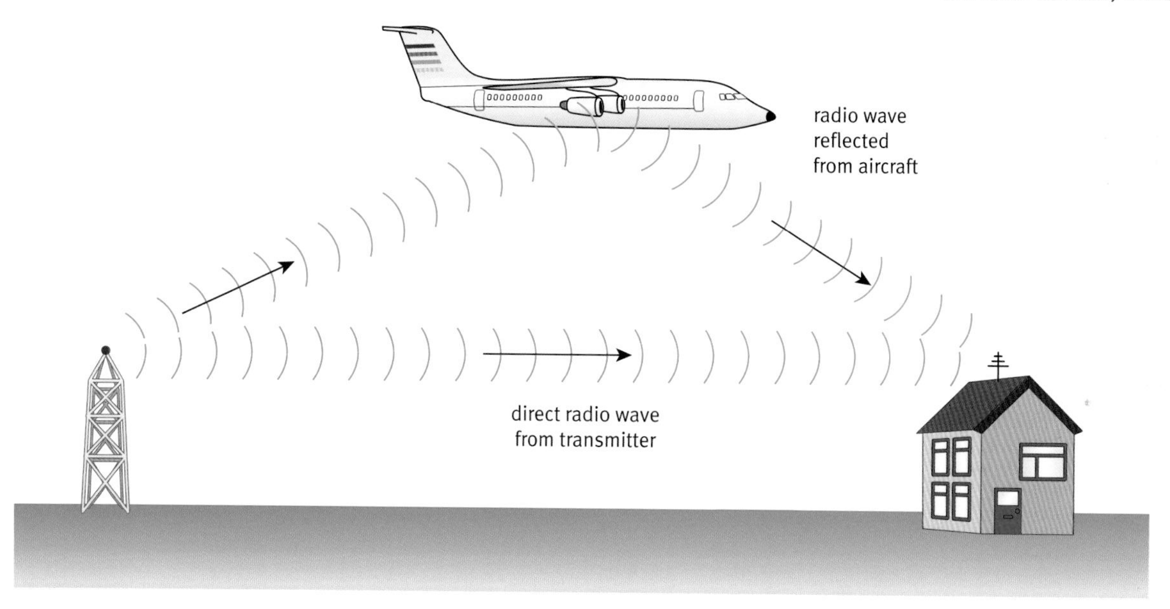

A low-flying aircraft can cause radio interference in homes nearby.

Questions

6 List two pieces of evidence that radio waves really are waves

7 A radio station is broadcasting on a frequency of 100 MHz.

- **a** How many radio waves does it send out every second?
- **b** Use the equation linking wavespeed, frequency, and wavelength to calculate the wavelength of the radio waves. (*Note*: First, you must write down the speed of radio waves in metres per second.)

8 **a** What is the ionosphere?

- **b** What effect does the ionosphere have on medium-wave (MW) radio waves? What effect does it have on microwaves?

Key words

infrared
microwaves
radio waves

Find out about:

- how information is carried by radio waves

I How radio works

Electric charges flowing backwards, forwards, backwards, forwards . . . many times a second is called an alternating current (a.c). When there is a.c. in a piece of wire, the wire acts as an aerial. Radio waves are produced. And when radio waves reach another aerial, they generate a.c. there. This is what makes radio communication possible.

Sending sounds by radio

The diagram below shows a simple radio system. When you speak into the microphone, radio waves carry the sound from one aerial to another, and a copy of your voice comes out of the loudspeaker. The incoming sound waves might have a frequency, typically, of around 1 kHz. This is the audio frequency (AF). The radio waves have a much higher frequency, for example 1 MHz. This is the radio frequency (RF).

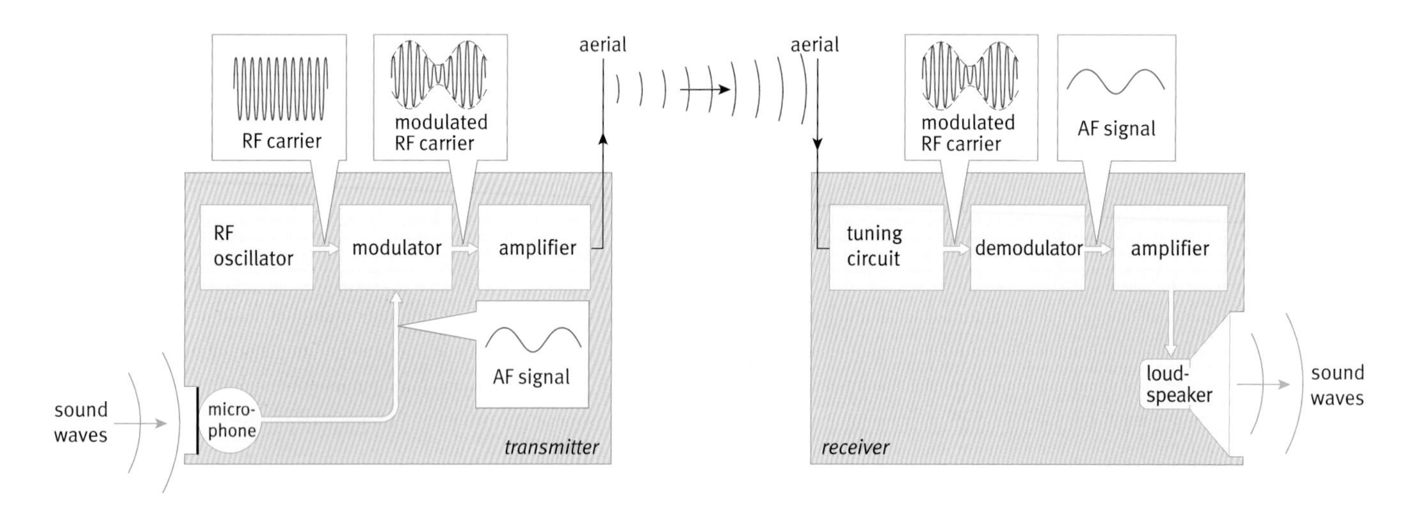

A simple radio system. When you speak into the microphone, a copy of the sound comes out of the loudspeaker.

In the microphone, the varying pressure from the sound waves produces a varying voltage – the AF signals.

The RF oscillator creates the high-frequency a.c. needed to produce the radio waves. The steady stream of waves is called the **carrier**. The AF signals are used to modulate (vary) the amplitude of the carrier so that the 'height' of the RF waves is a copy of the incoming sound waves. Before the modulated wave is sent to the aerial, its amplitude is boosted by an amplifier.

In the receiving aerial, the incoming radio waves generate electrical signals. This will be a mixture of the frequencies of all the different radio stations that are transmitting in the area. The tuning circuit selects the frequency of one RF carrier – the one carrying the radio station you want to listen to. The demodulator then removes the RF carrier, leaving only the AF signals. These are boosted by an amplifier and sent to the loudspeaker, which produces the sound.

TV pictures are also transmitted using radio waves. A TV camera scans each scene, breaking it down into hundreds of narrow strips. Information about how the brightness and colour changes along each strip is then used to modulate the carrier.

AM and FM

The system described above uses **amplitude modulation (AM)**: the amplitude (or height) of the radio frequency (RF) carrier wave is varied by the audio frequency (AF) signals. Another method is to use **frequency modulation (FM)**, where the frequency of the carrier is varied instead of the amplitude. One advantage of FM is that the signals are less affected by electrical interference, called **noise**. Noise causes extra unwanted variations in amplitude, but with FM, this has less effect.

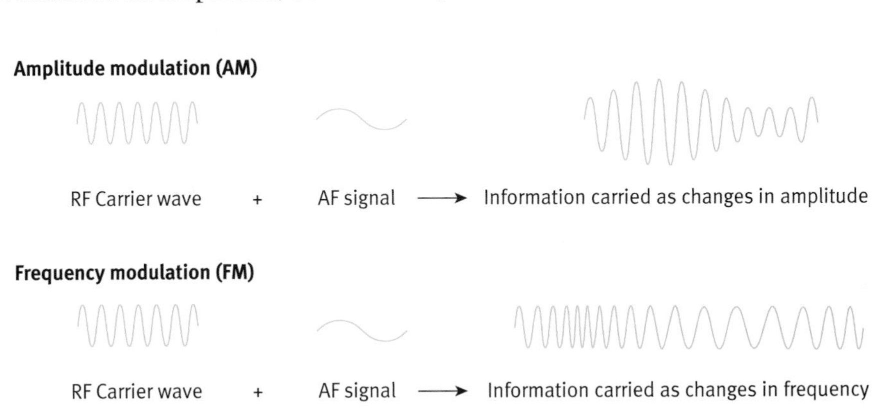

Two ways of modulating a radio frequency (RF) carrier wave, so that it carries an audio frequency (AF) signal.

Questions

1 The radio system in the diagram on the previous page has to deal with two frequencies, called RF and AF.

a Which of these is the frequency of the sound waves?

b Which is the frequency of the carrier?

2 Why does a radio receiver need a tuning circuit?

3 **a** In an AM radio system, what does 'modulating the carrier' mean?

b How would this be different in an FM system?

c What is the advantage of using FM rather than AM?

Key words

carrier
amplitude modulation (AM)
frequency modulation (FM)
noise

Find out about:
- how a digital signal differs from an analogue one
- the advantages of using digital signals for communications

J Going digital

Radio, TV, and telephone are all forms of telecommunication – ways of transmitting (sending) information long distances. The information may be sounds, pictures, text, or numbers, and it can be sent using wires, radio waves, or light.

Analogue and digital signals

When sound waves enter a microphone, a varying voltage is generated. The graph on the right shows how the voltage might change during one fraction of a second. Continuous variations like this are called **analogue signals**. In this case, the voltage rises and falls in the same way as the pressure in the sound waves.

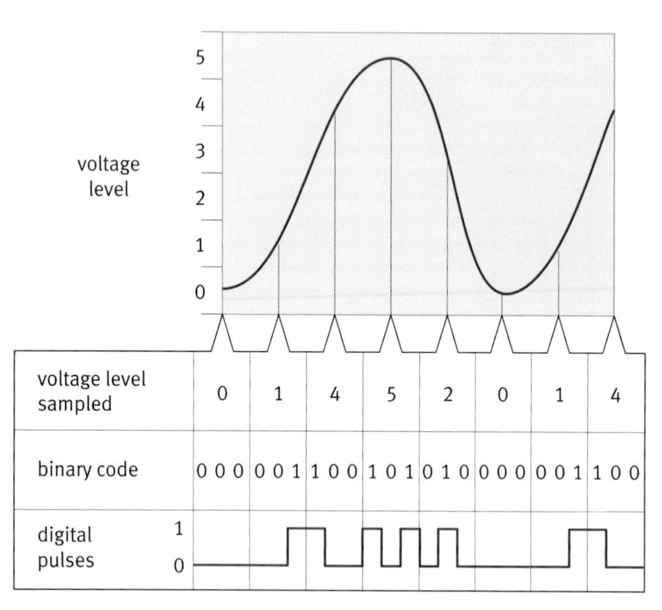

How an analogue signal can be converted into digital pulses. (Real systems use more levels and a much faster sampling rate.)

The table under the graph shows how this changing voltage can be converted into **digital signals**. These are signals represented by numbers. The voltage is sampled electronically many times per second. In effect, the height of the graph is measured repeatedly. Then the measurements are coded into binary (numbers using only 0s and 1s) – a **digital code**. This is transmitted as a series of pulses (no pulse = 0; pulse = 1). At the receiving end, the digital signals are **decoded** (turned back into analogue signals).

Real systems use more sampling levels than in this diagram, much faster sampling rates, and further stages of coding in order to work more efficiently.

Questions

1 A meter in which a pointer moves along a scale is called an analogue meter.

a Why do you think it is called this?

b What would you expect to see on a digital meter?

2 In a digital system, what does 'sampling' mean?

3 An analogue signal is converted into the following binary code:

100 010 001 010
100 100 010

Using the examples in the diagram above as a guide,

a sketch a graph showing the digital pulses

b sketch a graph to show what the analogue signal might have looked like

Advantages of digital transmission

For transmitting information such as sounds and pictures, digital systems have several advantages over analogue ones:

- Digital signals can be handled by microprocessors (as in computers).
- Digital signals can carry more information every second than analogue ones.
- Digital signals can be delivered with no loss of quality. In other words, the sequence of 0s and 1s does not change. Analogue signals lose quality, which cannot be restored.

Here is the reason for the last point. All signals get weaker as they travel along. Noise (interference) also gets added in. But with digital signals, these effects can be corrected. Even when the incoming signals have added noise, it is still possible to tell the 0s from the 1s, as these are the only values the signal can have. It would take a very large amount of noise before it became difficult to tell the 0s and 1s apart! A regenerator can then be used to restore the pulses to their original quality. Analogue signals cannot be 'cleaned up' in this way. They can be amplified but, unfortunately, any noise is amplified as well.

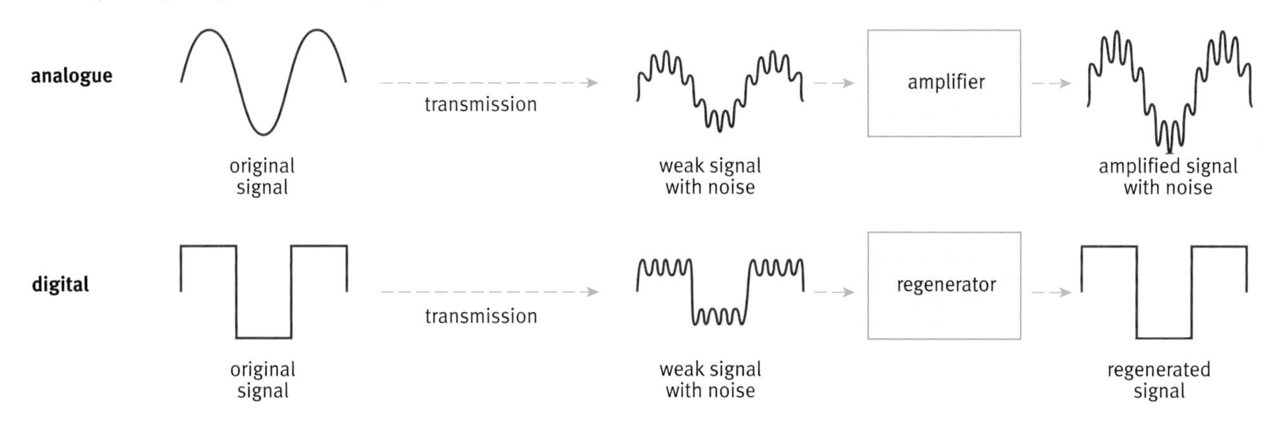

Digital signals can be 'cleaned up' by regenerators, but when analogue signals are amplified, noise gets amplified as well.

Digital recording

A CD (compact disc) contains a metal layer with millions of tiny bumps on it, arranged in a spiral track. When you play a CD, the disc is rotated and laser light is reflected from the bumps. The reflected pulses – the light signals – are turned into electrical signals and then decoded to produce the sound. DVDs (digital versatile discs) use the same idea, although they have to store much more information in order to create full-colour TV pictures.

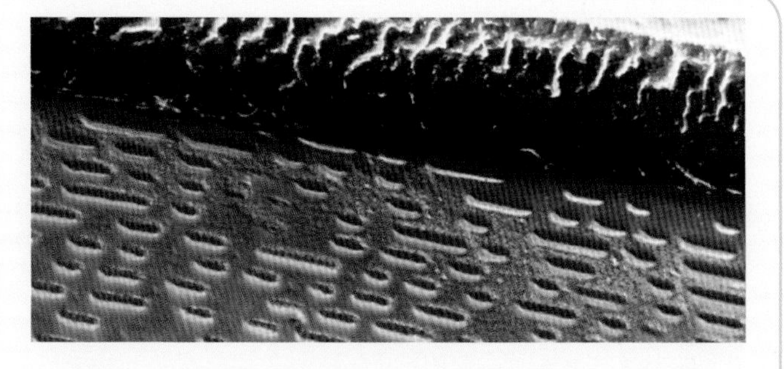

Magnified surface of a CD, showing the bumps on it (×2800)

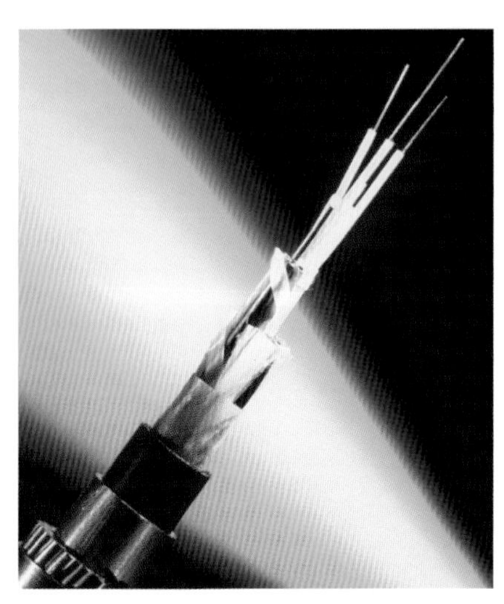

Optical fibres from inside a cable. Most telephone calls between Britain and America are carried by a cable like this, lying along the bottom of the Atlantic Ocean.

Transmitting through optical fibres

For long-distance transmission, many telephone and cable-TV networks use optical fibres rather than radio waves, to carry the signal (see page 158). These long, thin strands of glass or clear plastic carry digital signals in the form of very brief pulses of light (or infrared). In effect, a light beam is being switched on and off very rapidly to produce a digital code of 1s and 0s. At the transmitting end, electrical signals are changed into light signals by an LED (light-emitting diode) or a diode-laser. At the receiving end, the light signals are picked up by a photodiode and changed back into electrical signals.

Optical fibre cables are thinner and lighter than electric cables. They can also carry many more signals. With electrical cables, the signal gets weaker quite quickly, so amplifiers are needed at regular intervals to boost it again. With optical fibres, these 'booster stations' can be much farther apart. Optical fibres are not affected by electrical interference and they cannot be 'tapped' (people cannot listen in to others' messages).

Digital radio

DAB (Digital Audio Broadcasting) can handle more stations than the older analogue AM and FM systems and give high-quality, interference-free reception. With the older systems, reflections from hills and buildings cause interference. DAB actually uses these 'multipath' signals to improve quality! Its transmitters send out radio waves in pulses. In the receiver, a microchip does some clever processing on the various incoming signals, and uses them to produce a stronger signal.

DAB transmitters all use the same basic carrier frequency. However, this does not mean that you can only listen to one station. Digital signals from up to ten stations are split into groups in such a way that they can be transmitted alongside each other. The receiver unscrambles the different groups when they arrive. This system is called multiplexing. It is also used by telephone networks so that lots of calls can be sent together along the same cable.

A digital radio (DAB) receiver

Key words

analogue signals
digital signals
digital code
decoded

Questions

4 Why is it possible to remove noise from digital signals but not from analogue ones?

5 List as many reasons as you can why optical fibres are better than electric cables for long-distance communication.

K Radiation from space

Getting through

Only some types of electromagnetic wave can pass through the Earth's atmosphere. The graph below shows the percentage of radiation of each frequency reaching the ground. Radiation of many frequencies is completely absorbed by the atmosphere.

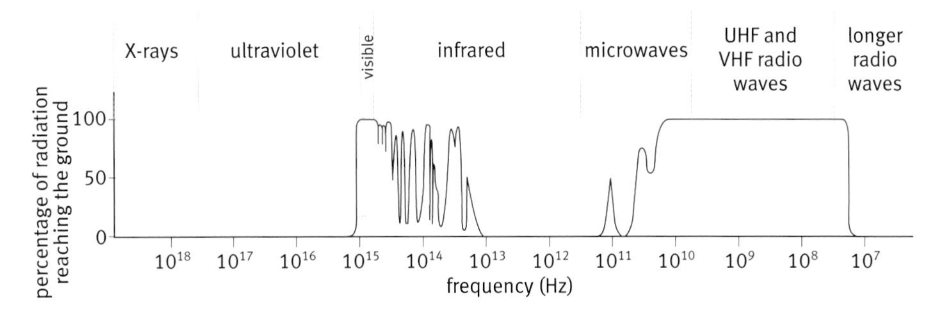

Graph showing the percentage of radiation reaching the earth's surface across the electromagnetic spectrum

One of the two main bands that can pass through the atmosphere is visible light (the band of frequencies that our eyes can detect.)

The only other large band of radiations that get through the atmosphere is lower-frequency microwaves and UHF and VHF radio waves. This is why radio telescopes have been developed.

Listening for aliens

As well as observing natural radio emissions from outer space, radio telescopes can be used to search for extraterrestrial intelligence. We on Earth have to hope that intelligent life is capable of sending out radio signals. The search for these signals is called the SETI project (the Search for ExtraTerrestrial Intelligence). It has been going on for over 40 years. If we ever do detect signals that we think come from extraterrestrials, having conversations with them would be almost impossible because of the time delay. Our radio signals would take nearly four years to reach even the nearest star. Then we would have to wait another four years for a reply.

Nowadays astronomers also use telescopes that detect infrared and ultraviolet, but the telescope itself has to be in space, outside the Earth's atmosphere. Information collected is then relayed back to Earth using radio signals which can pass through the atmosphere.

Find out about:

- which electomagnetic radiation can get through the atmosphere
- how astronomers make use of this

The big bang, coming to a screen near you!

Scientists think that the Universe began about 14 billion years ago with a huge explosion called the big bang. Radiation from the big bang is still reaching us from every direction in space – it is called the **microwave background.** If you turn on a TV and look at a channel that hasn't been tuned in, you will probably see lots of dancing spots on the screen. Much of this 'snow' is due to the microwave background.
You are seeing the remnants of the big bang!

Radio telescopes like this one are mainly used to detect the natural radio waves coming from distant stars and galaxies. However, they can also be used to search for signals from intelligent extraterrestrials.

Key words

microwave background

Summary

This module has introduced you to some basic ideas about waves, and how they behave. Light and other members of the electromagnetic spectrum behave like waves. Several types of electromagnetic wave (including radio, microwaves, infrared, and visible light) are used nowadays for communications.

Waves

- A wave is a disturbance moving through a medium. Parts of the medium move to and fro as the wave passes, but the medium does not move as a whole in the direction of the wave.
- A wave carries energy and information through the medium.

Wave characteristics

- The amplitude of a wave is the maximum disturbance of each particle of the medium as the wave passes.
- The frequency is the number of waves produced every second by the source (which is also the number of waves passing any point in the medium every second).
- The wave speed is the speed at which each wave crest moves through the medium.
- Amplitude and frequency depend on the source; wave speed depends on the medium.
- The wavelength is the distance from one wave crest to the next.
- For all waves:
  'wave speed = frequency × wavelength'.
  In any given medium, the wave speed is fixed. So the bigger the frequency, the smaller the wavelength.

Wave properties

- Four ways in which all waves behave are: reflection, refraction, diffraction, and interference.
- The last two are characteristic of waves. They provide evidence that a radiation behaves like a wave.

Light waves

- Light behaves like a wave, showing diffraction and interference, if you look carefully.
- Scientists believe that light consists of vibrating electric and magnetic fields – an electromagnetic wave.

Electromagnetic spectrum

- The electromagnetic 'family' of waves all travel through vacuum with a wave speed of 300 000 km per second.
- Electromagnetic waves have very different properties, depending on their frequency. In order of increasing frequency, they are: radio, microwaves, infrared, visible light, ultraviolet, X-rays, gamma rays.
- The three highest-frequency types are ionizing: they can cause chemical changes in materials that absorb them.

Communications

- Electromagnetic waves can be used to carry information. This is 'coded' on to a carrier wave, as changes in amplitude or frequency (analogue signals), or by pulsing the beam on and off very rapidly (digital signals).
- Digital signals can be communicated more accurately, with less unwanted 'noise'. H

Questions

1 Copy this table and complete it by writing either *gets bigger*, *gets smaller*, or *stays the same* in each cell:

	amplitude	frequency	wavelength	wave speed
a waves on the surface of a pond, as they travel away from the disturbance that caused them				
b a wave on a long spring, when the end is moved up and down more rapidly				
c water waves being diffracted as they pass through a gap in a barrier				
d a light wave as it moves from air into a clear plastic block				
e radio waves as they travel from a satellite to a receiving dish on the ground				

2 Sketch diagrams to show what happens to plane water waves in each of the following situations:

a hitting a straight barrier, at an angle of 45°

b moving from shallow into deeper water

c hitting an obstacle whose width is about 4 wavelengths

3 Describe what you would see as two sets of water waves, of exactly the same frequency and travelling across the same water surface at right angles to each other, met and crossed. Sketch a diagram if it helps you explain. Which wave property is involved here?

4 Make a list of all the pieces of evidence you would use to convince someone that the radiations of the electromagnetic spectrum behave like waves.

5 Make a table with four columns. In the left-hand column, list the following types of electromagnetic radiation:

radio waves, microwaves, infrared, light, ultraviolet, X-rays, gamma rays.

Complete the other three columns to show, for each type of radiation:

main sources, ways of detecting, main uses.

6 All electromagnetic waves travel at a speed of 300 000 km/s in a vacuum (and at much the same speed in air).

a How long does it take for a radio signal to travel from Britain to New Zealand (18 000 km)?

b How long does it take for a radio message to travel to Earth from an astronaut on the Moon (320 000 km)?

c How long does it take for light from the Sun to reach Earth (150 000 000 km)?

d When the *Voyager 2* probe passed Neptune, it was approximately 4 500 000 000 km from Earth. It sent back the photos it took as radio signals. How long did they take to reach us?

e The nearest star (after the Sun) is 4.35 light-years away, i.e. the distance light travels in 4.35 years. How many kilometres is this?

7 Optical fibres are increasingly important for communications. Explain why a light ray travels along an optical fibre, following any bends in it, and not escaping through the sides. Include a diagram to show what is happening.

Why study the Universe?

Physics offers an important way of looking at the world. Studying the whole Universe helps us know *where* the Earth is and *when* it is too because, it turns out, the Universe has a history. Though it may seem odd, explaining what happens in stars requires an understanding of matter at the microscopic scale, right down to the smallest sub-atomic particles. Everything in the physical world is made from a few basic building blocks.

The science

In the 1830s, the French philosopher Auguste Comte suggested that there were certain things that we could never know. As an example, he gave the chemical composition of the stars. By 1860, two years after his death, physicists had interpreted the spectrum of starlight and identified the elements present. In this module, you will look at how physicists have gradually extended our understanding of stars and galaxies.

Physics in action

In making their observations, astronomers use many different instruments – think of space probes, new telescopes, computing, and the Internet. Better instruments and new, collaborative observatories have played an essential part in extending astronomers' understanding of what is 'out there'. Physics enables such new technologies. New technologies enable new science.

Observing the Universe

Find out about:

- how astronomical instruments are used
- how observations of the night sky are interpreted
- how a star like the Sun functions
- a theory of the life history of stars
- how astronomers work together

Topic 1 Observatories and telescopes

Astronomy is the oldest science in the world. Ancient civilizations – for example, the Chinese, Babylonians, Egyptians, Greeks, and Mayans – practised naked-eye astronomy long before the invention of telescopes in the 17th century. They built costly observatories for both practical and religious reasons.

Everything beyond the Earth moves across the sky, from horizon to horizon. Calendars and clocks were based on cycles in these movements: day and night, the phases of the Moon, and seasonal changes in the Sun's path. Long-distance travellers, on sea and on land, navigated using the positions of familiar stars.

Today there are plenty of opportunities to visit astronomical observatories and see some of the great variety of telescopes that have been invented down the years, for a variety of purposes.

The Monument, London

The Monument in London was built to commemorate the Great Fire of 1666. It is a cylinder 202 feet (67m) tall – and it is a telescope! It was designed by Christopher Wren and Robert Hooke as an 'azimuthal telescope'. The observer lies down and looks up through the central tube at the sky above. As stars pass directly overhead, they can be timed from night to night, giving accurate measurements of their positions.

Greenwich Observatory, London

Jodrell Bank, Cheshire

The Royal Observatory Edinburgh

Greenwich Observatory in south-east London was built by the English Navy in the 17th century. They had suffered a number of defeats at the hands of the Dutch, because the enemy were able to navigate better. Dutch astronomy was in advance of the English, giving them more accurate charts of the night sky, which were used to work out a ship's position.

Jodrell Bank in Cheshire is famous for its radio telescopes. At the Visitor Centre, you can see the operators at work, day and night, mapping distant radio sources far beyond our galaxy.

The Royal Observatory Edinburgh is a world-renowned centre for astronomical instrumentation. Here, electronic cameras are designed and built, and images are processed electronically. The staff also work on the systems that control the direction in which a telescope is pointing.

The observatory at Knighton in mid-Wales is part of the Spaceguard Foundation. It keeps an eye out for Near-Earth Objects that might threaten our very existence.

The observatory at Knighton, mid-Wales

Find out about:

- the variety of telescopes
- the visual images produced by telescopes

1A What is a telescope?

In the autumn of 1609, Galileo made his first observations of the Moon using a **telescope**. He was not the first person to use a telescope to look at the night sky. But the observations Galileo made, and his interpretation of them, had repercussions down the centuries. He changed the way people see the Universe.

Telescopes make things visible which cannot be seen with the naked eye. Galileo's telescope allowed him to see craters on the Moon, and the moons of Jupiter.

The Effelsberg radio telescope in Germany. With a diameter of 100 m, it gathers radio waves from distant objects in the Universe, including galaxies other than our own. The dish is scanned across the sky to generate an image.

This reflecting telescope is at the Calar Alto observatory, over 2000 m above sea-level in southern Spain. Light passes through the ring (diameter 3.5 m) and reflects off the curved, shiny mirror at the back.

This astronomer is using a refracting telescope, which uses lenses to focus light.

An artist's impression of the X-ray Multi-Mirror (XMM) Newton telescope. X-rays enter the telescope at the near end and are reflected by mirrors onto detectors at the far end.

A new golden age of astronomy began in the 1930s, with the accidental discovery of radio waves coming from beyond the Earth. Astronomers realized that objects in space do not only produce visible light. Since then, many new kinds of telescope were invented. Each of them requires a suitable detector and a method of focusing the radiation. Astronomers now gather radiation from across the whole of the electromagnetic spectrum.

These pages show a number of telescopes, some more recognizable than others.

Key words
- telescope
- refraction
- reflection

Jocelyn Bell and Anthony Hewish were the discoverers of pulsars. These are distant objects which send out radio waves which vary with an extremely regular pulse. The wires in the field form an aerial which is their 'telescope'. Altogether, there are 1000 posts spaced out over 4.5 acres.

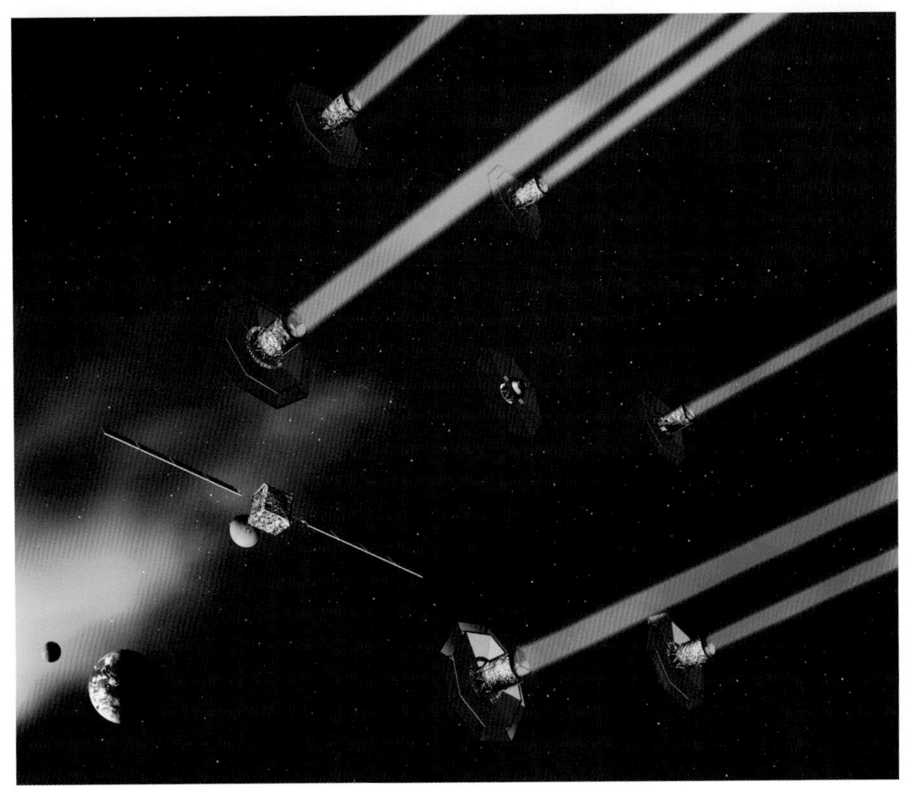

These are the Darwin infrared space telescopes which the European Space Agency proposes to launch in 2014. They will gather infrared radiation from planets orbiting stars other than the Sun.

Questions

1 Different telescopes make use of different types of electromagnetic radiation. List the telescopes shown on these two pages, together with the radiation which each gathers.

2 Telescopes 'make things visible which cannot be seen with the naked eye'. Is that true for all of the telescopes shown here?

This engraving of the Moon, as observed with a telescope, was made by Claude Mellan in 1635, just 25 years after Galileo's first lunar observations.

Seeing is believing

People like to see things with their own eyes. You feel you know something truly exists if you can see it. It is better still if you can touch it, but of course no one can touch most of the objects that astronomers study.

The first telescopes used visible light, so they produced images that people could see. Later, astronomers designed telescopes to gather radio waves, infrared radiation – in fact, every type of electromagnetic radiation, even gamma rays. The human eye cannot detect these, but ways were found to make them into pictures, using computers.

You probably think of the landscape as nearby trees and hills. But the landscape we live in extends well beyond the horizon, and out into space. A leap of human imagination can make that 'cosmic' landscape feel real too.

Radiation from objects in space reaches a telescope after a very long journey. As astronomers look at ever more distant objects, the light they gather set off ever further back in the past. In recent decades, astronomers have collected an enormous number of observations. Interpretation of the raw data is difficult. Yet, by putting together observations from different wavelengths, it is possible for them to build up a picture of the cosmic landscape, and its history.

Some astronomical images are shown on this page.

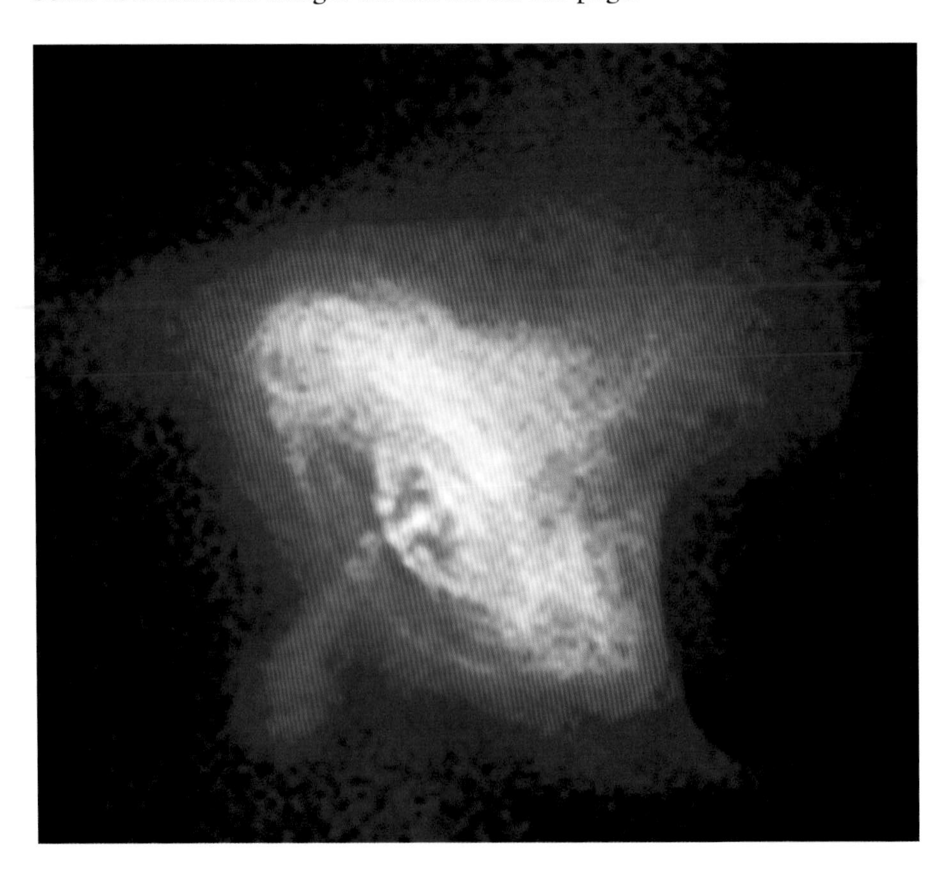

This image was made by an X-ray telescope. Although the human eye cannot detect X-rays, their energies can be measured and converted into an image. This is the Crab Nebula, the remains of a star that exploded in 1054 AD. At its centre is a pulsar.

Hearing pulsars

Jocelyn Bell, an Irish astronomer, was one of the discoverers of pulsars, strange sources of regularly pulsing radio waves. Her telescope was a giant radio aerial, and the 'image' it produced was a long paper chart. In October 1967, she was involved in making a survey of radio sources in the sky.

> Six or eight weeks after starting the survey I became aware that on occasions there was a bit of 'scruff' on the records, which did not look exactly like man-made interference. Furthermore, I realized that this scruff had been seen before on the same part of the records – from the same patch of sky.
>
> Whatever the source was, we decided that it deserved closer inspection, and that this would involve making faster chart recordings. As the chart flowed under the pen, I could see that the signal was a series of pulses, and my suspicion that they were equally spaced was confirmed as soon as I got the chart off the recorder. They were 1.3 seconds apart.

Jocelyn Bell with part of one of the charts produced by her radio 'telescope', showing the trace produced by a pulsar.

Electrical signals from the radio aerial can be played through a loudspeaker, so that the regular pulses can be heard. When first discovered, astronomers considered the possibility that they were a signal from some extraterrestrial civilization. Many other pulsing sources have since been identified, and astronomers now think that these signals come from rapidly spinning neutron stars.

Neutron stars are just one among the amazing variety of extreme and often violent objects that astronomers have discovered 'out there'. Later, you will learn how the classification of stars led to an understanding of how stars work, and how they change.

Questions

3 Look at the image of the Moon on the opposite page, and think about how the Moon appears to the naked eye when you see it in the night sky. What extra features has the artist who made this image been able to identify by using a telescope?

4 Look at the image of the Crab Nebula on the opposite page. How can you tell where the X-ray source is strongest?

Find out about:

- how converging lenses focus parallel rays
- the focal length and power of a lens

1B Describing lenses

The earliest and simplest telescopes used lenses. To understand these telescopes, you need first to understand what lenses do.

The focal length of a converging lens

The picture below shows how you can make a miniature image of a distant object using a single **converging lens**. If you position the screen correctly, you will see a small, inverted image of a distant scene.

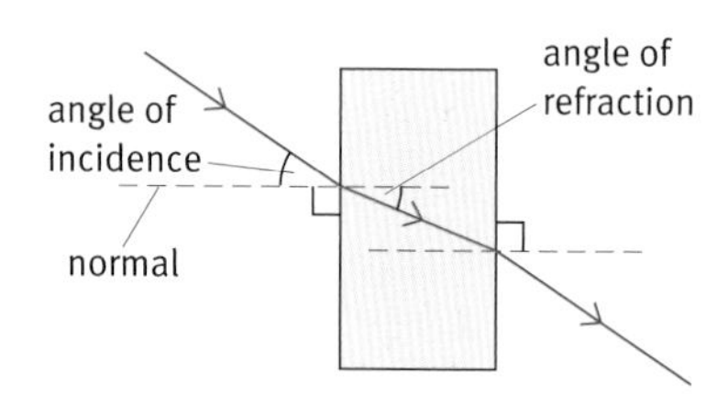

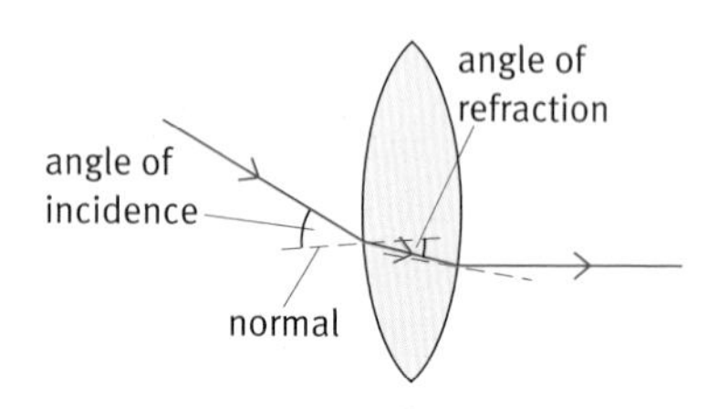

A ray of light bends as it passes at an angle from one material to another. This effect is called **refraction**. Angles are conventionally measured from the **normal,** an imaginary line drawn perpendicular to the boundary at the point where a ray strikes.

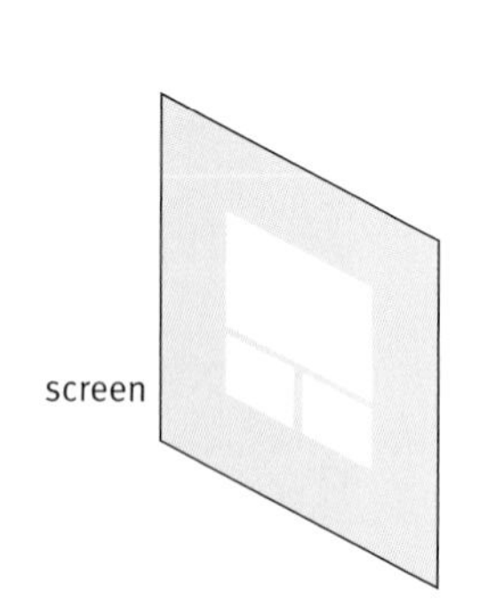

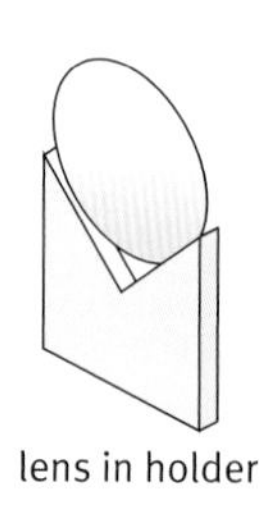

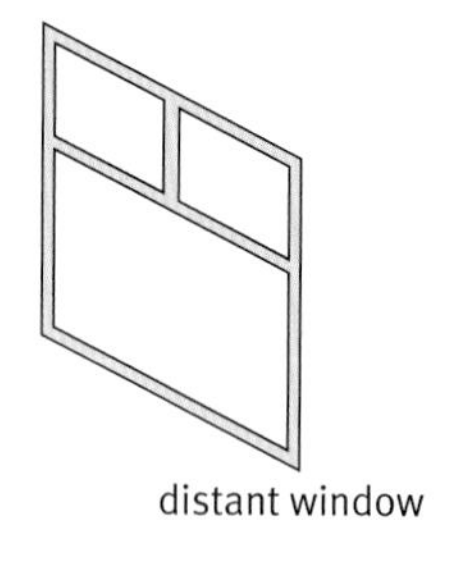

Using a converging lens to make an image.

Rays of light enter the lens. Because of the lens shape, they are refracted (they change direction), first on entering the lens and again on leaving.

A ray diagram shows this. A horizontal line passing through the centre of the lens is called the **principal axis**. Rays of light parallel to the axis are all refracted so that they meet at a point. This is why converging lenses are so called: they cause parallel rays of light to converge.

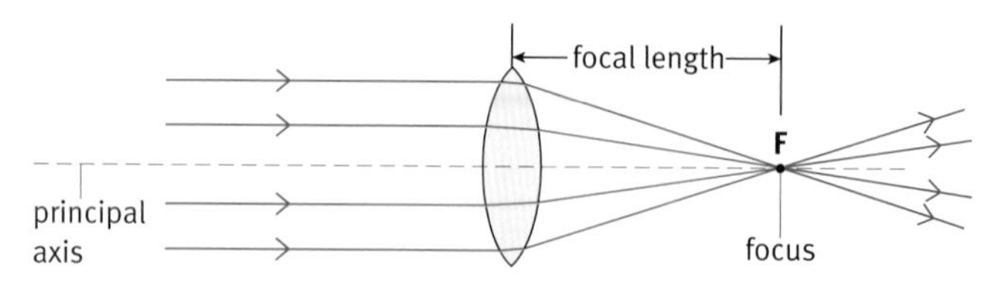

How a converging lens focuses parallel rays of light.

Parallel rays at an angle to the principal axis converge at a different point.

Lenses are cleverly designed. The surface must be curved in just the right way in order for the rays to meet at a point, the **focus,** F. When early astronomers made their own telescopes, they had to grind lenses from blocks of glass. If the glass was uneven, or if the surface was not smooth or was of the wrong shape, the telescope would give a blurred, poorly focused image. It is said that much of Galileo's success was achieved because he made high-quality lenses so that he could see details that other observers could not.

Key words

refraction
converging lens
focus
focal length
principal axis
power
dioptre

The distance from the centre of the lens to the focus is called the **focal length** of the lens. The longer the focal length of a lens, the larger (actual physical size) will be the real image that the lens produces of a distant object.

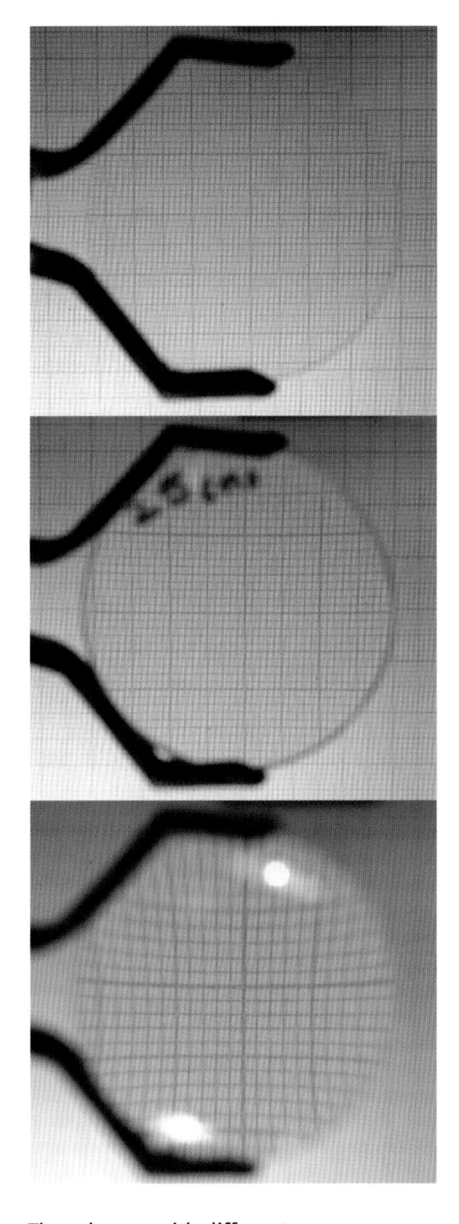

Three lenses with different focal lengths.

Estimating the focal length

You can compare lenses simply by looking at them.

- A lens with a long focal length is thin; its surfaces are not very strongly curved.
- A lens with a short focal length is fatter; its surfaces are more strongly curved.

To estimate the focal length of a converging lens, stand next to the wall on the opposite side of the room to the window. Hold up the lens and use it to focus an image of the window on the wall. Measure the distance from the lens to the wall – this will give you a good estimate of the focal length.

The power of a lens

A fat lens (one with a short focal length) bends the rays of light more. Its **power** is greater. So short focal length equals high power, and long focal length equals low power. As one increases, the other decreases.

Here is the equation used to calculate power when you know the focal length:

$$\text{power (in } \textbf{dioptres}\text{)} = \frac{1}{\text{focal length (in metres)}}$$

So if a lens has a focal length of 0.5 m, its power is

$$\frac{1}{0.5} = 2 \text{ dioptres}$$

If you look at the reading glasses sold in chemists' shops, or at an optician's prescription, you will see the power of the lens quoted in dioptres.

Questions

1 a When a ray of light passes from air into glass, which way does it bend: towards the normal or away from the normal?

b Which way does a ray bend when passing from glass into air?

2 Look at the three lenses in the photographs on this page. Put them in order, starting with the one with the shortest focal length. Which one has the greatest power?

3 The focal length of a lens is measured between which two points?

4 A lens has a focal length of 20 cm. What is its power?

5 A pair of reading glasses has lenses labelled +1.5 D (D stands for dioptres). What is their focal length?

Find out about:

- the lenses used in telescopes
- the magnification produced by a telescope
- how bigger apertures give brighter images

1C Refracting telescopes

Magnifying glass

Any converging lens can act as a magnifying glass. Simply hold it above a small object, look through, and see a magnified view. The image on the retina of your eye is larger than the object itself.

Lenses can produce reduced or magnified images.

Just two lenses

Telescopes evolved from the lenses used for correcting poor eyesight. These were in use before 1300, though you had to be quite well off to afford a pair of spectacles in those days. The lenses were biconvex; that is, they were convex (bulging outwards) on both sides. The word 'lens' is Latin for 'lentil', which has the same shape.

Such lenses work as magnifying glasses, producing an enlarged image when you look through them at a small object. For medieval scholars, whose eyesight began to fail in middle age, spectacles meant that they could go on working for another 20 or 30 years.

Seeing further

The Dutch inventor Hans Lippershey is credited with putting two converging lenses together to make a telescope. The picture below illustrates the legend that it was in fact his children who, in 1608, held up

Hans Lippershey and his children. Their play with lenses led to the invention of the telescope.

two lenses and noticed that the weathercock on a distant building looked bigger and closer. Lippershey tested their observation, and went on to offer his invention to the Dutch military.

In fact, Lippershey failed to get a patent on his device. Other 'inventors' challenged his claim, and the Dutch government decided that the principle of the telescope was too easy to copy, so that a patent could not be granted.

Telescopes rapidly became a fashionable item, sold widely across western Europe by travelling salesmen. That is how one came in to Galileo's hands in Padua, near Venice, in 1609.

Key words
refractor
eyepiece lens
objective lens

Refracting telescopes

A telescope that uses lenses to gather and focus light is called a refracting telescope, or **refractor**.

A refractor has two lenses:

- The **eyepiece lens** is the one next to your eye.
- The **objective lens** is the one closest to the object you are observing.

The same terminology is used for microscopes.

In the sort of telescope that Lippershey invented, the eyepiece lens is fatter than the objective lens. Its surfaces are more strongly curved.

This helps to explain why the telescope was not invented earlier, despite the fact that lenses had been in use in spectacles for three centuries. Spectacle lenses are not very strong, and they tend to be rather similar to each other. In a telescope, to achieve a degree of magnification, you need to have two different lenses, of different powers.

Questions

1 Look at the photograph on the opposite page.

a Which lens is a converging lens? How can you tell?

b What sort of image is produced by a diverging lens?

2 Why do you think Lippershey offered his invention to the Dutch military?

3 What would you expect to see if you made a telescope using two identical lenses?

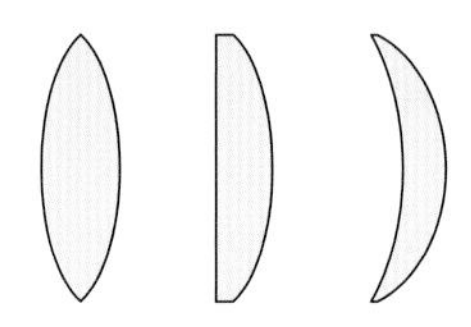

All converging lenses are fatter in the middle than at the edges.

DIY telescope

You can make a telescope using almost any pair of converging lenses. As the illustration on the left shows, this means any lens that is fatter in the middle than at the edges.

Sometimes converging lenses are described as 'convex', but this can be misleading. You can see from the picture that some converging lenses are convex on one side and concave on the other.

Eyepiece and objective

Two converging lenses, mounted in a line, are enough to make a simple telescope.

- The fatter (more curved) lens is the eyepiece lens.
- The thinner lens is the objective lens.

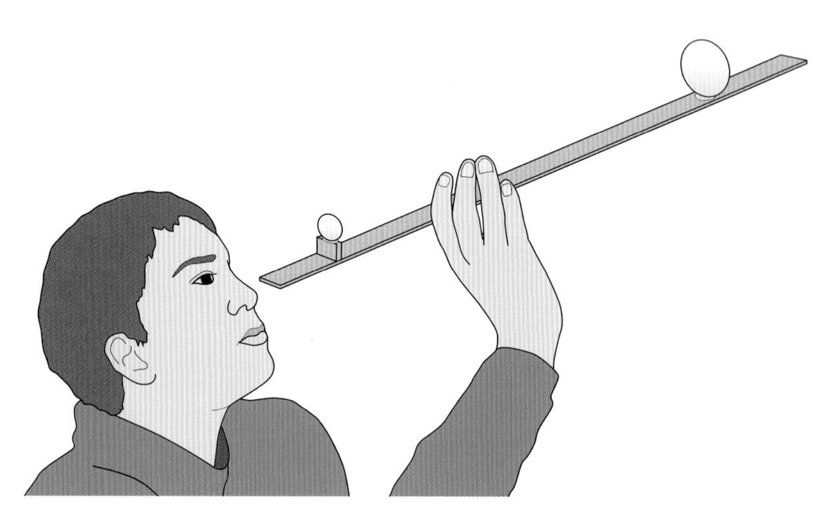

Looking through the eyepiece, you should see an **inverted** (upside-down) **image** of a distant object.

Top: A telescope that uses two converging lenses, like Lippershey's. *Bottom:* This is what you can see through it.

Getting focused

You will only see a clear image if the two lenses are the correct distance apart. That is why optical instruments (such as telescopes, binoculars, and microscopes) are designed so that you can adjust the separation of the lenses. Adjusting the focus means altering the separation of the lenses until you see a clear, unblurred image.

Focusing depends on how far away the object is that you are looking at. Birdwatchers buy binoculars that will focus on a bird just a few metres away, or at a distance of hundreds of metres. Astronomers are only interested in much more distant objects. In fact, astronomical objects are all sufficiently far away that they can be described as being 'at infinity'.

Galileo's telescope was of a different design, with a diverging lens as the eyepiece. This has the advantage that the image seen through the telescope is the right way up. Although telescopes of this type are no longer made, the Galilean telescope is the basis of the opera glasses found in some theatres.

A **diverging lens** makes parallel rays of light diverge. Lenses that are fatter at the edges than in the middle are diverging lenses.

Key words
focusing
inverted image
diverging lens

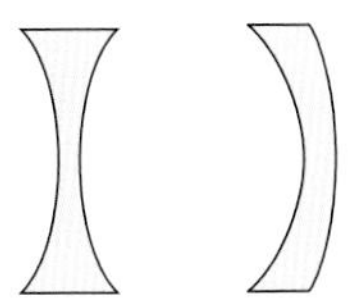
diverging lenses

Diverging lenses are thinner in the middle than at the edges.

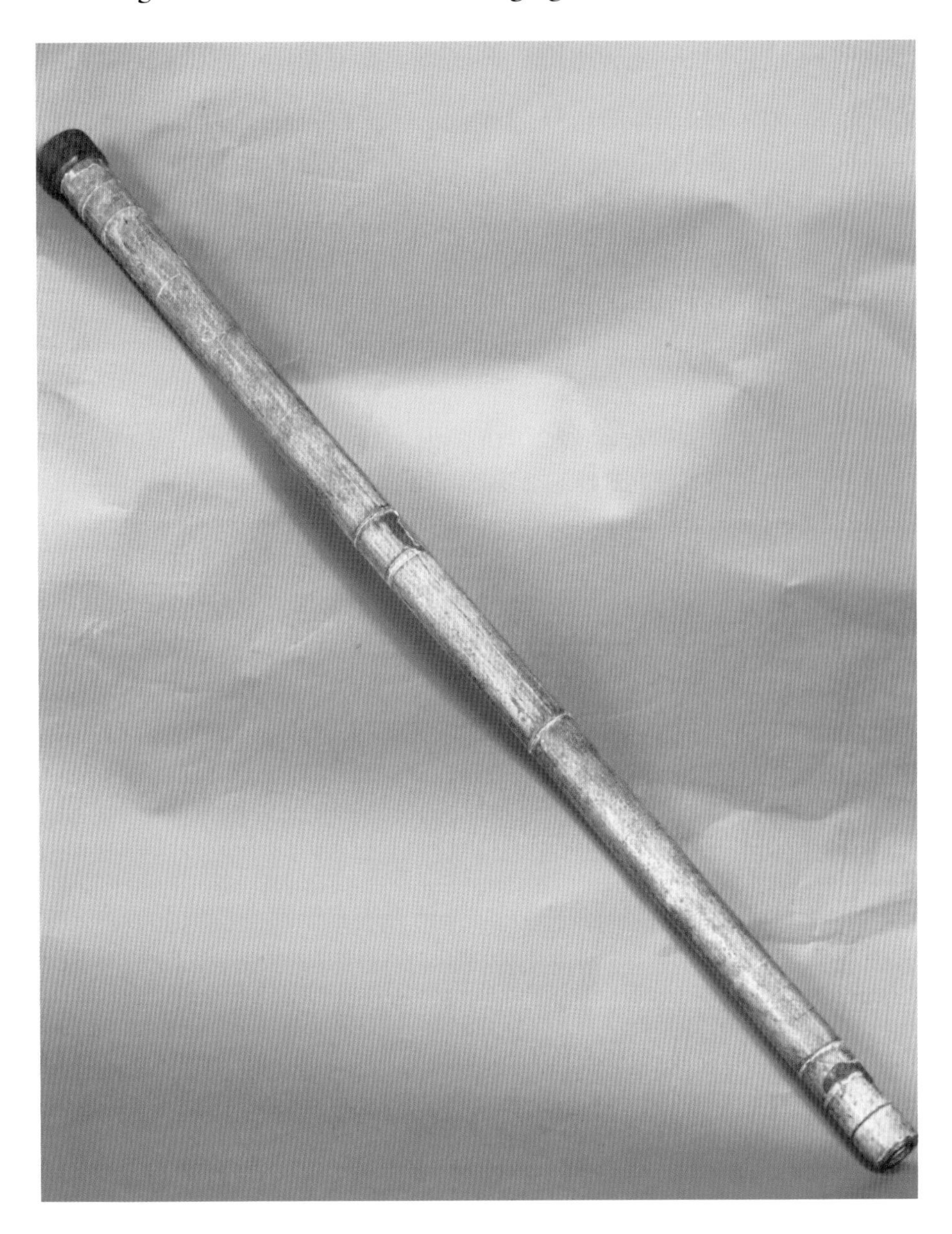

This is a replica of Galileo's telescope. His best telescopes had a magnification of 30 times.

Questions

4 Name four optical instruments mentioned on these pages.

5 What word describes the fact that the image seen through a simple telescope is the wrong way up?

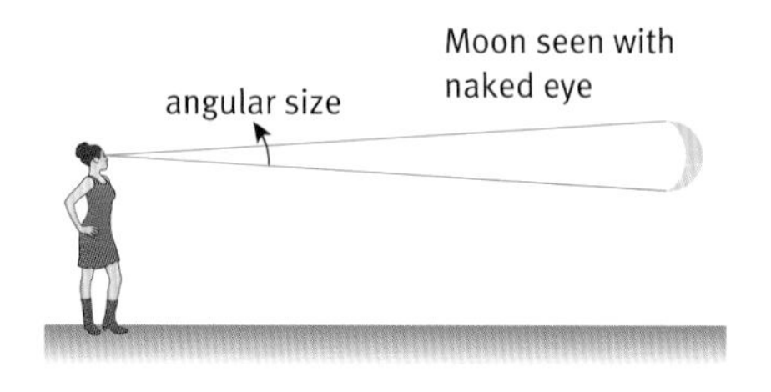

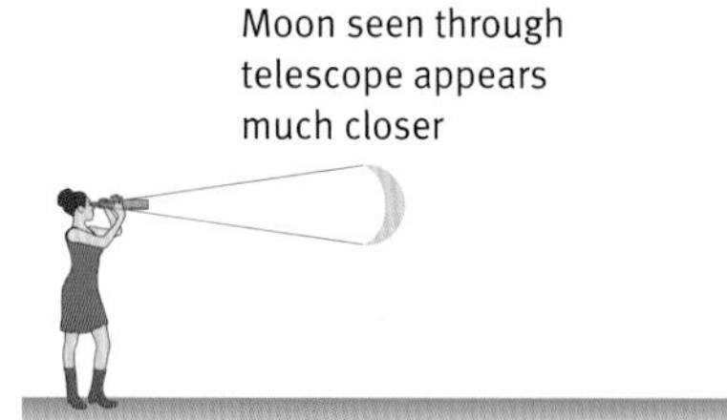

A telescope increases the angular size of the Moon.

H Magnification of a telescope

When you look at the Moon, it appears quite small in the sky. Its image on the retina of your eye is quite small. Look at the Moon through a telescope, and it looks enormous. The telescope produces a greatly enlarged image on your retina.

Suppose your telescope is labelled 50×. This says that its **magnification** is 50. There is more than one way of thinking about this:

- The telescope makes the Moon appear 50 times as close.
- The telescope makes the Moon's angular size seem 50 times as great.

To the naked eye, the Moon has an angular size of about half a degree (0.5°). With the telescope, this is increased to 25°. The telescope has an **angular magnification** of 50.

A telescope does not make a distant star look any bigger – it remains a point of light. However, a telescope spreads out a group of stars, by magnifying the angles between them. This makes it possible to see two stars that are close together as separate objects.

Calculating magnification

The magnification produced by a refracting telescope depends on the lenses from which it is made:

$$\text{magnification} = \frac{\text{focal length of objective lens}}{\text{focal length of eyepiece lens}}$$

Suppose you choose two lenses with focal lengths 50 cm and 5 cm:

$$\text{magnification} = \frac{50\,\text{cm}}{5\,\text{cm}} = 10$$

Aperture and image brightness

A telescope cannot make stars look bigger, because they are too far away. But there is something important the telescope can do – it makes stars look brighter. Dim stars look bright, and stars that are too faint to see come into view.

Without a telescope, you can see up to 3000 individual stars in the night sky; a small telescope can increase this by a factor of at least 10.

So a telescope is better than the naked eye for seeing dim stars. The reason is that the telescope gathers more light than the eye. Think about the objective lens of a telescope. It may be, say, 10 centimetres across. Its aperture is 10 cm. Compare this with the pupil of your eye. On a dark night, the **aperture** of your eye (the 'pupil') might be 5 mm across.

This amateur astronomer has two telescopes with 7- and 11-inch apertures.

Key words
magnification
angular magnification
aperture

Nocturnal creatures such as owls and mice have large pupils to make the most of the available night light (and not for star-gazing). The lenses of a telescope gather light from a star and concentrate the rays before they enter your eye.

The amount of light gathered by a telescope depends on the collecting area of its objective. To see the faintest and most distant objects in the Universe, astronomers require telescopes with very large collecting areas.

Questions

6 Look at the equation for magnification. Use it to explain why a telescope made with two identical converging lenses will be useless.

7 Calculate the magnification provided by a telescope made from lenses with focal lengths 1 m and 5 cm.

8 You are asked to make a telescope with as large a magnification as possible. You have a box full of lenses. How would you choose the two most suitable lenses?

(9) Show that four telescopes of diameter 8 m gather as much light as one of diameter 16 m.

Here is one way to gather a lot of light – build four identical telescopes. These quadruplets are part of the European Southern Observatory in Chile. Each has an aperture of 8 m, so when combined they equal a single telescope with an aperture of 16 m.

Find out about:

- the advantages of reflecting telescopes over refractors
- how mirrors collect light in reflecting telescopes
- how reducing diffraction gives better resolution

1D Reflecting telescopes

Mirrors that focus light

The first telescopes used lenses to focus light. But from ancient times people knew that mirrors too could focus light. The mirror must be curved; a curved mirror with the correct shape is described as 'parabolic'.

When parallel rays of light reach a parabolic mirror parallel to its axis, they reflect so that they meet at the focus.

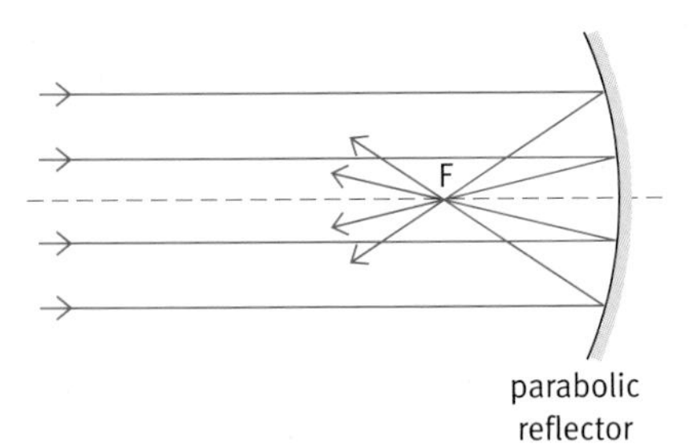

A parabolic reflector is correctly shaped to ensure that every ray parallel to the axis is reflected to the focus.

Mirror v. lens

A simple converging lens focuses different colours (frequencies) of light at slightly different points. Used as the objective of a refracting telescope, this lens will produce an unclear image.

The first design for a reflecting telescope, or **reflector**, was proposed in 1636, not long after Galileo's refracting telescope. It used a parabolic mirror as the telescope objective. One advantage of using a mirror is that it reflects rays of all colours in exactly the same way.

Mirrors have other advantages. Astronomers were soon designing optical telescopes with objectives that had larger diameters, to gather more light. You can make very large parabolic mirrors, but you cannot make very large lenses.

- The largest objective lens possible has a diameter of about 1 m. Any larger and the lens would sag and change shape under its own weight, making it useless for focusing light.
- It is very difficult to ensure that the glass of a large-diameter lens is uniform in composition all the way through.
- It is difficult, but perfectly possible to make a mirror 10 m in diameter. Its weight can be supported from the back, as well as the sides.

For these reasons, the largest professional astronomical telescopes use mirrors as their objectives.

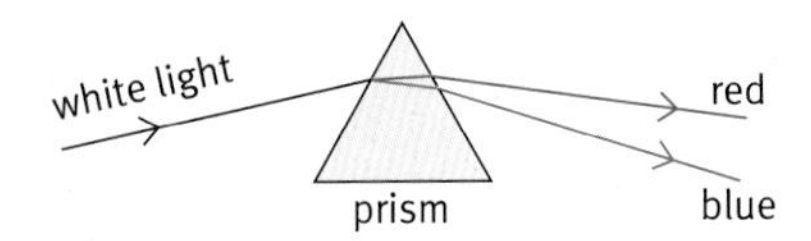

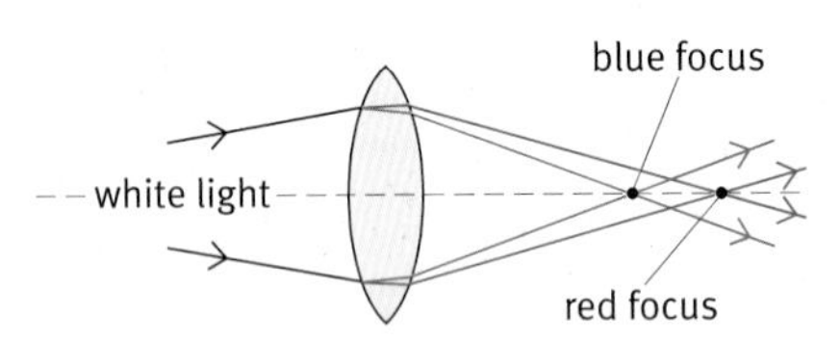

The fact that glass refracts light of different colours by different amounts is useful for prisms but is a problem for simple lenses.

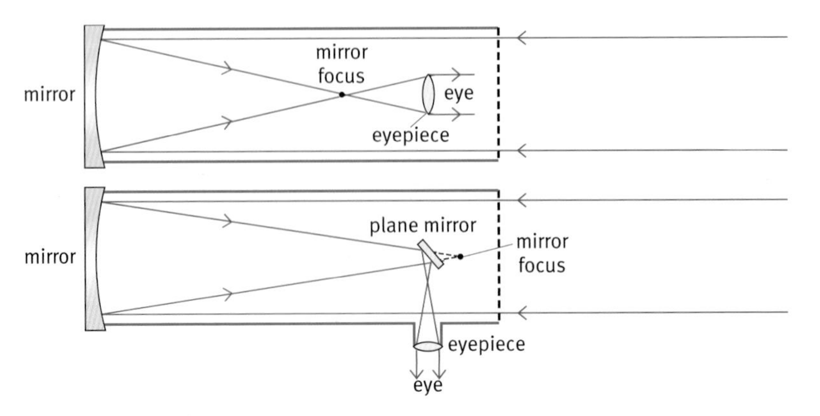

A problem with a reflecting telescope is where to place the observer. In the top diagram, the observer must be inside the telescope. In a different design, a small plane mirror close to the focus of the objective reflects light out of the telescope to an external eyepiece. Many other solutions are in common use.

Diffraction effects

A telescope with a big aperture gathers more light. There is another advantage of a big aperture: it makes it easier to see two stars that are close together. Here's why:

Recall from Module P6 *The wave model of radiation* (page 155) that when light (or any other kind of radiation) passes through an aperture, it tends to spread out. This effect is called **diffraction**. The result is that when you look at a star through a telescope, you see a slightly blurred blob of light rather than a perfect spot.

If you look at two stars that are side-by-side in the sky, their blobs may overlap and you will not be able to resolve (distinguish) them. They will look like a single blob of light.

A large aperture causes less diffraction, so a telescope with a larger aperture has a greater **resolving power**.

Wavelength and diffraction

It is easier to understand how diffraction works by thinking of the radiation passing through an aperture in the form of waves. The effect is greatest when the aperture is similar to the wavelength of the waves.

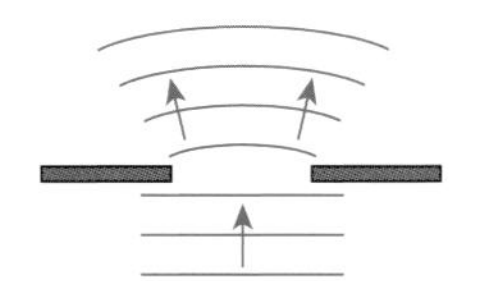

Waves spread out as they pass through the aperture.

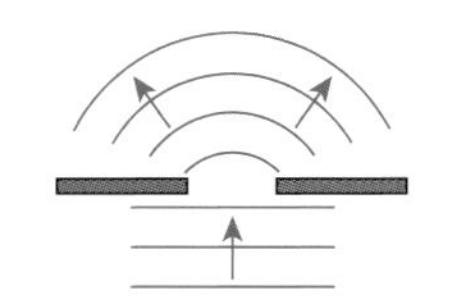

A narrower aperture has more effect.

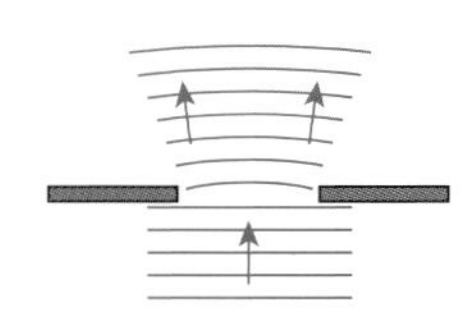

A smaller wavelength gives less diffraction.

To avoid diffraction effects, astronomers design telescopes with apertures much larger than the wavelength of the radiation they want to gather. This explains why radio telescopes, for example, are so large – the wavelength of the radio waves may be several metres (see the photo of a radio telescope on page 180).

Key words

- reflector
- diffraction
- resolving power

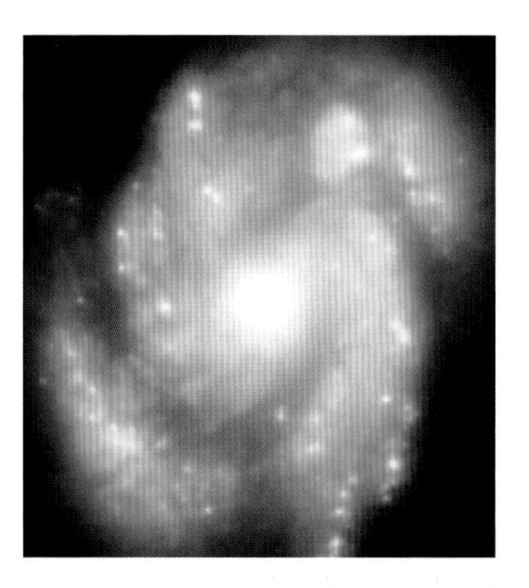

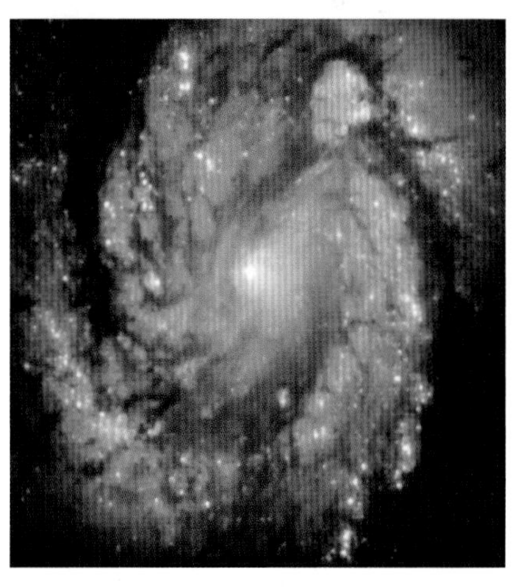

Two views of a region of the night sky, taken with telescopes having different resolving powers.

Questions

1. In a reflecting telescope, is the mirror the objective or the eyepiece?
2. Look at the two photos of the night sky (above). In which are the stars better resolved? Which telescope had the larger aperture?
3. An astronomer finds that she can scarcely resolve (separate) two stars. Which of these will improve the situation?
   - using a telescope with a smaller aperture
   - looking at light of shorter wavelength

Find out about:

- engineering of telescopes
- how data is collected in radio telescopes

1E Bigger and better telescopes

Telescope mountings

Very small telescopes can be hand-held. But with a magnification greater than about 6×, a telescope needs support. Otherwise you see nothing more than a blurred movement.

Telescopes must be moveable, so that they can be pointed to different parts of the night sky. Rotation around two axes at right angles to each other enables a telescope to point in any direction. Telescope supports are called mountings.

The tripod holds this amateur telescope in a fixed position.

Amateur astronomers will use a tripod to support their telescope. The largest of professional telescopes can have masses as big as 100 000 kilograms. Their bearings must have low friction, and the weight of the telescope itself must be balanced carefully, if the telescope is to be easily moved and able to maintain its alignment. This represents an enormous technical challenge.

Temperature changes

Telescope designers have many other interesting problems to solve. Think of the locations of some telescopes – on mountaintops and in deserts. Temperatures fall from day to night, sometimes quite dramatically. The telescope structure will expand as its temperature rises and contract as it cools. Clever engineering gets round these problems too.

Radio reflectors

The 'dish' of a radio telescope is a giant reflector. It is parabolic in shape and needs a wide aperture because the radio waves that it gathers have a much longer wavelength than light. The famous radio telescope at the Jodrell Bank Observatory in Cheshire has a diameter of 76 m.

Radio telescopes at Jodrell Bank. The 76 m dish is on the right.

The radio wave detector is placed at the focus of the dish, where the reflected radio waves are most concentrated. As the dish is steered about, it gathers radio waves from different points in the sky. A picture is gradually built up from the signals received.

The reflecting surface of the dish must be smooth. Any irregularities would reflect radio waves at the wrong angle, and this would contribute to fuzziness of the image.

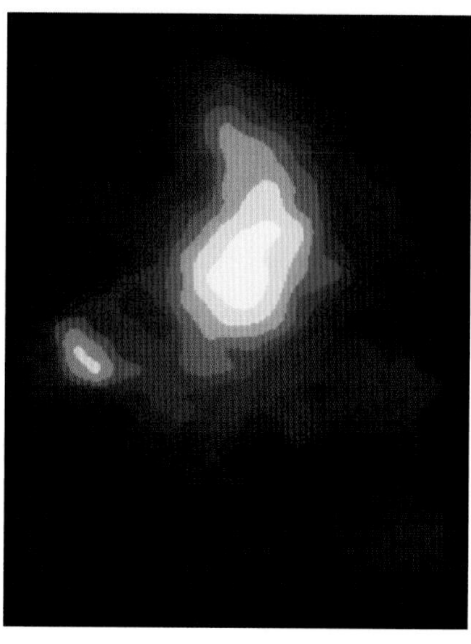

This radio telescope image shows a cluster of galaxies (the M81 cluster). It was made by tuning the Jodrell Bank telescope to the wavelength of radio waves given out by hydrogen gas (21 cm). The most intense waves come from two large galaxies (white areas), but you can also see that the whole cluster of galaxies is embedded in a giant cloud of hydrogen gas.

Here is another way to achieve a large aperture – use several reflectors, then combine their data using sophisticated electronics and fast computing. This is the Hanbury Brown radio telescope in Australia. Six reflectors move along a 6 km track.

Questions

1. Look at the radio telescope image of the M81 cluster of galaxies. What can be seen in this image that would not be seen using an optical (light) telescope?
2. Compare the Hanbury Brown telescope (this page) with the Calar Alto Observatory telescope (page 181). What similarities and differences can you identify?

1F Ray diagrams for telescopes

Find out about:

- the parallel nature of light from a distant object
- interpreting ray diagrams for mirrors and lenses

To an astronomer, most objects of interest are far off, 'at infinity'. This means that all of the rays of light coming from a distant star can be regarded as being parallel to one another.

Reflectors

The **ray diagram** on the left shows what happens when parallel rays of light reach a parabolic mirror.

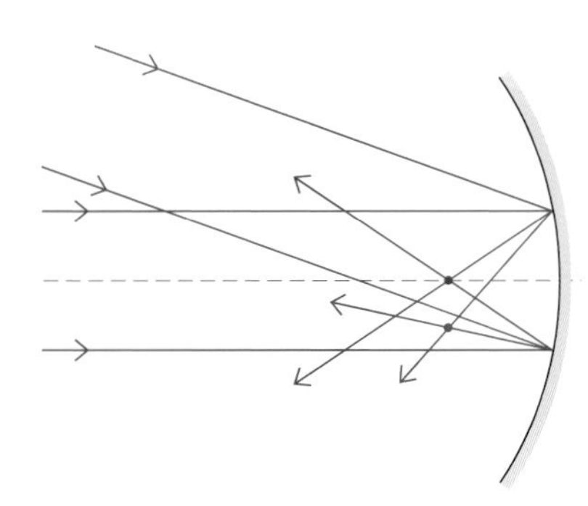

Rays parallel to the axis of the reflector are reflected to the focus. Parallel rays from another direction are focused at a different point.

This diagram is simple to construct, because each ray has to obey the law of reflection:

angle of incidence = angle of reflection

The photograph below shows the giant radio telescope at Arecibo, in Puerto Rico. Hanging above the dish is the detector, which is moved around to collect reflected waves coming from different directions in space.

The Arecibo radio telescope is built into a natural crater. It cannot be steered about.

Refractors

A ray diagram can also be used to represent the way in which a converging lens focuses parallel rays of light. Rays parallel to the axis meet at the focus. Parallel rays from another direction are focused at a point above or below the focus. (See the diagram on the left.)

Parallel rays that are not parallel to the principal axis are focused at a point below the focus. All of the light from a star which strikes the lens goes to the same point. The star's image position can be located by drawing just two rays.

Extended object

A distant star is so far away that it appears as a point of light, even through a powerful telescope. But a galaxy looks bigger than this, and is described as being an **extended object**. The Sun, Moon, and planets are extended objects too. The ray diagram on the right shows how a real image is formed of an extended object. It also shows why the image seen in a refracting telescope is inverted.

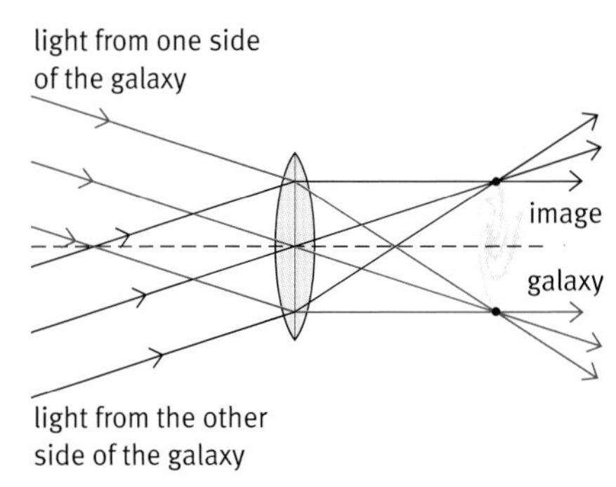

An objective lens is shown gathering light from two sides of a galaxy. Note that the image of the galaxy is inverted.

A refracting telescope

Using what you have learned about lenses can give an explanation of how a telescope made of two converging lenses works. The diagram below shows what happens.

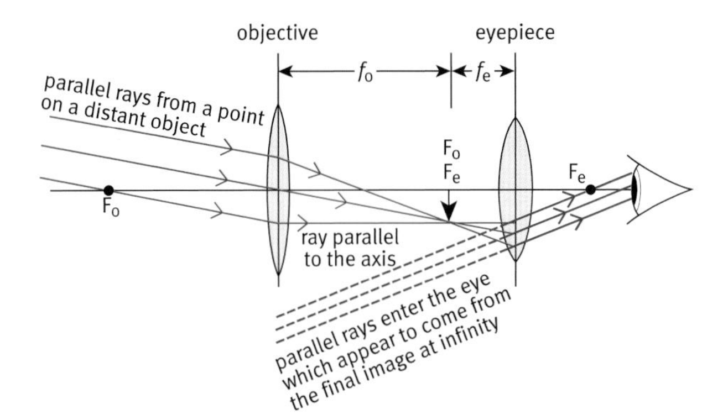

How a telescope made of two converging lenses works.

- Parallel rays of light enter the objective lens from a point on a distant object.
- Each set of parallel rays is focused by the objective lens, so a real image is formed. A weak lens is used for the objective because it produces a larger image of the distant object.
- The eyepiece is a magnifying glass, which you use to look at this real image. A strong lens is used for the eyepiece because it magnifies the image more.

From the diagram, you can see that the angle between the rays from the eyepiece and the principal axis is larger than the angle between the rays from the object and the principal axis. This means that any extended object, for example the Moon, which looks small to the naked eye, will look much larger through the telescope.

The diagram is useful in another way. It shows that, when the telescope is adjusted to focus on a distant object, point F is the focal point of both lenses. So the lenses must be separated by a distance equal to the sum of their focal lengths.

distance between lenses =
focal length of objective + focal length of eyepiece

Questions

1 Sketch a curved mirror. Draw two rays of light striking the mirror parallel to the axis, and show how the reflected rays cross. Where the rays strike the mirror, draw the normal to the surface of the mirror and mark the angles of incidence and reflection.

2 If you look through a converging lens at a distant scene, you see an inverted image. Explain why this is so, using a diagram to support your explanation.

Key words

ray diagram
extended object

Find out about:

- the effects of the atmosphere on light from stars
- the need to reduce light pollution

1G Windows in the atmosphere

Some telescopes are built on high mountains. Others are carried on spacecraft in orbit around the Earth. This is partly because some types of electromagnetic radiation are absorbed by the atmosphere (see the diagram on page 173).

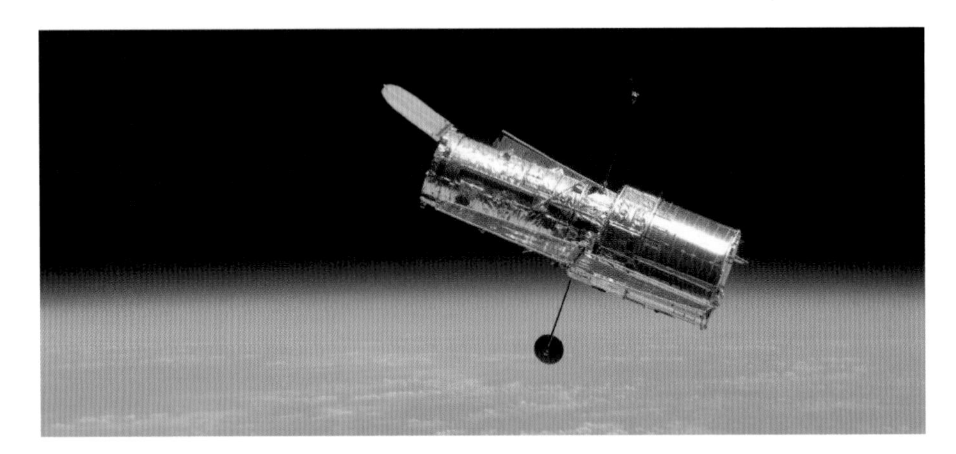

The *Hubble Space Telescope* orbits above the Earth, avoiding the effects of atmospheric absorption and refraction of light. As well as gathering visible light, Hubble also gathers ultraviolet and infrared radiation.

- Visible light, microwaves, and radio waves can pass through the atmosphere without being significantly absorbed.
- Other radiation, including X-rays, gamma rays, and much infrared, is absorbed.

Our eyes, and those of other creatures, have evolved to make use of the wavelengths that reach the surface of the Earth.

Twinkle, twinkle

If you look up at the stars in the night sky, you are looking up through the atmosphere. Even when the sky appears clear, the stars twinkle. This shows an important feature of the atmosphere.

The stars, of course, shine more or less steadily. Their light may travel across space, uninterrupted, for millions of years. It is only on the last few seconds of its journey to your eyes that things go wrong. The twinkling, or scintillation, is caused when starlight passes through the atmosphere.

The atmosphere is not uniform. Some areas are more dense, and some areas are less dense. As a ray of light passes through areas of different densities, it is refracted and changes direction. The atmosphere is in constant motion (because of convection currents and winds) so areas of different density move around. This causes a ray to be refracted in different directions, and is the cause of scintillation.

Some astronomical telescopes record images electronically. Computer software allows them to reduce or remove completely the effects of scintillation from the images they produce.

Dark skies, please

A telescope sees stars against a black background. However, many astronomers find that the sky they are looking at is brightened by light pollution. Much of this is light that shines upwards from street lamps and domestic lighting. Scattered by the atmosphere, the light enters any nearby telescope.

Another consequence is that it has become difficult to see the stars at night from urban locations. People today are less aware of the changing nature of the night sky, something that was common knowledge for our ancestors.

In 2006, the city authorities in Rome decided that enough was enough. To reduce light pollution and to save energy, it was decided to switch off many of its 170 000 street lights, thereby cutting its lighting bill by 40%. Illumination of its ancient monuments has been dimmed, as well as lights in shop and hotel windows.

The Campaign for Dark Skies, supported by amateur and professional astronomers, campaigns to reduce light pollution. But it is not only light radiation that affects astronomers. Radio waves used for broadcasting and mobile phones can interfere with the work of radio telescopes, so certain ranges of frequencies must be left clear for astronomical observations.

These two maps show the amount of light pollution across the United Kingdom in 1993 and 2000. The red areas indicate where most light is emitted, the dark-blue areas where the least is emitted. You can see that the red areas have grown over the seven years between surveys. But the biggest change is in the countryside. Here the light pollution, although at lower levels than in towns and cities, has increased across large areas of England

1993

2000

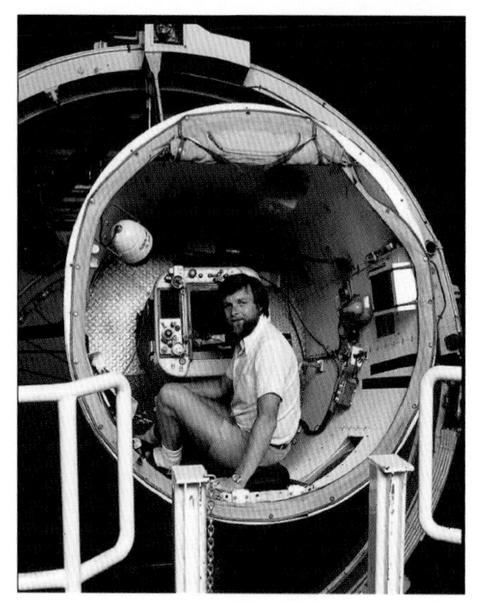

David Malin as a young scientist.

Images of the stars

A large aperture gathers more light, so that fainter stars can be seen. But there is another way to see faint stars – collect their light over a long period of time. There are two ways to do this:

- use photographic film or plates, or
- use an electronic device to record the light.

David Malin is an astronomer based in Sydney, Australia. He has made a name for himself by producing stunning images of the night sky.

'Light passing down the telescope is passed through three filters in turn – red, green and blue – to produce three separate images on black-and-white photographic plates. These are then used in the enlarger to create a single coloured image.

'The photographs below compare the results of this technique with a photo of the same area of the sky photographed on the standard colour film use by astronomers.'

The photo on the left was produced from three separate colour images. The one on the right used standard colour film.

Going electronic

Photographic images of stars can give great images, but most astronomers today use electronic imaging systems. These are based on charge coupled devices (CCDs), which are the devices used in digital cameras to record images. Electronic imaging has a number of advantages:

- Images are stored digitally.
- They can be merged and processed by computers.
- They can be sent around the world, for use by colleagues working elsewhere.

However, photographic plates can be scanned to turn them into digital images.

A barred spiral galaxy, photographed using a CCD.

Processing digital images

Astronomical images often suffer from unwanted speckles called noise, and from poor contrast. They can be improved using some of these techniques:

- adding images together but retaining only the parts, like stars, that are the same in each image
- sharpening images to accentuate the edges of astronomical objects
- using false colour to highlight information in a single picture

These techniques are based on methods that were developed to improve photographic plates. However, astronomers can now apply them more quickly and with more control using their desktop computers rather than a darkroom.

Topic 2

Mapping the heavens

Telescopes have revealed many unexpected features of the Universe, and they continue to do so today. However, many fundamental ideas about the Solar System and the stars beyond were developed before the invention of the telescope.

Look, no lenses! This sailor is demonstrating the use of an **astrolabe**, an astronomical instrument invented over 2000 years ago. The instrument measures the angle of a star above the horizon. It was used for navigation, astronomy, astrology, and telling the time.

Day-time astronomy

You do not have to stay up all night to make valid astronomical observations. You can see the Sun cross the sky every day from East to West, moving at a steady rate. That is an observation that any scientific theory of the Universe must account for. You may also have noticed that the Moon follows a similar path, sometimes by day and sometimes by night.

The Sun sets – a time-lapse photograph.

Around the pole

The stars also move across the night sky. Their movement is imperceptible, but it is revealed by long-exposure photography. The photo also shows that the stars appear to rotate about a point in the sky directly above one of the Earth's poles.

This photograph shows the motion of the stars across the sky. The exposure time was 10.5 hours.

Eclipses

Sometimes it is rare and unusual events that reveal something important. When the Sun passes behind the Moon in a total eclipse, you can see the Sun's gaseous corona.

A total eclipse of the Sun. The outer atmosphere or corona becomes visible. Its appearance changes from one eclipse to the next.

Key words
astrolabe

Find out about:

- phases of the Moon
- the difference between sidereal and solar days
- the motion of the planets against the background of fixed stars

The Moon is shown here at intervals of three days. The Sun is off to the right. Notice that it is the half of the Moon facing the Sun which is lit up.

2A Naked-eye astronomy

The spinning observatory

The Sun and Moon move across the sky in similar but slightly different ways:

- The Sun appears to travel across the sky once every 24 h (on average).
- The Moon moves very slightly slower, reappearing every 24 h 49 m.

People are, of course, deceived by their senses. The Sun is not moving round the Earth. It is the Earth that is spinning on its axis. That is why the Sun rises and sets every day, and why we experience day and night.

The situation with the Moon is more complex. The spinning of the Earth makes the Moon cross the sky. But the Moon is also slowly orbiting the Earth, from West to East. One complete orbit takes about 28 days.

Even without taking into account the fact that the Earth is orbiting the Sun, you can use these ideas to explain the changing **phases of the Moon**.

The Moon's phases

At any time, half of the Moon is lit up by the Sun's rays, just like the Earth. The view from the Earth depends on where the Moon is around its orbit.

- When the Moon is on the opposite side of the Earth to the Sun, an observer on the Earth can see the whole of its illuminated side. This is a full Moon.
- When the Moon is in the direction of the Sun, the side that is in darkness faces the Earth. This is a new Moon.

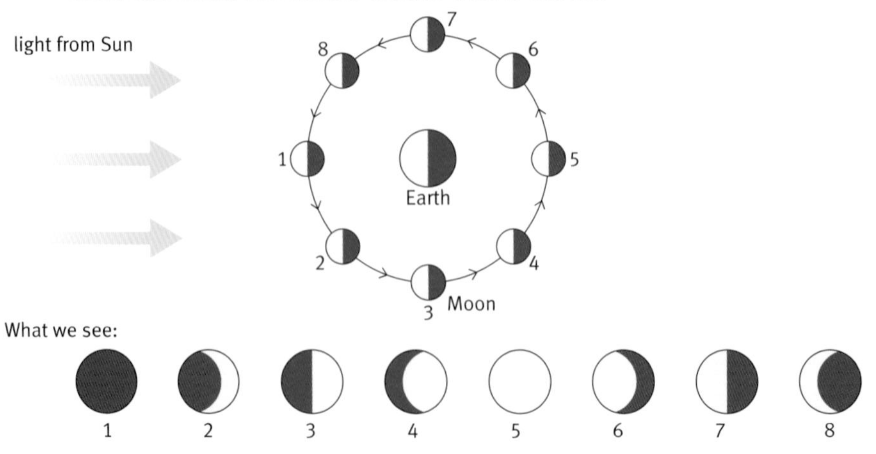

The phase of the Moon changes as it orbits the Earth.

The spinning, orbiting observatory

Earth-bound observers see the sky from a rotating planet. That is why the stars appear to move across the night sky. Their apparent motion is slightly different from the Sun's:

- The stars appear to travel across the sky once every 23 h 56 m.

That is 4 minutes less than the time taken by the Sun. The difference arises from the fact that the Earth orbits the Sun, once every year.

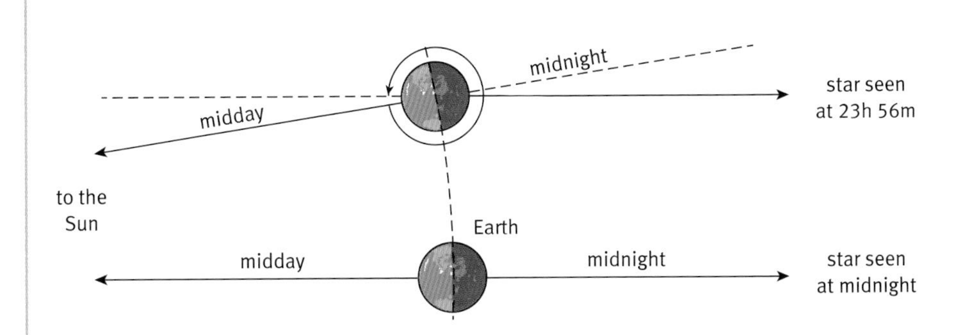

Imagine looking up at a bright star in the sky. 23 h 56 m later, it is back in the same position. This tells you that the Earth must have turned through 360° in this time, and you are facing in the same direction in space.

The Sun not behave like the stars. Repeat the above observation, this time looking at the Sun. After 23 h 56 m, the Earth has turned through 360°, but the Sun has not quite reached the same position in the sky. The diagram shows that, in the course of a day, the Earth has moved a short distance around its orbit. Now it must turn a little more (4 minutes' worth) for the Sun to appear in the same direction as the day before.

Days are measured by the Sun. The average time it takes to cross the sky is 24 h, and this is called a **solar day**. We could choose to set our clocks by the stars (although this would be very inconvenient). Then a day would last 23 h 56 m; this is called a **sidereal day** ('sidereal' means 'related to the stars').

Key words

phases of the Moon
sidereal day
solar day

You can think of the Sun and stars as fixed. We view them from a spinning, orbiting planet.

Questions

1 Draw a diagram to show the relative positions of the Earth, Sun, and Moon when the Moon is at first quarter (half-illuminated, as seen from Earth).

2 a If the Earth orbited the Sun more quickly – in, say, 30 days – would the difference between sidereal and solar days be greater or less?

b Work out the time difference between sidereal and solar days.

3 Why would it be 'inconvenient' if we set our clocks according to sidereal time?

Mapping the heavens

The night sky changes through the year. You may have noticed that some newspapers publish monthly star charts to help you work out what stars are visible, and when. These charts are drawn using knowledge of the positions of the stars and planets in the sky.

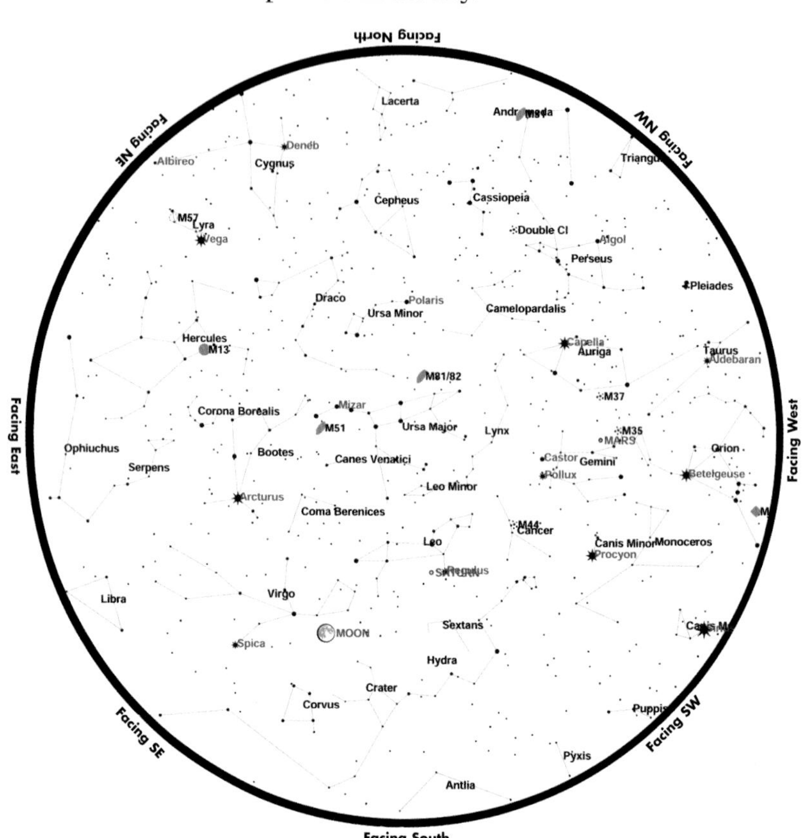

A star chart for March 2008.

Astronomers give the position of astronomical objects in the sky in terms of angles. Imagine standing in a field, looking at a star. Two angles are needed to give its position:

- Start by pointing at the horizon, due North of where you are. Turn westwards through an angle until you are pointing at the horizon, directly below the star. That gives you the first angle.
- Now move your arm upwards through an angle, until you are pointing directly at the star. That gives you the second angle.

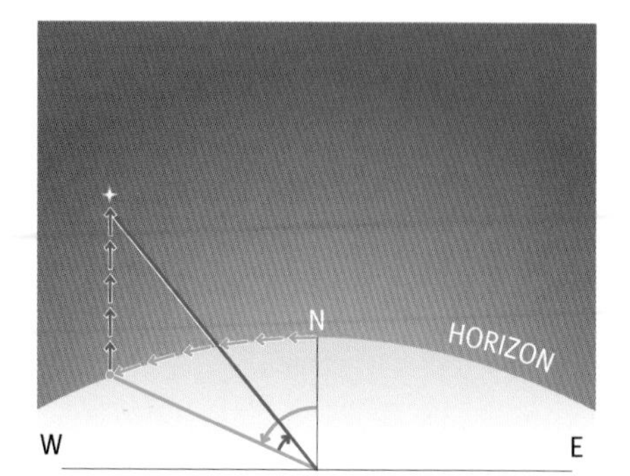

Two angles describe the position of a star.

Constellations

Another way astronomers identify particular stars is to state the **constellation** they are in. A constellation is a group of stars that form a pattern in the sky, and the names people use for them go back a very long time. Some are familiar from the signs of the zodiac. They have no real significance – the stars in a constellation may be vastly separated from each other in space, but it is convenient to use their names to identify areas of the sky.

Some constellations seen on a winter night are different from those of a summer night. This is because the Earth travels half way round its orbit in six months. You see the stars that are in the opposite direction to the Sun and so, after six months, you can see the opposite half of the sky.

Each day, a particular star rises 4 minutes earlier. After 6 months, those extra minutes add up to 12 hours, so that a star which is rising at dusk in June will be setting at dusk in December.

Heavenly wanderers

Five **planets** can be seen with the naked eye from Earth – Mercury, Venus, Mars, Jupiter, and Saturn. These were recognized as different from stars long, long ago, because they appear to move, very slowly, night by night, against the background of 'fixed stars'. The diagram on the right shows the changing positions in the sky of three planets at dawn in late 2007.

The most striking thing is that the planets generally move steadily in one direction across the background of stars, but, at times, they slow down and go into reverse. This is known as **retrograde motion**.

To explain the behaviour of planets, recall that both the Earth and the planets are orbiting the Sun. An observer looking towards Mars sees it against a backdrop of the fixed stars. Its position against this backdrop depends on where the Earth, and Mars, are in their orbits.

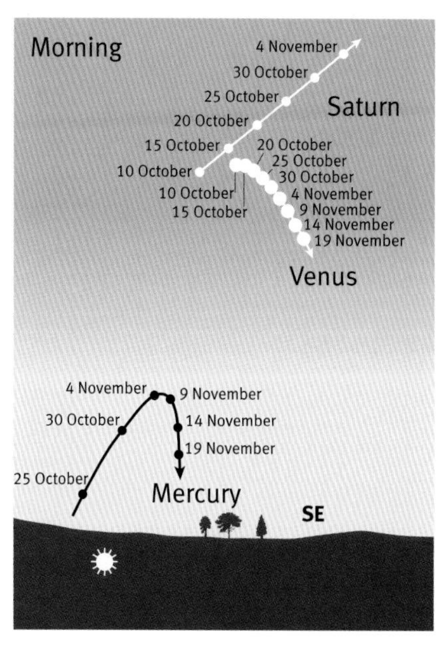

The pattern of movements of planets in the sky is different from that of stars. This diagram shows the pattern of movement of three planets just before sunrise.

Key words
- constellation
- planet
- retrograde motion

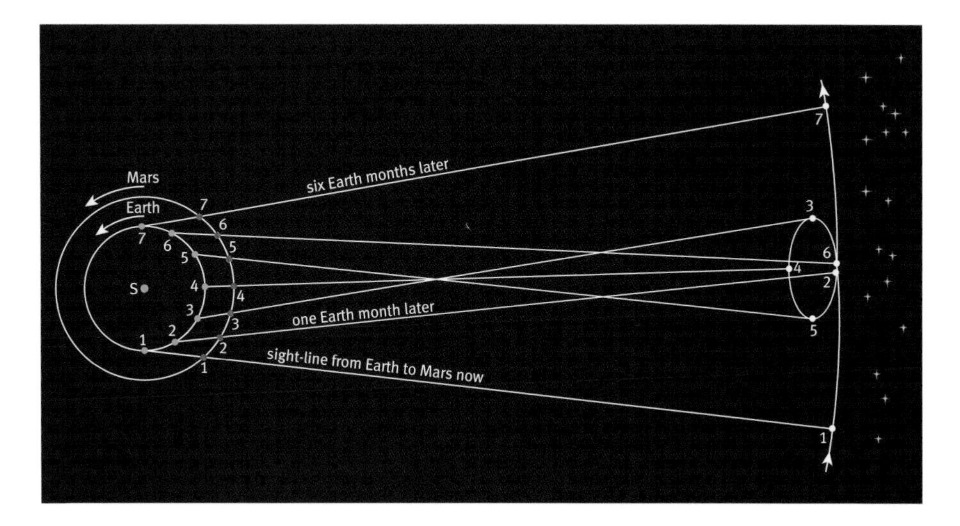

From months 1 to 3, Mars appears to move forwards. Then, for two months, it goes into reverse before moving forwards again.

Questions

4 a Draw a diagram to explain why you see some different constellations in winter and summer. The diagram on page 205 may help you.

b Use your diagram to explain why there are some stars that can never be seen from the UK, but that can be seen from places in the southern hemisphere.

5 Mercury is the closest planet to the Sun. It is only ever seen at dawn or dusk, close to the Sun in the sky. Draw a diagram to explain why.

Find out about:

- why solar and lunar eclipses happen
- the effect of the Moon's orbital tilt

2B Eclipses

Astronomers can predict when an eclipse of the Sun (a solar eclipse) will occur. A solar eclipse happens just a few times each year. And a total eclipse at any particular point on the Earth is a rare event.

Eclipses involve both the Sun and the Moon.

- In a **solar eclipse**, the Moon blocks the Sun's light.
- In a **lunar eclipse**, the Moon moves into the Earth's shadow.

The predictability of eclipses shows that they must be related to the regular motions of the Sun and Moon. Their rarity suggests that some special circumstances must arise if one is to occur.

The first way to explain a solar eclipse is to think of the Sun and Moon and their apparent motion across the sky. The Sun moves slightly faster across the sky than the Moon, and its path may take it behind the Moon. For us to see a total eclipse, the Sun must be travelling across the sky at the same height as the Moon. Any higher or lower and it will not be perfectly eclipsed.

The fact that the Moon precisely blocks the Sun is probably a coincidence. The Sun is 400 times the diameter of the Moon, and it is 400 times as far away.

Provided the Sun's path across the sky matches the Moon's, a total eclipse may be seen.

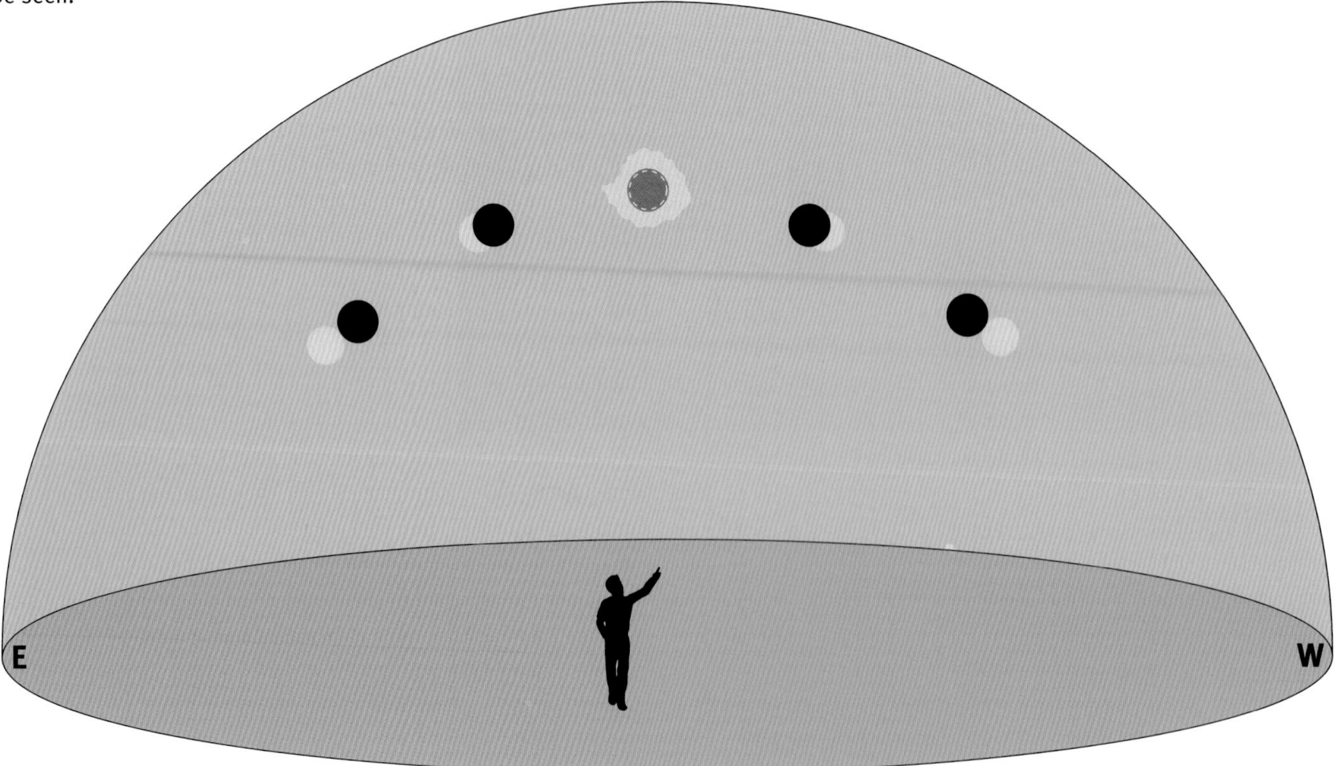

Umbra and penumbra

The diagram below shows a different way of explaining eclipses, both solar and lunar. Both the Earth and the Moon have shadows, areas where they block sunlight. Because the Sun is an extended source of light, these shadows do not have hard edges. There is a region of total darkness (the **umbra**), fringed by a region of partial darkness (the **penumbra**). The Earth's shadow is much bigger than the Moon's.

- If the Moon's umbra touches the surface of the Earth, a solar eclipse is seen from inside the area of contact.
- If the Moon passes into the Earth's umbra, a lunar eclipse is seen.

Key words
lunar eclipse
solar eclipse
umbra
penumbra

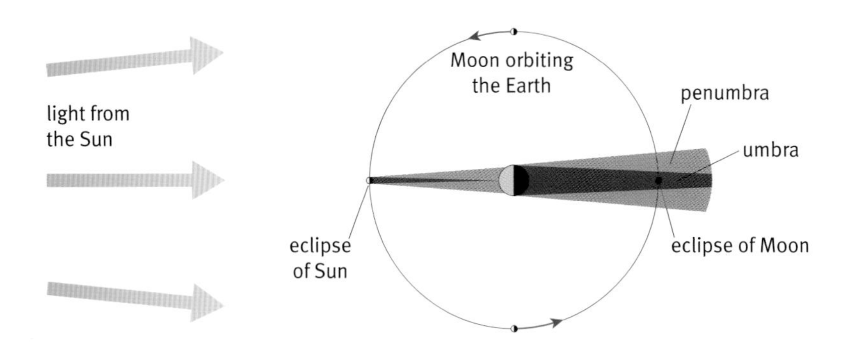

The umbra and penumbra for an eclipse of the Sun and an eclipse of the Moon.

Why the rarity?

The Moon orbits the Earth once a month, so you might expect to see a lunar eclipse every month, followed by a solar eclipse two weeks later. You do not – eclipses are much rarer than this. The reason is that the Moon's orbit is tilted relative to the plane of the Earth's orbit by about 5°. Usually Earth, Sun and the Moon are not in a line so no eclipse occurs.

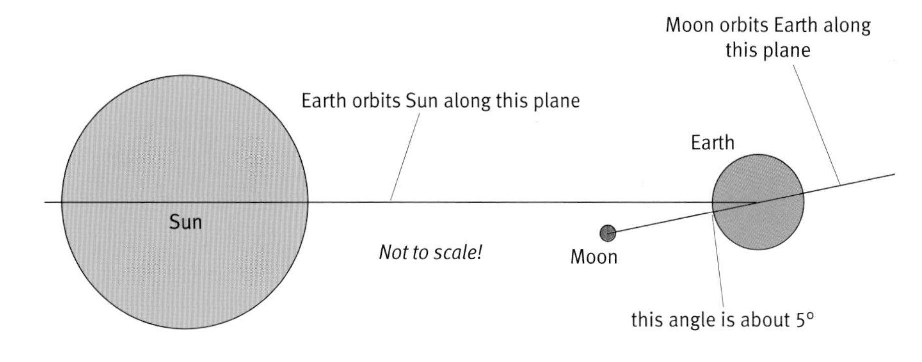

The Moon's orbit is tilted relative to the plane of the Earth's orbit around the Sun. The effect is exaggerated here.

Questions

1 At times, the Moon's orbit takes it further from Earth so that it looks smaller in the sky. Now, if it is in front of the Sun, the result is that we see a ring of bright sunlight around the black disc of the Moon. This is an annular eclipse. Construct a diagram like the one opposite to show why this happens.

2 What is the phase of the Moon at the time of

a a solar eclipse, and

b of a lunar eclipse?

3 Explain why a person on Earth is more likely to see a lunar eclipse than a solar eclipse.

Eclipse trips

Today, solar eclipses are big business. Thousands of people select their holiday dates to coincide with an eclipse. Tour operators organize plane-loads of eclipse spotters, and cruise liners sail along the track of the eclipse. Guest astronomers give lectures to interested audiences. And, provided the clouds hold off, hundreds of thousands of satisfied customers will get a view of a spectacular natural phenomenon.

The moment of total eclipse. For a few tens of seconds, the Moon blocks the Sun's bright disc and the solar corona is visible. Photograph by Fred Espenak.

Scientific expeditions

For centuries, astronomers have travelled to watch eclipses for scientific purposes. They have helped us to learn about the dimensions of the Solar System, and about the Sun and Moon. Take the question of the corona. For a long time, scientists had been unable to agree whether the corona was actually part of the Sun, or a halo of gas around the Moon, illuminated by sunlight during an eclipse.

The picture on the left shows a scientific expedition that travelled to India to observe and record the total solar eclipse of 12 December 1871. They took photographs from which it was possible, for the first time, to develop a scientific description of the Sun's corona (outer atmosphere).

Preparing to observe a solar eclipse in 1871.

A scientific purpose?

But is it worth studying eclipses today? Are there good scientific reasons to take tonnes of scientific equipment off to some distant land? One person who thinks so is Fred Espenak, an astrophysicist at NASA's Goddard Space Flight Centre. His interest is in the atmospheres of planets, moons, and the Sun.

Fred uses an infrared spectrometer to examine radiation coming from planetary atmospheres. A spectrometer is a device that splits light (or other radiation) into its different wavelengths, just as a prism or diffraction grating splits white light into the colours of the spectrum. By studying the wavelengths that are present, it is possible to deduce the chemical composition of the source of the radiation. One of Fred's experiments, to measure atmospheric flow, was carried on the Space Shuttle.

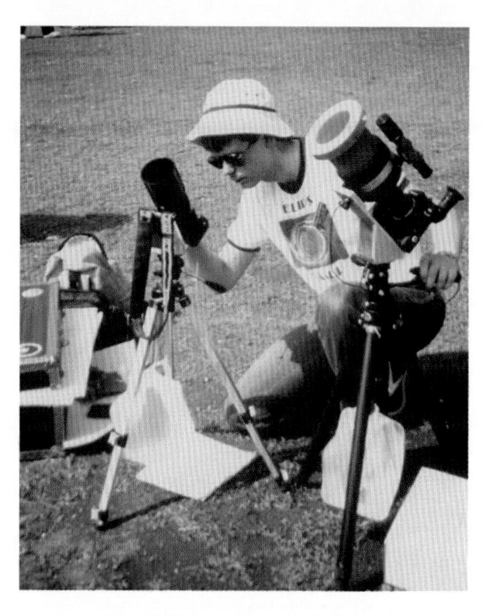

A young Fred Espenak, preparing to observe an eclipse in 1983. He is now a veteran of over 20 eclipse expeditions.

A solar eclipse is the only time that an earthbound observer can study the Sun's corona. This is a mysterious part of the Sun, extending far out into space. The mystery is its temperature. We see the surface of the Sun, which is hot, at about 5500 °C, but the corona is far hotter – perhaps 1.5 million degrees. Measurements during eclipses may help to explain how this thin gas become heated to such a high temperature.

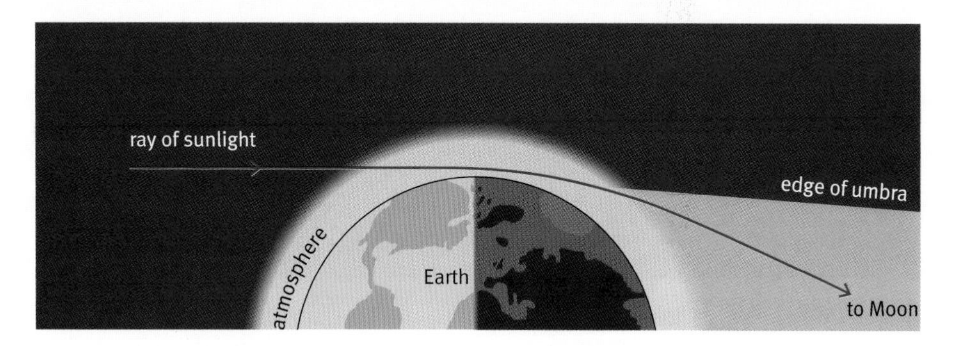

During an eclipse, light from the Sun is refracted as it passes through the Earth's atmosphere, and lights up the Moon.

And lunar eclipses? Fred's main interest is in examining the light that reaches the Moon through the Earth's atmosphere at this time. The quality of the light can be a good guide to the state of the Earth's atmosphere, indicating pollution from such causes as forest fires and volcanoes.

A composite image of the Moon moving in and out of eclipse. The central image shows the Moon lit up by sunlight that has been refracted through the Earth's atmosphere. Another photograph by Fred Espenak.

Find out about:

- parallax as an indication of distance
- parsecs and light-years as units of distance
- the observed and intrinsic brightness of a star

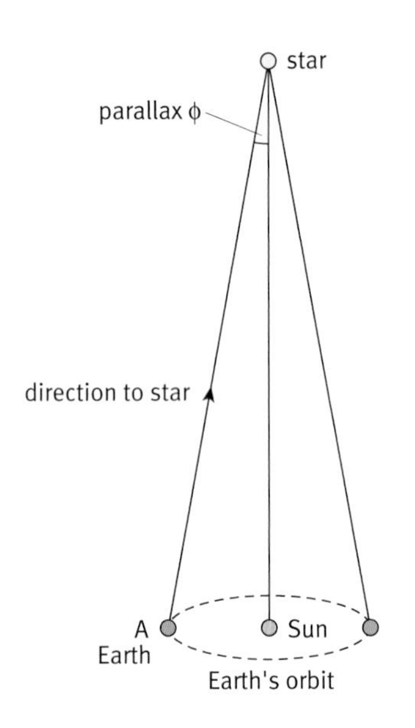

Defining the parallax angle.

These two photos illustrate the effect of parallax. They show the same view, but the photographer moved sideways before taking the second one. The closest object, the person on the bench, has moved furthest across the image.

2C Star distances

Parallax angles

The stars are far off. How can we measure their distances? One way is to use the idea of parallax.

Imagine looking across a city park in which there are a number of trees, scattered about. You take a photograph. Now take two steps to the right and take another photograph. Your photos will look very similar, but the *relative* positions of the trees will have changed slightly. Perhaps one tree that was hidden behind another has now come into view.

Now superimpose the photos one on top of the other and you will see that the closer trees will have shifted their positions in the picture more than those which are more distant. You have observed an effect of **parallax**.

Astronomers can see the same effect. As the Earth travels along its orbit round the Sun, some stars seem to shift their positions slightly against a background of fixed stars. This shifting of position against a fixed background is what astronomers call parallax, and it can be used to work out the distance of the star in question.

The diagram on the left shows how astronomers define the **parallax angle** of a star. They compare the direction of the star at an interval of six months. The parallax angle is *half* the angle moved by the star in this time. Equally, it is half the angle moved by the star on a star map.

From the diagram, you should be able to see that, the closer the star, the greater is its parallax angle.

The scale of things

In the Middle Ages, astronomers imagined that all of the stars were equally distant from the Earth. It was as if they were fixed in a giant crystal sphere, or perhaps pinholes in a black dome, letting through heaven's light. They could only believe this because the patterns of the stars do not change through the year – there is no obvious parallax effect.

They were, of course, wrong. But it is not surprising that they were wrong, because parallax angles are very small. The radius of the Earth's orbit is about 8 light-minutes, but the nearest star is about 4 light-years away – that's over 250 000 times as far.

Parallax angles are usually measured in fractions of a second of arc. There are:

- $360°$ in a full circle
- $60'$ (minutes) of arc in $1°$
- $60''$ (seconds) of arc in $1'$

So a second of arc is $\frac{1}{3600}$ of a degree.

Astronomers use a unit of distance based on this: the **parsec**.

- An object whose parallax angle is 1 second of arc is at a distance of 1 parsec.

Because a smaller angle means a bigger distance:

- An object whose parallax angle is 2 seconds of arc is at a distance of 0.5 parsec.

A parsec is about 3.1×10^{13} km. This is of a similar magnitude to a light-year, which is 9.5×10^{12} km. Typically, the distance between neighbouring stars in our galaxy is a few parsecs.

Key words

parallax
parallax angle
parsec

Questions

1. Draw a diagram to show that a star with a large parallax angle is closer than one with a small parallax angle.
2. Which is bigger, a parsec or a light-year?
3. How many light-years are there in a parsec?
4. If a star has a parallax angle of 0.25 seconds of arc, how far away is it (in parsecs)?

How bright is that star?

Measurements of parallax angles allow astronomers to measure the distance to a star. This only works for relatively nearby stars. But there are other methods of finding how far away a star is.

In the late 17th century, scientists were anxious to know just how big the Universe was. The Dutch physicist Christiaan Huygens devised a technique for measuring the distance of a star from Earth. He realized that, the more distant a star, the fainter its light would be. This is because the light from a star spreads outwards, and so, the more distant the observer, the smaller the amount of light that reaches him or her. So measuring the **observed brightness** of a star would give an indication of its distance.

Here is how Huygens set about putting his idea into practice:

- At night, he studied a star called Sirius, the brightest star in the sky and one of the Sun's closest neighbours.
- The next day, he placed a screen between himself and the bright disc of the Sun. He made a succession of smaller and smaller holes in the screen until he felt that the speck of light he saw was of the same brightness as Sirius.
- Then he calculated the fraction of the Sun's disc that was visible to him. It seemed that roughly 1/30 000 of the Sun's brightness equalled the brightness of Sirius. His calculation showed that Sirius was 27 664 times as distant as the Sun.

How wrong he was! Astronomers now know that Sirius is more than 500 000 times as distant as the Sun. This is where Huygens went wrong:

- First, there was subjectivity in his measurements. He had to judge when his two observations through the screen were the same.
- Second, his method assumed that Sirius and the Sun are identical stars, radiating energy at the same rate.
- Third, he had to assume that no light was absorbed between Sirius and his screen.

Huygens understood these problems with his method, but he was keen to find a method of estimating the distances of the stars.

Luminosity

Scientists now think that stars are not all the same. There are big ones and small ones, hot ones and cooler ones. A big, hot star radiates more energy every second than a small, cool star. To judge the **luminosity** of a star, you could imagine putting it next to the Sun. If you could see all stars at this distance, you would have a fair test of their luminosity.

So luminosity depends on two factors:

- Its **temperature**: a hotter star radiates more energy every second from each square metre of its surface.
- Its **size**: a bigger star has more surface that radiates energy.

Observed brightness (what an observer sees from Earth) depends on the star's distance from Earth, as well as the star's luminosity. Also, any dust or gas between Earth and the star may absorb some of its light.

Key words
luminosity
observed brightness

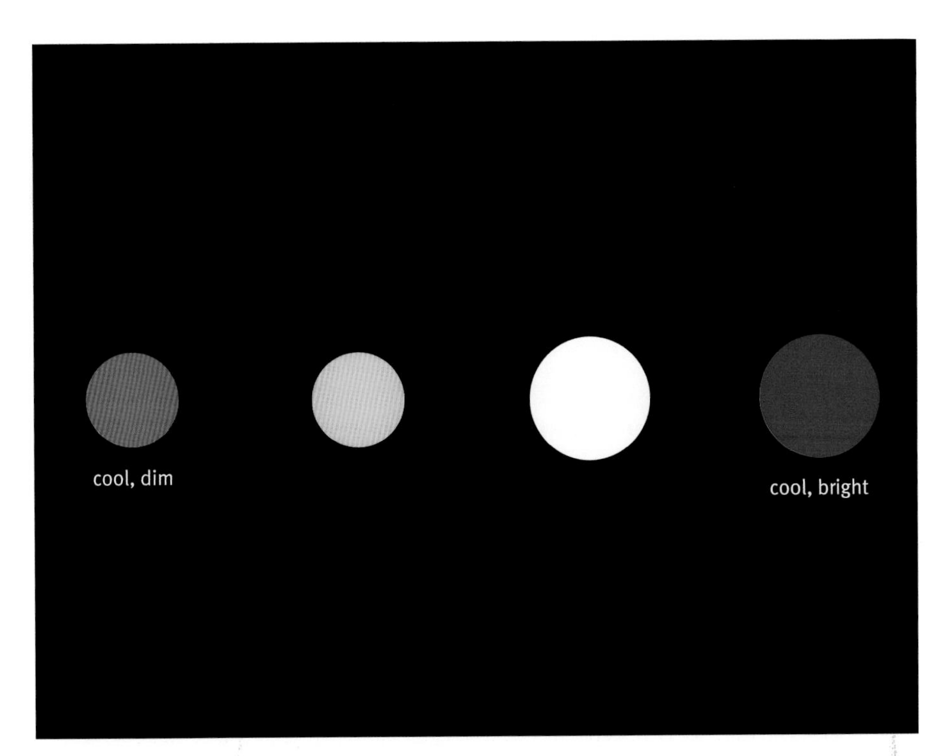

Stars differ in size and temperature, so they radiate different amounts of energy.

Questions

5 Look at the diagram on this page. Which of the stars shown has the greatest luminosity? Which has the least? Explain how you know.

6 Explain how two stars having the same observed brightness may have different luminosity.

Find out about:

- how colour is related to temperature
- how the spectrum of radiation from a hot object depends on its temperature

Stars come in different colours.

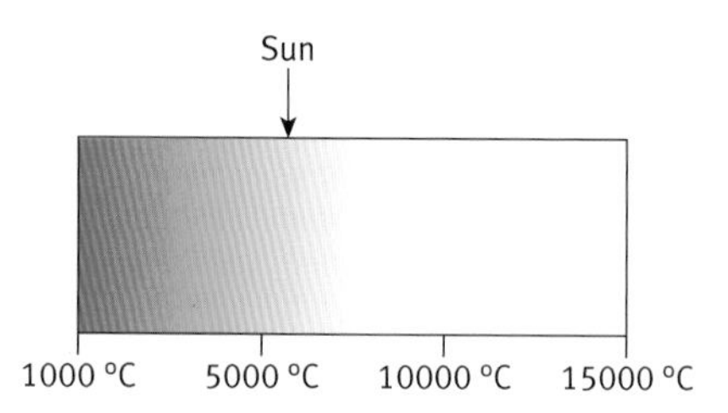

The colour of a star depends on its surface temperature.

2D Star temperatures

Colour and temperature

A star produces a continuous range of frequencies across the **electromagnetic spectrum**. Most of its radiation is in the infrared, visible and ultraviolet regions of the spectrum, but stars also produce radio waves and X-rays. If you look out at the stars at night, you may get a hint that they shine with different colours – some reddish, some yellow (like the Sun), others brilliantly white. It is more obvious if you look through binoculars or a telescope.

Colour is linked to temperature. Look back to the diagram on page 164 which compares the radiation from objects at different temperatures. Imagine heating a lump of metal in a flame. At first, it glows dull red. As it gets hotter, it glows orange, then yellow, then bluish white.

You might notice that these colours appear in the order of the spectrum of visible light. Red is the cool end of the spectrum, violet the hot end. For centuries, the pottery industry has measured the temperature inside a kiln by looking at the colour of the light coming from inside.

So the colour of a star gives a clue to its surface temperature.

Analysing starlight

At one time, astronomers judged the colours of stars and classified them accordingly. However, it is better to analyse stars, light using an instrument called a **spectrometer**. A spectrometer can be attached to a telescope so that it produces a spectrum, showing all of the frequencies that are present. The photographs show how a spectrometer turns the light from each star into a spectrum.

The Pleiades is a group of bright stars. With a spectrometer, the light from each star is broken up into a spectrum (above).

Comparing stars

Better still is to turn the spectrum of a star into a graph. This shows the intensity (energy radiated per unit area of a star's surface) for each frequency in the spectrum.

Key words
electromagnetic spectrum
spectrometer
peak frequency

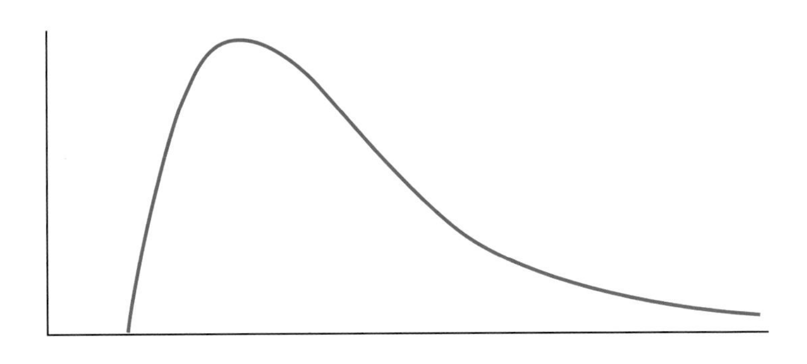

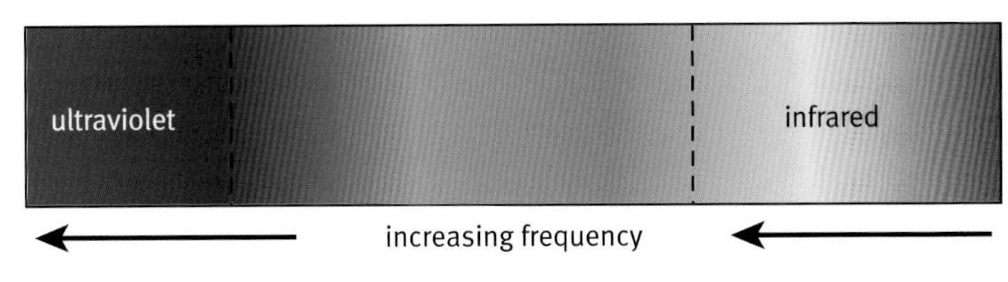

The graph of the spectrum provides information about intensity as well as showing which frequencies are present.

The next diagram, on the right, shows the results of comparing the spectra of hotter and cooler stars.

- For a hotter star, the area under the graph is greater; this shows that the luminosity of the star is greater.
- For a hotter star, the **peak frequency** is greater; it produces a greater proportion of radiation of higher frequencies.

These are not special rules for stars: they apply to any hot object.

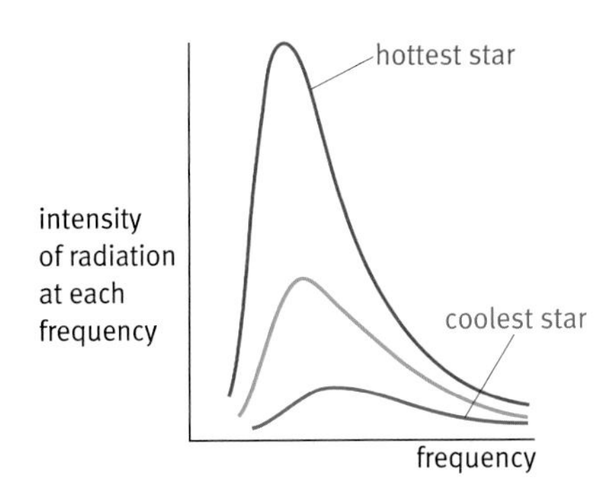

The spectra of hotter and cooler stars

Questions

1 Stars A and B are the same size, but star A is hotter than star B.

a Which star has greater luminosity?

b If you examined the spectra of these stars, which would have the greater peak frequency?

c Sketch graphs to show how these stars' spectra would differ.

Find out about:

- the brightness–period relationship for Cepheid variable stars
- how this helps to measure astronomical distances
- how we know about galaxies beyond our own

2E Galaxies

Cepheid variable stars

Life on Earth relies on the fact that the Sun burns at a steady rate – we would be in trouble if it got brighter and dimmer as the days went by. Most stars burn steadily like this. But, in 1784, a new type of star was discovered by an English astronomer called John Goodricke.

Goodricke noticed that a star called δ Cephei (δ = delta) varied in brightness. It went from bright to dim and back to bright again in an interval of a week or so, and this variation was very regular. The graph below shows modern measurements of the brightness of this star.

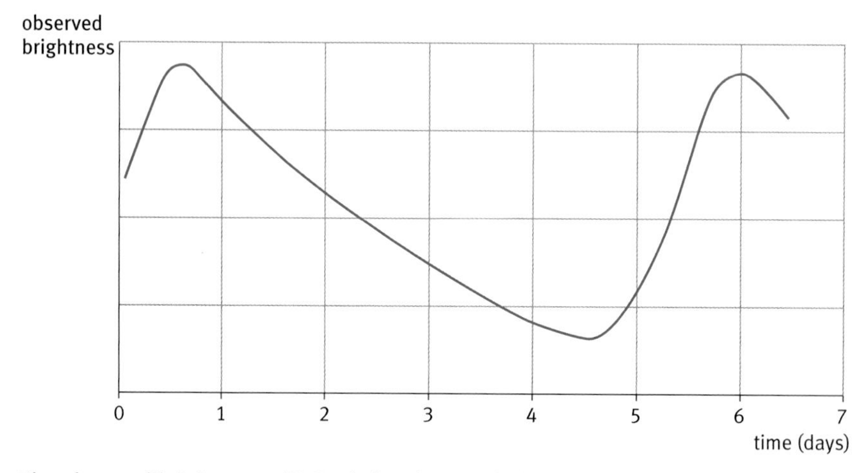

The observed brightness of δ Cephei varies regularly.

Many stars have been found that vary in this way, and they have been named **Cepheid variables**, or simply Cepheids. It is now thought that a star like this is expanding and contracting so that its temperature and luminosity vary. Its diameter may vary by as much as 30%.

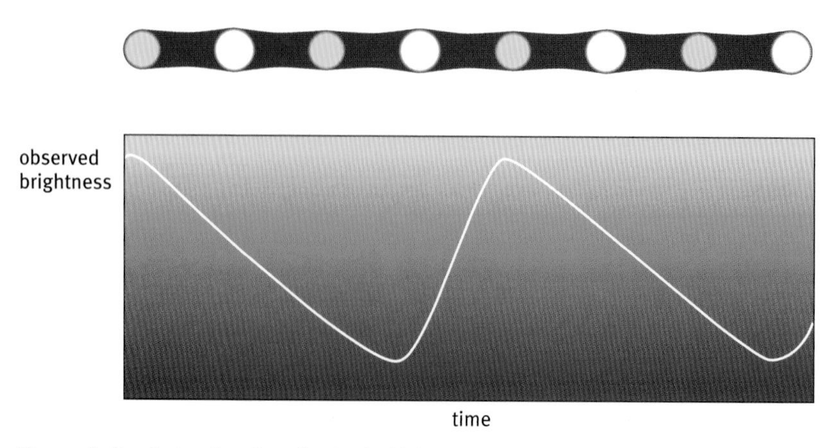

The variation in luminosity of a Cepheid is caused by its expansion and contraction.

Henrietta Leavitt

Key words
Cepheid variable

Henrietta Leavitt, whose work opened up a new method of measuring the Universe.

In the early years of the twentieth century, an American astronomer called Henrietta Leavitt made a very important discovery. She looked at Cepheids in a nearby group of galaxies (the Magellanic Clouds). She noticed that the brightest Cepheids varied with the longest periods, and drew a graph to represent this.

Because the stars she was studying were all at roughly the same distance, Leavitt realized that the stars which appeared brightest were also the ones with the greatest luminosity – they were not brighter simply because they were closer. So, by measuring the period of variation of a Cepheid, she could determine its luminosity.

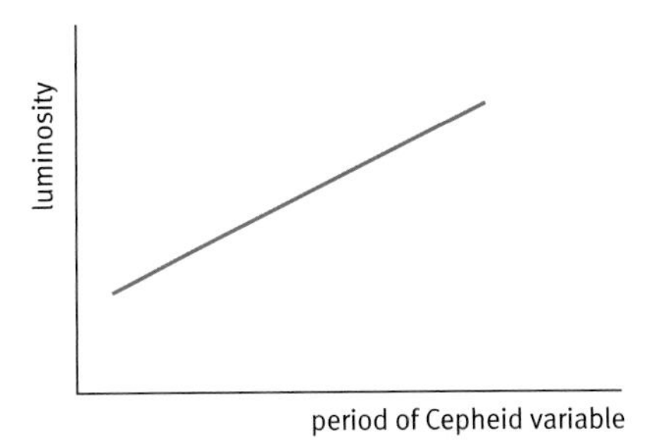

The luminosity of a Cepheid variable star is related to its period.

Measuring the Universe

Henrietta Leavitt had discovered a method of determining the distance to a star in a distant galaxy. This is how to do it:

- Look for a Cepheid variable in the galaxy of interest.
- Measure its observed brightness and its period of variation.
- From the period, determine its luminosity.
- Knowing both the luminosity and the intensity of its light at the telescope, calculate the distance of the star (and hence of the galaxy).

The same method can be used to find the distance to a Cepheid variable in our own galaxy.

Questions

1 From the graph on the opposite page, deduce the period of variation of δ Cephei.

2 Why did Henrietta Leavitt assume that the stars she was studying were all at roughly the same distance from Earth?

Galaxies: one or many?

Today, astronomers think that our Sun is just one of many thousands of millions in our galaxy, the Milky Way. They also believe that the Milky Way is just one galaxy among hundreds of billions in the Universe. But it took a long time to develop this picture.

Most of the stars seen with the naked eye are stars of the Milky Way. However, telescopes reveal many more stars, as well as some fuzzy-looking objects that look bigger than individual stars. These were originally called nebulae. ('Nebula' means 'cloud'.)

At the start of the 20th century, scientists generally thought that these nebulae were stars gradually forming from matter thinly scattered through space. If they appeared as fuzzy blobs, it seemed that they could not be very far away.

An American astronomer called Harlow Shapley set out to test this idea. He directed his telescope at a number of nebulae and measured their distances. He found that they seemed to form a spherical cloud whose centre was far from the Solar System. He guessed that each nebula was, in fact, a cluster of stars, and together these formed a sphere around the centre of the Milky Way galaxy.

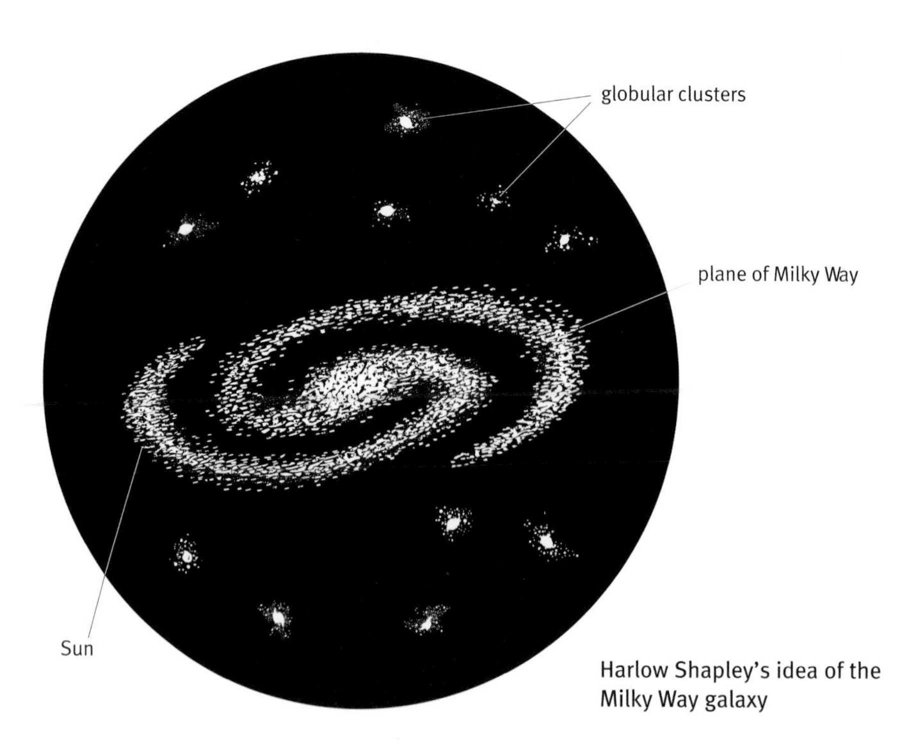

Harlow Shapley's idea of the Milky Way galaxy

Shapley presented his findings in 1920. His idea of our galaxy is a simple version of what is generally accepted today. The star clusters he was looking at are now known as **globular clusters**, and they orbit the centre of the Milky Way.

Is that all there is?

But Harlow Shapley was wrong about one thing. He claimed that the Milky Way galaxy was the entire Universe. This was challenged by another American, Heber Curtis. Curtis had been studying 'spiral nebulae' (rather than the globular clusters). He felt that these objects were very distant from the Milky Way, and possibly even were other objects on a similar scale to the Milky Way. In other words, they were galaxies in their own right.

The *Hubble Space Telescope* has given us a striking view of a Universe that contains many billions of galaxies.

The matter was decided in 1923, by exciting new results from Edwin Hubble (yet another American). Hubble was studying the Andromeda Nebula when he spotted a dim Cepheid variable star within it. He measured its period and its observed brightness, and deduced its surprising distance. Andromeda was almost one million light-years away, far greater than the dimensions of the Milky Way.

Hubble's striking result convinced astronomers that the Andromeda Nebula, at least, was a giant object outside of our galaxy. Today, the *Hubble Space Telescope* has revealed many more galaxies, scattered through space, at distances of up to 13 or 14 *thousand million* light-years. The distances between galaxies are measured in **megaparsecs** (Mpc). 1 Mpc = 1 million parsecs.

Questions

3 What force holds the globular clusters in their orbits around the centre of the galaxy?

4 Edwin Hubble observed a Cepheid in the Andromeda nebula. Why did it appear dim?

5 If there are 100 thousand million galaxies each with 100 thousand million stars, how many stars is that?

Key words

globular cluster
megaparsec

Mapping the Milky Way

In 2005, a brand new image of our Galaxy, the Milky Way, was published. From the image below, you can see that the Milky Way is a spiral galaxy, with several arms emerging from the centre. It is described as a barred spiral, because there appears to be a denser region of stars, a 'stellar bar', crossing the centre of the galaxy.

The Milky Way

Here is part of the press release that accompanied the publication of this image:

With the help of NASA's *Spitzer Space Telescope*, astronomers have conducted the most comprehensive structural analysis of our galaxy and have found tantalizing new evidence that the Milky Way is much different from your ordinary spiral galaxy.

The survey using the orbiting infrared telescope provides the fine details of a long central bar feature that distinguishes the Milky Way from more pedestrian spiral galaxies.

'This is the best evidence ever for this long central bar in our galaxy,' says Ed Churchwell, a UW-Madison professor of astronomy and a senior author of a paper describing the new work in an upcoming edition of *Astrophysical Journal Letters*, a leading astronomy journal.

Using the orbiting infrared telescope, the group of astronomers surveyed some 30 million stars in the plane of the galaxy in an effort to build a detailed portrait of the inner regions of the Milky Way. The task, according to Churchwell, is like trying to describe the boundaries of a forest from a vantage point deep within the woods: 'This is hard to do from within the galaxy.'

Spitzer's capabilities, however, helped the astronomers cut through obscuring clouds of interstellar dust to gather infrared starlight from tens of millions of stars at the center of the galaxy. The new survey gives the most detailed picture to date of the inner regions of the Milky Way.

> 'We're observing at wavelengths where the galaxy is more transparent, and we're bringing tens of millions of objects into the equation,' says Robert Benjamin, professor of physics at the University of Wisconsin.
>
> It shows a bar, consisting of relatively old and red stars, spanning the center of the galaxy roughly 27 000 light-years in length – 7 000 light years longer than previously believed. It also shows that the bar is oriented at about a 45-degree angle relative to a line joining the Sun and the center of the Galaxy.

The press release mentions a new discovery – that our Galaxy has a bar at the centre – as well as two of the big problems with mapping the Galaxy: it is difficult to make a map from inside the Galaxy, and there are clouds of dust and gas that obscure the view.

An earlier attempt

In 1785, William Herschel attempted to determine the shape of the galaxy. Looking through his telescope, he counted all the stars he could see in a particular direction. Then he moved his telescope round a little and counted again. Once he had completed a complete circle, he could draw out a map of a slice through the Milky Way.

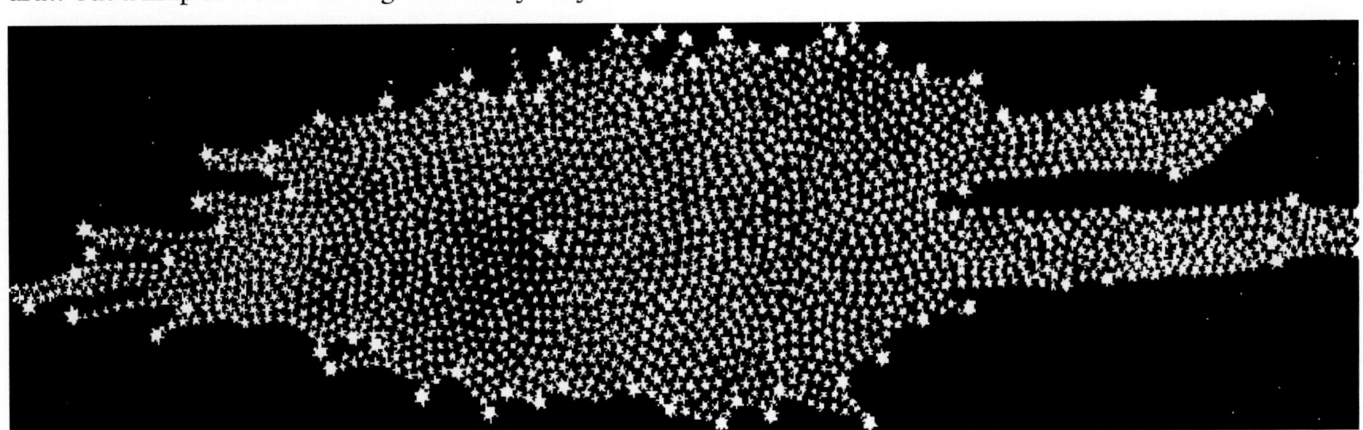

William Herschel's map of a slice through the Milky Way. The Sun is shown near the centre. The more stars seen in a particular direction, the greater the distance to the edge in that direction.

He knew that he was making the following assumptions:

- that his telescope could detect all the stars in the direction he was looking, and
- that he could see to the far end of the galaxy

Herschel himself discovered his first assumption to be incorrect when he built a bigger telescope. Astronomers now know that his second assumption too was incorrect. Dust in the galaxy makes it difficult to see stars in the packed centre of the galaxy.

Today, astronomers use infrared telescopes because infrared radiation is less affected by dust and so gives a clearer view. And it was William Herschel who, in 1800, discovered the existence of infrared radiation.

Find out about:

- the discovery of the recession of galaxies
- evidence for the expansion of the Universe

2F The changing Universe

Moving galaxies

Edwin Hubble was fortunate to be working at a time (the 1920s) when the significance of Cepheid variables had been realized. They could be used as a 'measuring stick' to find the distance to other galaxies. At the same time, he was able to use some of the largest telescopes of his day, reflectors with diameters up to 200 inches (5 metres).

He conducted a survey of galaxies, objects that had not previously been seen, let alone understood, until these powerful instruments became available. In his book *The Realm of the Nebulae,* he described what it was like to see individual stars in other galaxies:

> The observer looks out through the swarm of stars which surrounds him, past the borders and across empty space, to find another stellar system . . . The brightest objects in the nebula can be seen individually, and among them the observer recognizes various types that are well known in his own stellar system. The apparent faintness of these familiar objects indicates the distance of the nebula – a distance so great that light requires seven hundred thousand years to make the journey.

Edwin Hubble using the 48-inch telescope at the Mount Palomar observatory.

Redshift

Hubble used Henrietta Leavitt's discovery to determine the distance of many galaxies. At the same time, he made a dramatic discovery of his own. This was that the galaxies all appeared to be receding (moving away) from us. He deduced this by looking at the spectra of stars in the galaxies. The light was shifted towards the red end of the spectrum, a so-called redshift.

It turned out that, the more distant the galaxy, the greater its **speed of recession** – another linear relationship. Hubble's graph shows that, although his data points are scattered about, the general trend is clear.

Key words

speed of recession
Hubble constant

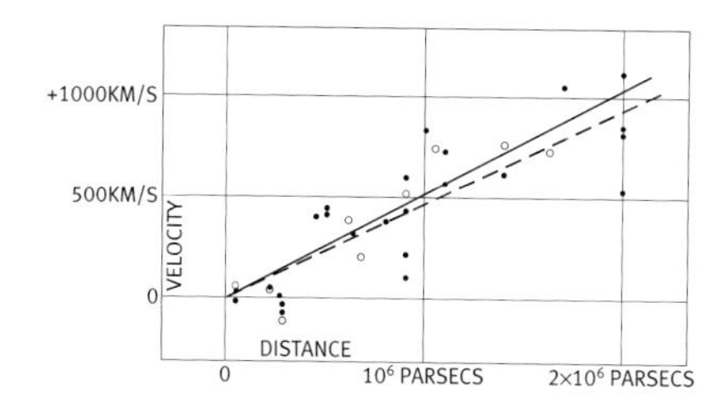

Edwin Hubble's graph, relating the speed of recession of a galaxy to its distance.

The Hubble constant

Hubble's finding can be written in the form of an equation:

$$\text{speed of recession} = \text{Hubble constant} \times \text{distance}$$

The quantity called the **Hubble constant** shows how speed of recession is related to distance. His first value (from the graph) was about 500 km/s per megaparsec. In other words, a galaxy at a distance of 1 Mpc would be moving at a speed of 500 km/s. A galaxy at twice this distance would have twice this speed.

Other astronomers too began measuring the Hubble constant, using many more distant galaxies. Hubble's first value was clearly too high. For decades, the measurement uncertainties remained high and so there were disputes about the correct value. By 2001, the accepted value of the Hubble constant was 72 ± 8 km/s per Mpc.

Back to the big bang

The fact that the galaxies are moving apart led to two important ideas:

- The Universe itself may be expanding, and may have been much smaller in the past.
- The Universe may have started by exploding outwards from a single point – the big bang,

Edwin Hubble's discovery of the moving galaxies was to lead to one of the most extraordinary scientific ideas ever.

Questions

1 Calculate the speed of recession of a galaxy which is at a distance of 100 Mpc, if the Hubble constant is 70 km/s per Mpc.

2 Using the same value of the Hubble constant, calculate the distance of a galaxy whose speed of recession is 2000 km/s.

3 A galaxy lies at a distance of 40 Mpc from Earth. Measurements show its speed of recession is 3000 km/s. What value does this suggest for the Hubble constant?

Topic 3 Inside stars

The Sun is our star. By understanding the Sun better, astronomers hope to be able to make more sense of the variety of stars they see in the night sky.

The Sun is much closer than any other star, so it is the easiest star to gather detailed scientific data on. Scientists study its surface and analyse the radiation coming from it. They turn their telescopes on it, and send spacecraft to make measurements from close up.

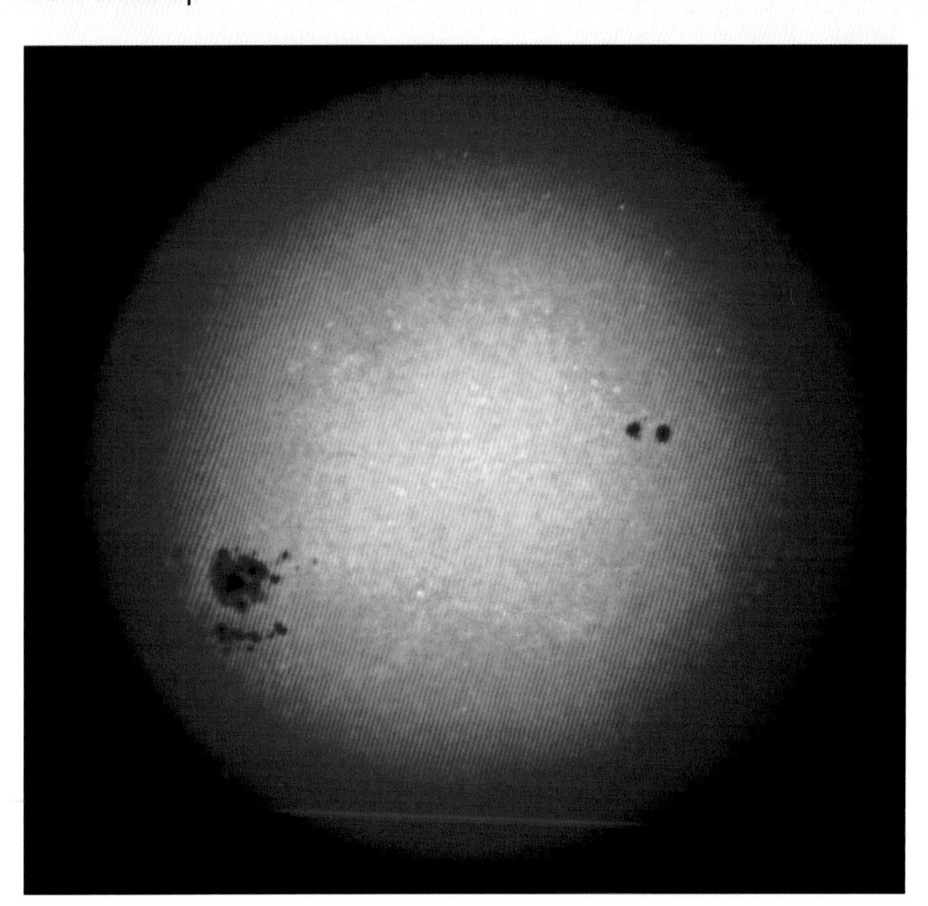

The surface of the Sun is quite dramatic when seen from close up. These are sunspots, cooler areas of the surface. They are still very hot, but perhaps 1000 °C cooler than the average surface temperature of about 5500 °C.

An energy source

What makes the Sun work, how it keeps pouring out energy, day after day, year after year, millennium after millennium – for billions of years, and why it burns so steadily are questions that puzzled scientists for centuries.

Some suggestions were that it was powered by volcanoes, that it was burning coal, that it used the energy of comets that fell into it, or simply the energy of the Sun itself as it collapsed inwards under the pull of its own gravity.

A 17th century view – the Sun as an enormous lump of coal.

Solar telescope

The McMath–Pierce solar telescope stands on a hill-top in Arizona, USA. The astronomers who use it study large images of the Sun and analyse the light gathered by the telescope. In doing this, they are applying some of the latest technology to repeat some historic experiments first performed two centuries ago.

The McMath–Pierce solar telescope. The tower on the right is 30 m high and carries the reflector that sends a beam of sunlight down the sloping shaft to an underground observation chamber.

The image of the Sun produced by the solar telescope is about 80 cm across. Some of the light passes through the hole in the table into a spectrometer for detailed analysis.

Here you can see the 2.1 m reflector, which sends a beam of light down the 152 m focusing channel. The reflector is automatically rotated to track the movement of the Sun across the sky.

Find out about:

- absorption and emission spectra as 'fingerprints' of elements
- electron energy levels and photons

3A The composition of stars

The mystery of the Sun

It might seem impossible to find out what the Sun is made of. But it turns out you can do this by examining the light it gives out. If sunlight is passed through a prism or diffraction grating, it is split into a spectrum, from red to violet.

In 1802, William Woolaston noticed that the spectrum of sunlight had a strange feature – there were black lines, showing that some wavelengths were missing from the continuous spectrum. These lines are now called Fraunhofer lines, after Joseph von Fraunhofer who made many measurements of their wavelengths. He could not, however, explain their origin.

Here is how you can make a spectrum. Light is passed through a narrow slit, so that you start with a tall, narrow strip of white light. This is then spread out into its different wavelengths by the prism. The photograph below shows a spectrum of sunlight.

The spectrum of sunlight, showing the dark Fraunhofer lines.

The colours of the elements

Before he looked at sunlight, Fraunhofer had been studying the light given out by different chemicals when they burn. He knew that sodium burns with a yellow flame. When he looked at the spectrum of light from a sodium flame, Fraunhofer saw that it consisted of just a few coloured lines, rather than a continuous band from red to violet. He could measure the wavelengths of the different colours that made up the spectrum of sodium.

A spectrum like this is called an **emission spectrum**, because you are looking at the light emitted by a chemical. Today, we know that each element has a different pattern of lines in its emission spectrum, and that this can be used to identify the elements present.

Making sense of sunlight

The dark Fraunhofer lines in the spectrum of sunlight were not explained until 1859. They are caused by the absorption of some colours, rather than emission. To understand what is going on, you have to think of the structure of the Sun.

- The interior of the Sun produces white light, with all wavelengths present.
- As this light passes through the Sun's atmosphere, some wavelengths are absorbed by atoms of elements (including sodium) that are present there.

As a result, the light that reaches us from the Sun is missing some wavelengths, which correspond to elements in the Sun's atmosphere.

This is an example of an **absorption spectrum**. The wavelengths of the absorption lines reveal which elements have been doing the absorbing. From this, astronomers can identify the elements present in the Sun and in the most distant stars.

Key words

emission spectrum
absorption spectrum

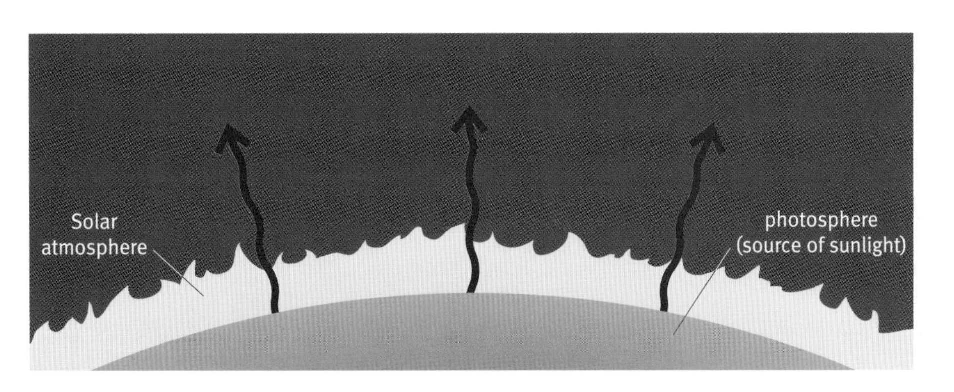

Light from the interior of the Sun must pass through its atmosphere before it reaches us.

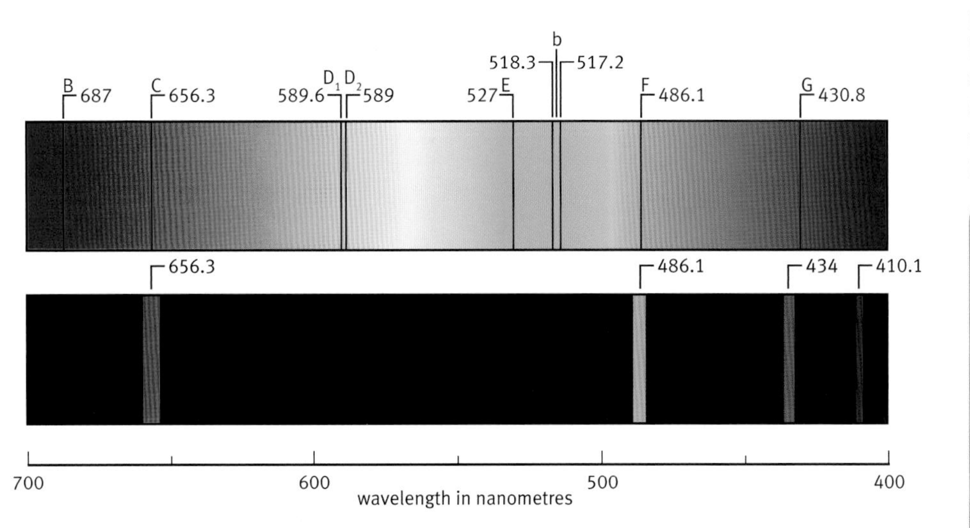

Lines in the emission spectrum of hydrogen (*bottom*) correspond to dark absorption bands in the Sun's spectrum (*top*). The numbers are the wavelengths of lines in the hydrogen spectrum in nanometres.

Questions

1 'Light is a messenger from the stars'. Explain how this statement is true.

2 Look at the emission spectrum of hydrogen. What colours are the main emission lines in this spectrum? Which is the strongest (most intense) line?

Understanding line spectra

The line spectrum of an element is different from that of every other element – it can be thought of as the 'fingerprint' of the element. In 1868, two scientists used this fact to discover a new element: helium. Norman Lockyer (English) and Jules César Janssen (French) took the opportunity of an eclipse of the Sun to look at the spectrum of light coming from the edge of the Sun. Janssen noticed a line in the spectrum that he had not seen before. He sent his observation to Lockyer, who realized that the line did not correspond to any known element. He guessed that some other element was present in the Sun, and he named it 'helium' after the Greek name for the Sun, 'Helios'.

Emitting light

To understand why different elements have different emission spectra, you need to know how atoms emit light. The light is emitted when electrons in atoms lose energy – the energy they lose is carried away by light.

That is a simple version of what happens, and it does not explain why only certain wavelengths appear in the spectrum. Here is a deeper explanation:

- The electrons in an atom can only have certain values of energy. Scientists think of them as occupying points on a 'ladder' of **energy levels**.
- When an electron drops from one energy level to another, it loses energy.
- As it does so, it emits a single **photon** of light – that is, a packet of energy. The energy of the photon is equal to the difference between the two energy levels.

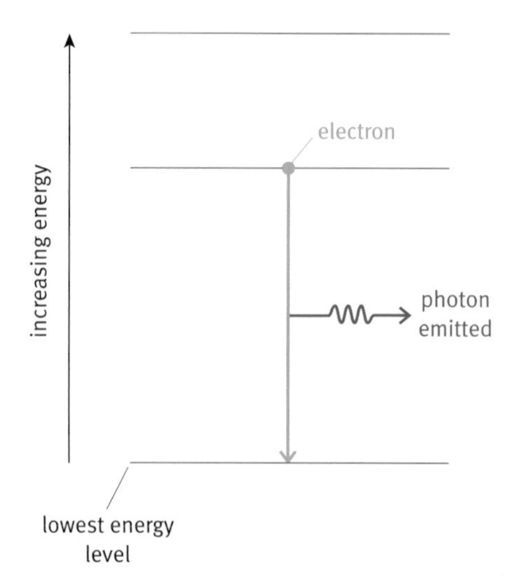

An electron gives out a single photon of light as it drops from one energy level to another.

The greater the energy gap, the greater is the energy of the photon. High-energy photons correspond to high-frequency, short-wavelength light.

In the simplified diagram of energy levels shown here, you can see that only three photon energies are possible. The most energetic photon comes from an electron which has dropped from the top level to the bottom level.

Key words
energy level
photon

Absorbing light

The same model can explain absorption spectra. The dark lines come about when electrons absorb energy from white light.

- White light consists of photons with all possible values of energy.
- An electron in a low energy level can only absorb a photon whose energy is just right to lift it up to a higher energy level. When it absorbs such a photon, it jumps to the higher level.
- The white light is now missing photons that have been absorbed because their energies corresponded to the spacings in the ladder of energy levels. The 'missing' photons correspond to the dark lines in an absorption spectrum.

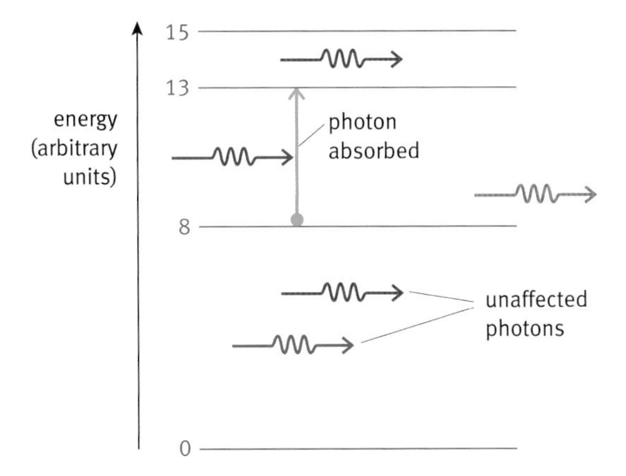

An electron jumps from one energy level to another when the atom absorbs a single photon of light.

The diagram above shows values of energy for each level (in arbitrary units). You can see that:

- Photons of energies 2, 7, and 8 would be absorbed.
- Photons of energies 1, 6, and 9.3 would not be absorbed.

A photon with sufficient energy can ionize an atom. An electron gains so much energy that it escapes the attractive force of the nucleus, and leaves the atom. As a result, the atom is left with net positive charge.

Questions

3 What sub-atomic particles are associated with the emission and absorption of light?

4 For the energy level diagram on this page:

a Explain how photons of energies 2, 5, and 8 would be absorbed.

b Give other energy values that would also be absorbed.

c Explain how an atom could emit a photon of energy 13 units.

Find out about:

- models of the atom
- how alpha particle scattering reveals the existence of the atomic nucleus

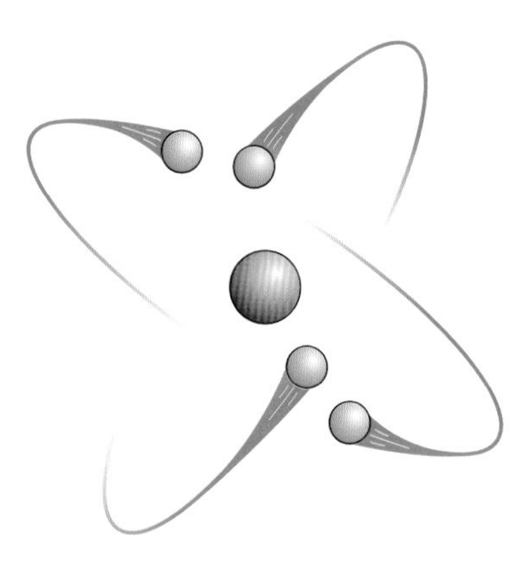

The 'solar system' model of the atom.

3B Atoms and nuclei

How do scientists know about the structure of atoms, the diagram of an atom as a miniature 'solar system', with the **nucleus** at the centre and electrons whizzing round like miniature planets, has become very familiar. It is often used just to suggest that something is 'scientific'.

The 'solar system' model of the atom dates back to 1910, and an experiment thought up by Ernest Rutherford. Scientists were beginning to understand radioactivity, and were experimenting with radiation. Rutherford realized that alpha and beta particles were smaller than atoms, and so they might be useful tools for probing the structure of atoms. So he designed a suitable experiment, and it was carried out by his assistants, Hans Geiger and Ernest Marsden.

Here is how to do it:

- Start with a metal foil. Use gold, because it can be rolled out very thin, to a thickness of just a few atoms.
- Direct a source of alpha radiation at the foil. Do this in a vacuum chamber, so that the alpha particles are not absorbed by air.
- Watch for flashes of light as the alpha particles strike the detecting material around the outside of the chamber.
- Work all night, counting the flashes at different angles, to see how much the alpha radiation is deflected.

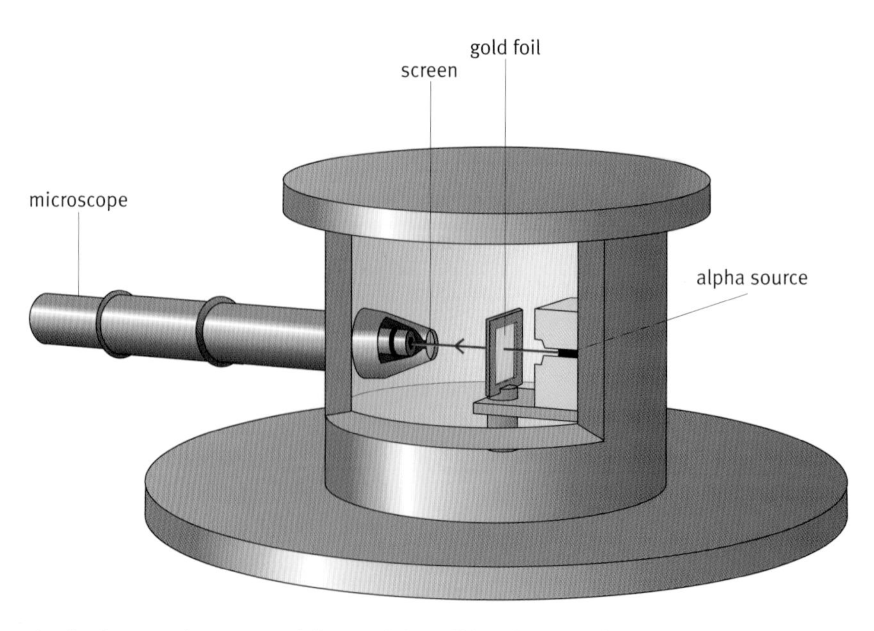

Rutherford's experiment. An alpha particle striking the scintillating material gives a tiny flash of light.

Results and interpretation

This is what Geiger and Marsden observed:

- Most of the alpha particles passed straight through the gold foil, deflected by no more than a few degrees.
- A small fraction of the alpha particles were actually reflected back towards the direction from which they had come.

And here is what Rutherford said:

'It was as if, on firing a bullet at a sheet of tissue paper, the bullet were to bounce back at you!'

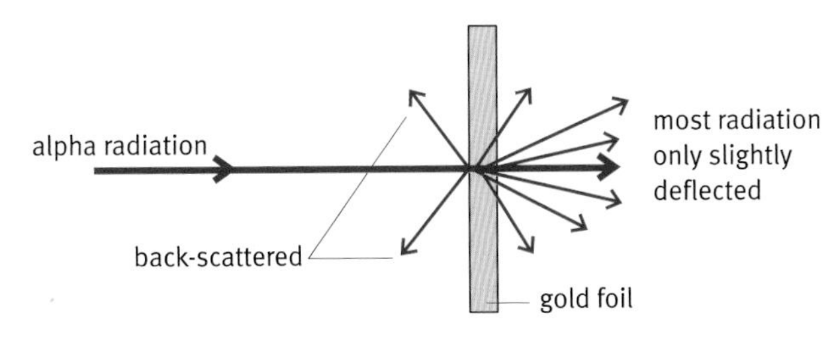

Only alpha particles passing close to a nucleus are significantly deflected.

In fact, less than 1 alpha particle in 8000 was back-scattered (deflected through an angle greater than 90°), but it still needed an explanation.

Rutherford realized that there must be something with positive charge that was repelling the alpha particles (which also have positive charge). And it must also have a lot of mass, or the alpha particles would just push it out of the way.

This 'something' is the nucleus of a gold atom. It contains all of the positive charge within the atom, and most of the mass. Note that, at this time, scientists had no knowledge of protons and neutrons, the particles that make up the nucleus.

Rutherford's analysis of his data showed that the nucleus was very tiny, because most alpha particles flew straight past without being affected by it. The diameter of the nucleus of an atom is roughly a hundred-thousandth of the diameter of the atom.

Key words

nucleus

Questions

1 What charge do the following have:

- **a** the atomic nucleus?
- **b** alpha radiation?
- **c** electrons?

2 Put these in order, from least mass to greatest: gold atom, alpha particle, gold nucleus, electron.

3 Make a prediction: Geiger and Marsden repeated their experiment using a thicker gold foil. Would more alpha particles be reflected back towards the source, or fewer? Explain your answer.

Small particles, big science

At the time of Rutherford's experiment, no-one knew about protons and neutrons and the role they played in the atom. But once the neutron had been discovered in 1932, the modern picture of the atom was established. These ideas explained a lot – the pattern of the periodic table, for example – so physicists had been able to explain most of chemistry!

It seemed that all matter was made of just three particles: protons, neutrons, and electrons. It was very satisfying to think that all of matter could be boiled down to just three particles, but that idea did not last long. Physicists studying cosmic radiation from space discovered something that was different, particles with masses in between those of protons and electrons. As so often happens, a new discovery caused an attractive, simple scientific theory to collapse like a pack of cards. But at least that left plenty of scope for new ideas and experiments.

Probing the very small

Rutherford's experiment showed scientists how to investigate something as small as an atomic nucleus. You need a lot of atoms (the gold foil) and something of a similar size to the nucleus (the alpha particle) to act as a probe.

Alpha particles move fast – at about 10 million m/s. That means they have enough momentum to get close to an atomic nucleus and feel its effects. But when scientists began to suspect that protons and neutrons were not fundamental after all, they wanted to look *inside* the particles of the nucleus.

CERN nestles at the foot of the Alps, straddling the border of Switzerland and France. The circle shows the location of the 27 km tunnel containing the LHC.

To look inside protons and neutrons, it takes even more energetic particles than alpha particles, and to get these, large particle accelerators must be built. Big particle accelerators are amongst the biggest and most complex machines ever made, and they are very expensive. An example is the Large Hadron Collider (LHC) at CERN, Geneva.

Today, most physicists think that there are even more fundamental particles called **quarks**. It takes three quarks to make a proton or a neutron, and just two to make some of the other particles that have been discovered from studies of cosmic rays and in other experiments.

An international project

CERN is the European Centre for Nuclear Research. It is funded by 20 countries because, for most, just one accelerator would be too expensive. By cooperating, their scientists get to use all of the facilities at CERN, including the LHC. The LHC is in the form of a circular tunnel, 27 km in circumference.

- The accelerator produces two beams of high-energy protons.
- The beams travel around the tunnel in opposite directions.
- At one point, the paths of the two beams cross and protons collide.
- Detectors record the debris from the collision, and scientists analyse these records.

Key words
quarks

Back to the beginning

The point of all this is that the protons CERN's scientists are using have energies that are millions of times greater than Rutherford's alpha particles, so they hope that the LHC will tell them more about the fundamental nature of matter.

At the same time, there is a link with astronomy. In the early history of the Universe, shortly after the Big Bang, all matter and energy was compressed into a tiny volume so that particles such as protons had very high energies – just like those in the LHC. So the LHC is effectively recreating the conditions that existed at that time.

Just some of the people involved in the LHC project. This team manages and coordinates the experimental areas.

Inside the LHC tunnel. These are some of the giant magnets used by the LHC accelerator.

Find out about:

- the processes of fusion and fission
- the attractive and repulsive forces between nuclear particles

3C Nuclear fusion

Knowing about the structure of the atom can help us to understand how the Sun releases energy. The Sun is made mostly of hydrogen. Hydrogen burns with oxygen to make water, but there is very little oxygen in the Sun, so that cannot be the source of its energy. To understand how hydrogen on its own can have fuelled the Sun for 4.5 billion years, you need to look inside the atom and inside the nucleus.

Recall from Module P3 *Radioactive materials* that atomic nuclei are made of two particles:

- **protons**, which have positive charge, and
- **neutrons**, which are electrically neutral

These particles have similar masses and account for most of the mass of an atom (because electrons have very little mass).

A balance of forces

The nucleus of an atom is made up of protons and neutrons. This tells us something important – protons and neutrons are happy to stick together. There must be an **attractive force** that holds protons and neutrons together, and that can even hold two protons together despite the fact that they repel each other because of their positive charges. This force is called the **strong nuclear force**.

The strong nuclear force has a short range. It only acts when two nucleons (protons or neutrons) are very close together. In a nucleus, the particles are separated by just the right distance so that the strong nuclear force is balanced by the electrical (electrostatic) force.

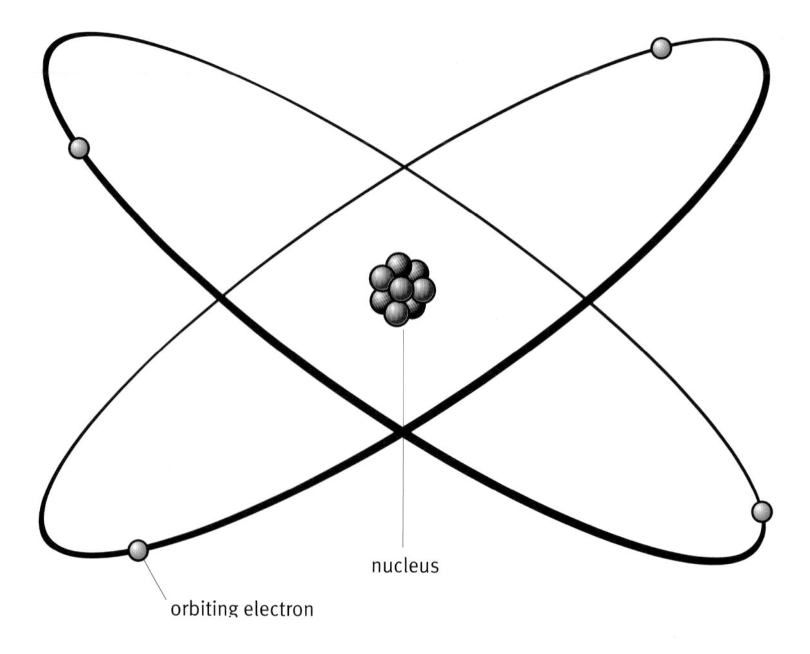

In this diagram, protons are shown as red spheres and neutrons as black.

Joining up

In the process of **nuclear fusion**, the nuclei of two hydrogen atoms join together and energy is released. The diagram below right shows one way in which this happens. Note that the nuclei are of two different isotopes of hydrogen – both are hydrogen, because they have just one proton in the nucleus.

Picture bringing two atoms close together: their nuclei repel each other, because of the electrical (electrostatic) force. They will not fuse together. Push hard enough: they come close enough for the attractive force to take over, and the nuclei fuse. Energy is released.

You would have to do a lot of work to push two nuclei together, but you would get a lot more energy out when they fused.

You can think of fusion as the opposite of nuclear fission, the process used in nuclear power stations.

- **Fission**: a large nucleus splits, releasing energy and a few neutrons, and forming two medium-sized nuclei.
- **Fusion**: two small nuclei join together, releasing energy and forming one bigger nucleus.

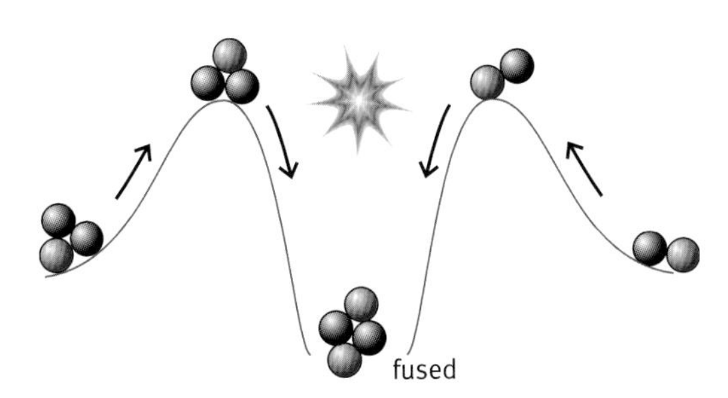

Pushing two hydrogen nuclei together: the 'hills' represent the repulsive force between them. The deep 'valley' represents the stable state they reach when they fuse together.

In the stars

It is tricky to force two nuclei to fuse in a laboratory on Earth, but it happens all of the time in stars. Because the interior of a star is very hot – millions of degrees – the particles within it are very energetic. They are moving very fast. There is a small chance, when any two particles collide, that their momentum will be enough to overcome the repulsion between them, and to bring them within the short range of the strong nuclear force. They will then fuse. Each time this happens, energy is released.

The Sun was almost 100% hydrogen when it formed. Today, it is 71% hydrogen and 27% helium.

Key words

fusion
fission
proton
neutron
strong nuclear force

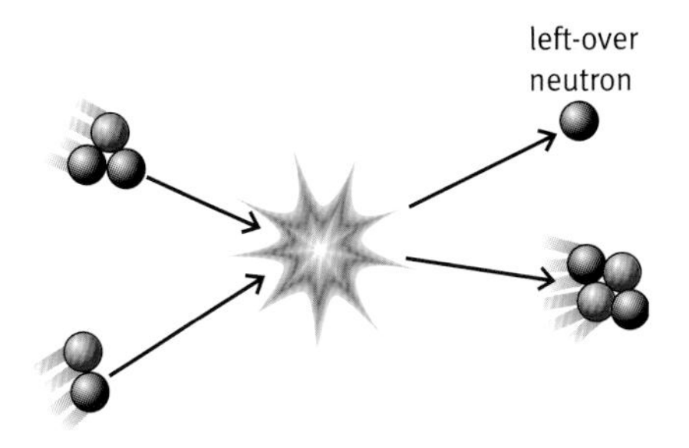

Fusion

Questions

1 a What attractive force acts between particles in the nucleus? And what repulsive force?

b Which of these two forces has the greater range?

2 The Sun contains 1% oxygen nuclei. There are 8 protons in an oxygen nucleus. Explain why these are less likely to fuse together than hydrogen nuclei.

3 Use information on these two pages to make an estimate of the lifespan of the Sun. Explain the assumptions you have made.

Find out about:

- how the pressure, volume, and temperature of a gas are related
- how to convert between temperature scales

3D How gases behave

Great balls of fire

The Sun is mostly hydrogen and helium – two gases. As you have seen, it releases energy by fusing hydrogen nuclei to make helium nuclei. This goes on in the core of the Sun and most other stars.

So a star is a giant ball of hot gas. To understand better how stars work – how they get hot, the forces that hold them together – you need to learn about gases in general.

A steam engine works by allowing a gas (steam) to expand so that it pushes a piston.

Describing a gas

Think of a balloon. You blow it up, so it is filled with air. How can you describe the state of that air? What are its properties?

Volume – the amount of space the gas occupies, in m^3.

Mass – the amount of matter, in kg.

Pressure – the force the gas exerts per unit area on the walls of its container, in Pa ($= N/m^2$).

Temperature – how hot the gas is, in °C (or K – see page 241).

These are all measurable quantities that a physicist would use to describe the gas. Understanding how these properties change is important. For example, the engine of a car relies on the pressure of an expanding gas to provide the motive force that makes the car go.

Pressure and volume

Now picture squashing the balloon. You are trying to decrease its volume. Its pressure resists you. It is easier to understand what is happening by pressing on a gas syringe.

volume

pressure

Increasing the pressure on a gas reduces its volume.

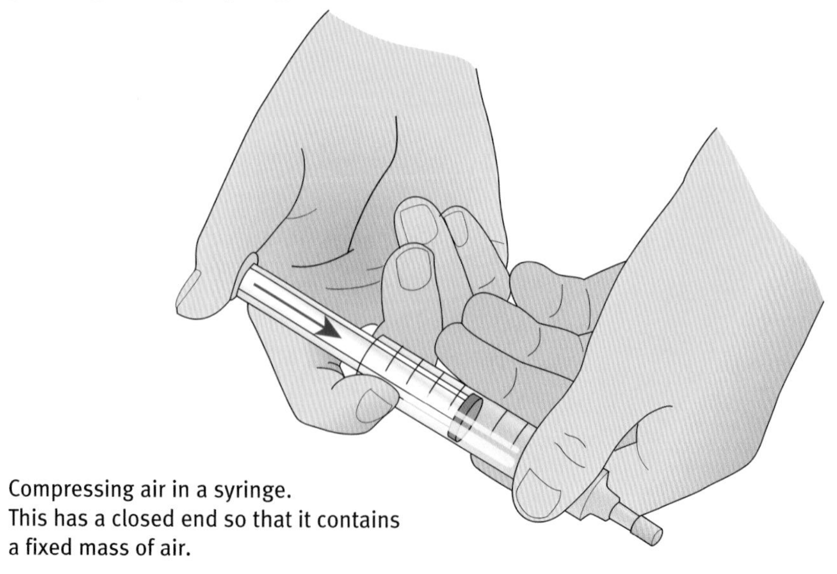

Compressing air in a syringe. This has a closed end so that it contains a fixed mass of air.

- It is easy to push the plunger in a little.
- The more you push on the plunger, the harder it gets to move it further.

This shows that the pressure of the air is increasing as you reduce its volume. If you reduce the force with which you press on the plunger, the air will push back and expand. The graph on the previous page shows this relationship:

- When the volume of a gas is reduced, its pressure increases.

Key words
volume
mass
pressure
temperature
kinetic model

Explaining pressure

The connection between the pressure and the volume of a gas was worked out before anyone was sure that gases were made of particles. Yet you can use the **kinetic model of matter** to explain these findings. In this model, a gas consists of particles (atoms or molecules) that move around freely, and most of the volume of the gas is empty space.

- The particles of a gas move around freely. At room temperature, they have speeds around 450 m/s.
- As they move around, they bump into the walls of their container – see the first diagram on the right. (They also bump into each other.)
- Each collision with the walls causes a tiny force. Together, billions of collisions produce gas pressure.

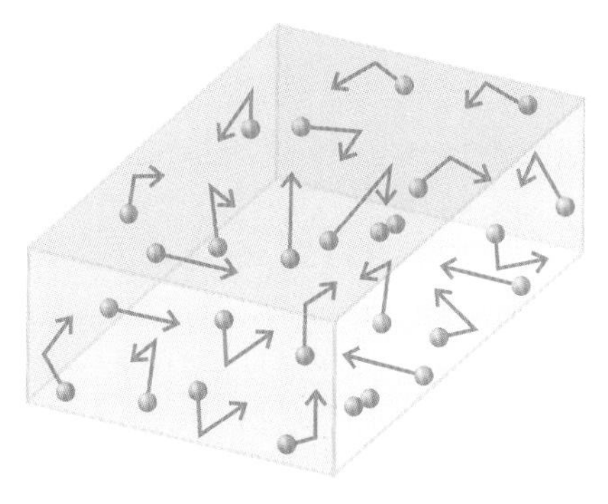

Particles of a gas collide with the walls of its container. This causes pressure.

Now think about what happens if the same gas is compressed into a smaller volume – see the diagram on the right. The collisions with the walls will be more frequent, and so the pressure will be greater.

Questions

1 Imagine that a fixed mass of gas is compressed into half its original volume. The temperature remains constant. Which of the following statements are correct?

A The pressure of the gas will increase.
B The average separation of the particles of the gas will decrease.
C A particle of the gas will strike the walls of the container with greater force.
D A particle of the gas will strike the walls more frequently.

As cold as it gets

Now think about what happens when a gas gets cold. Blow up a balloon and put it in the freezer – it starts to shrink. Its pressure and volume have both decreased.

In an experiment to investigate this, it is best not to change one factor (temperature) and then to allow two others (pressure and volume) to change. So experiments are designed to control one factor while the other is allowed to change. The picture below shows how a fixed volume of air (in a rigid flask) can be heated to change its temperature. The gauge shows how the pressure of the gas changes.

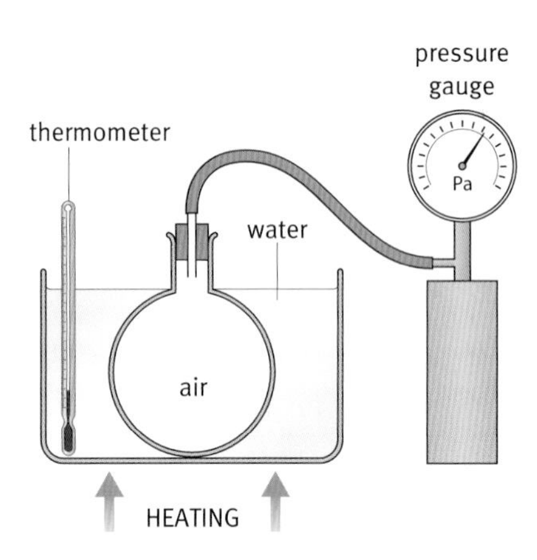

Because the flask is rigid, the volume of the air inside does not change as it is heated or cooled.

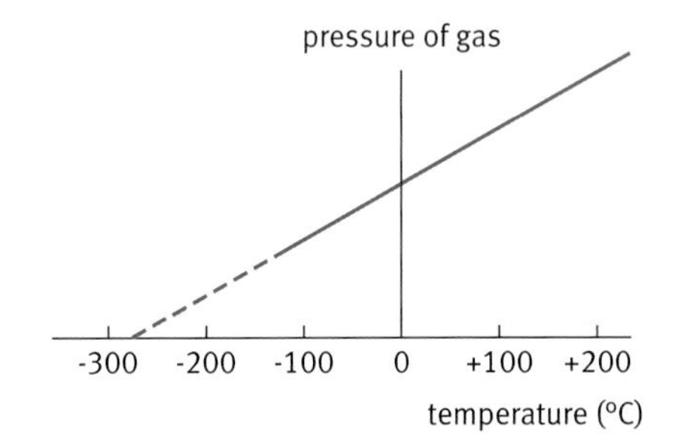

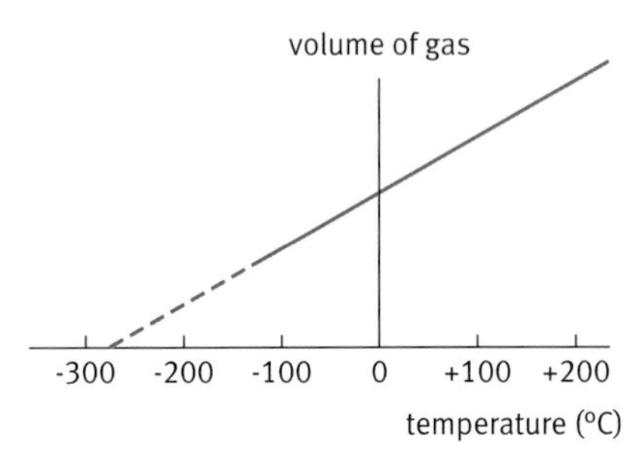

As a gas is cooled, its pressure and volume decrease.

Heating up, cooling down

The graphs on the left show the results of experiments like this. Think about the effects of cooling down a fixed mass of gas.

- Fixed volume of gas: as the gas is cooled, its pressure decreases steadily.
- Fixed pressure of gas: as the gas is cooled, its volume decreases steadily.

Both of these graphs show the same pattern: the pressure and volume of the gas decrease as the temperature decreases, and both seem to be heading for a value of zero at a temperature well below 0 °C. Whatever gas is used, the graph heads for roughly the same temperature.

The point where a graph like this reaches zero is known as the **absolute zero** of temperature. In practice, all gases condense to form a liquid before they reach this point.

$$\text{absolute zero} = -273\,°\text{C}$$

Temperature scales

In everyday life, we use the Celsius scale of temperature. This has its zero, 0 °C, at the temperature of pure melting ice. You can also define a scale that has its zero at absolute zero, the **Kelvin scale**. The individual divisions on the scale (degrees) are the same as on the Celsius scale, but the starting point is much lower. Because nothing can be colder than absolute zero, there are no negative temperatures on the Kelvin Scale.

Temperatures on this scale are given in kelvin (K).

Here is how to convert from one scale to the other:

- temperature in K = temperature in °C + 273
- temperature in °C = temperature in K – 273

So, for example, suppose your body temperature is 37 °C. What is this in K?

- temperature in K = 37 °C + 273 = 310 K

A kinetic explanation

When a gas is cooled down, the particles of the gas lose energy, so they move more slowly.

- If the volume of the gas is fixed, each particle takes longer to reach a wall. So particles strike the walls less frequently. They also strike with less force. So the pressure decreases for these two reasons.
- If the pressure is to remain constant, the volume of the gas must decrease to compensate for the fact that the collisions are weaker and less frequent.

Eventually, you can picture the particles of the gas losing all of their kinetic energy, so that they do not collide with the walls at all. There is no pressure. This is absolute zero.

Key words
absolute zero
Kelvin scale

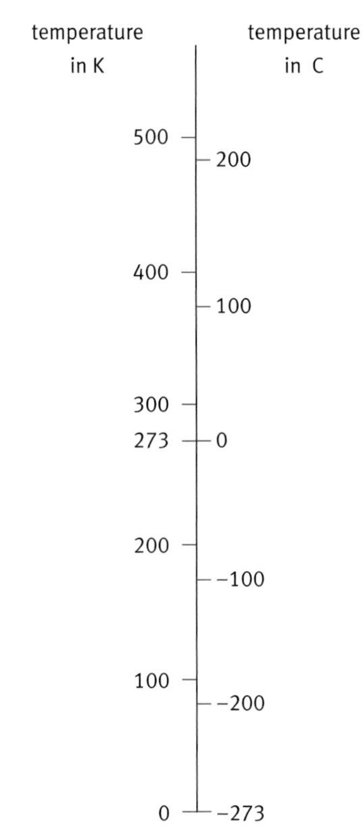

Comparing the Celsius and Kelvin scales of temperature.

Questions

2 What are the values of the following temperatures on the Kelvin scale?
- **a** 0 °C
- **b** 100 °C
- **c** −100 °C

3 What are the values of the following temperatures on the Celsius Scale?
- **a** 0 K
- **b** 200 K
- **c** 300 K.

4 The surface temperature of the Sun is roughly 5800 K. What is this in °C, equally roughly?

Weather from the Sun

- January 1997, USA. Millions of satellite TV viewers lose their picture as their satellite is destroyed.
- March 1989, eastern Canada. Six million people shiver as their power grid trips out.

What links these two events? Could they happen again?

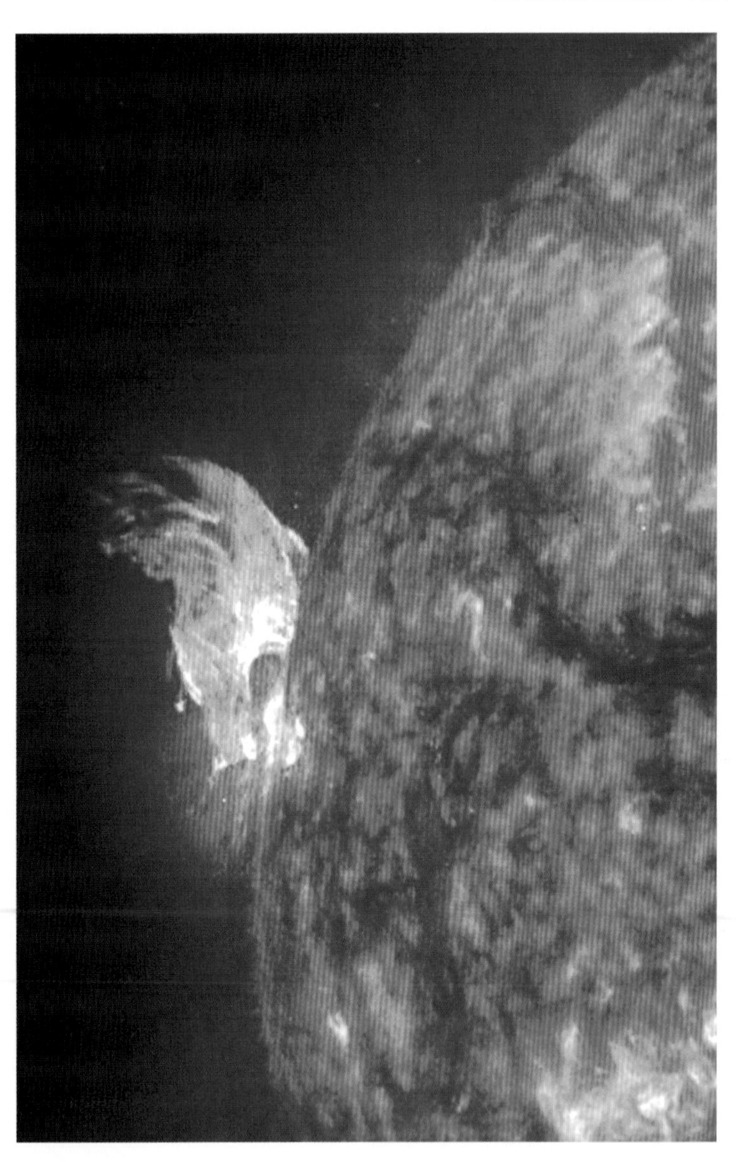

A coronal mass ejection bursting out from the surface of the Sun. (Photograph taken using ultraviolet light.)

Both these dramatic events were caused by the Sun. From Earth, the Sun appears to burn brightly and steadily, day by day. When astronauts first travelled above the atmosphere, they took photographs that revealed a different side to it. The photos showed giant bubbles of gas bursting out of the Sun and flying out into space. These are now known as coronal mass ejections (CMEs) and may weigh as much as 100 billion tonnes.

Heading this way

A coronal mass ejection can travel through space at speeds of around 1.5 million km/h, so they take about 4 days to travel the 150 million km to Earth. This is fast, but much slower than light, which takes just 500 seconds to reach us from the Sun.

CMEs consist of electrically charged particles. When one reaches Earth, it can produce dramatic aurora effects in the night sky (the aurora borealis). It can also have some more drastic effects:

- Satellites' electrical systems may be destroyed and radio communications disrupted.
- Power systems may be damaged by the associated magnetic field.

The realization that events on the Sun could have such serious impacts on the technologies people depend on here on Earth led scientists and engineers to wonder whether it is possible to forecast the 'weather' coming our way from the Sun.

Scopes in space

The answer was the *SOHO* space observatory, launched in December 1995. (*SOHO* stands for Solar and Heliospheric Observatory.) This carried an instrument called a coronagraph, designed and built by Professor George Simnett and colleagues at Birmingham University. The corona is the Sun's outer atmosphere; monitoring the corona with the coronagraph allows us to see when a CME bursts out of the Sun, and to track it. If it is heading our way, a warning forecast is issued.

You cannot usually see the Sun's corona, except when there is a total eclipse of the Sun. Then the Moon blocks the Sun's bright disc and the glowing corona becomes visible. A coronagraph is a telescope inside which a circular disc is fixed. Like a permanent eclipse, this blocks the Sun and allows a camera at the back of the telescope to photograph the corona.

The coronagraph has other uses. Because it blocks out the disc of the Sun, it has made it possible to spot over 1000 previously unobserved comets, mostly as they plunged into the Sun. (Even in the early 20th century, many scientists believed that the Sun gained its energy from the kinetic energy of comets that fell into it.)

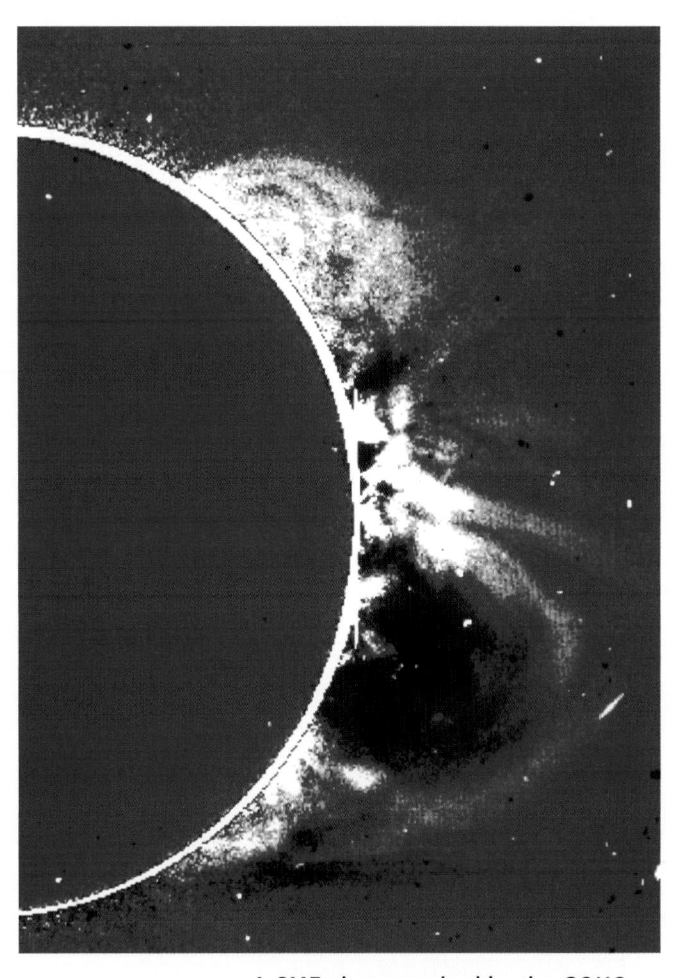

A CME photographed by the *SOHO* coronagraph. You can see the eclipsing disc clearly on the left.

Warning on the web

Websites today carry space 'weather forecasts' so that precautions can be taken where necessary. For example, astronauts on the *International Space Station* may need to turn their craft so that they are shielded from incoming particles from the Sun. Other spacecraft may fold up their solar panels to avoid them being damaged.

Topic 4 The lives of stars

The photograph at bottom left shows a set of prominent stars in the southern sky, forming the shape of a cross. This is the Southern Cross, which Catholic sailors regarded as very significant when the Spanish and Portuguese Empires expanded in the 15th and 16th centuries. It now appears on the Australian flag. Once this particular group of stars has been pointed out to you, it is hard *not* to recognize it whenever it is above the horizon.

Humans seem to impose a sense of order whenever they can. In the night sky, the signs of the zodiac provide an example of this. The picture at bottom right shows the pattern of stars that make up the constellation of Leo (the Lion). To see the lion, you need to be told which stars form its legs, tail, head, and so on. From the picture, you can see that this is not at all obvious.

The picture came from a book published in Italy in the 13th century. It was a new Latin translation of a book called *The Pattern of Fixed Stars*, written in Arabic in 964 AD. But these patterns go back much further in history, to at least 2500 years ago. And they are still in use today. Astronomers name stars according to the constellation they are in. Astrologers use the movements of planets across constellations to tell fortunes.

The Southern Cross, photographed from Cuba.

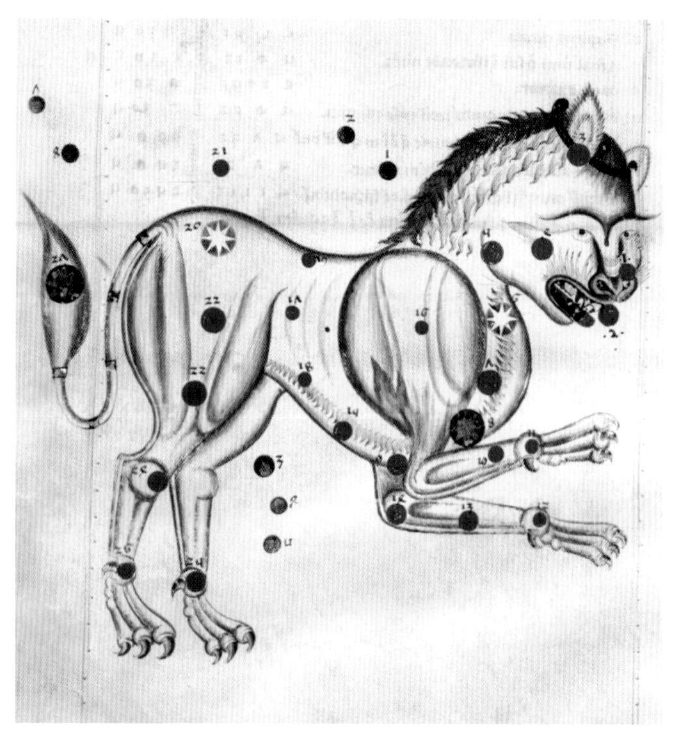

The constellation of Leo the Lion.

Unchanging Universe?

The human history of constellations tells us something striking: the pattern of stars in the night sky has changed very little over thousands of years. In the Middle Ages, astronomers generally thought that the heavens were unchanging. In that way, the sky was quite different from our everyday, Earth-bound experience. Different laws applied 'up there' from 'down here'.

Then, in 1572, a Danish astronomer called Tycho Brahe made a surprising discovery – what he took to be a new star, or 'nova'. It appeared in the constellation of Cassiopeia, almost overhead. It was visible for 18 months or so, after which it gradually faded away. Now we know that what Tycho saw was not a new star, but an old one, exploding towards the end of its life. It was what is called a supernova, and it showed that stars really can change.

Today, astronomers keep a watch for supernovas. They are rare events in our galaxy, but they can also be spotted in other galaxies. They are useful, because one type (Type 1a) seems to explode with a standard brightness. This means that they can be used for measuring distances – the dimmer the supernova, the more distant it is. Combining this with measurements of redshift provides a way of gathering more data for estimates of the expansion of the Universe.

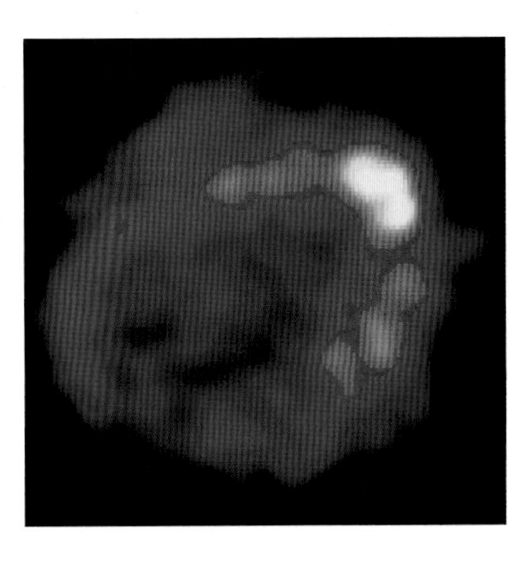

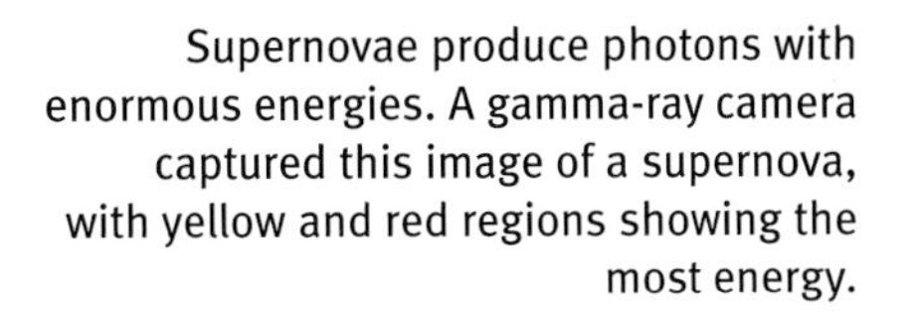

Supernovae produce photons with enormous energies. A gamma-ray camera captured this image of a supernova, with yellow and red regions showing the most energy.

Find out about:

- mapping stars on to the Hertzsprung–Russell diagram
- the structure of the Sun, and how energy leaves the Sun

4A Stars change

Looking at the night sky, you see stars of different brightnesses and colours. By the beginning of the 20th century, astronomers had worked out how to make sense of this:

- Stars might be dim because they were a long way off. Knowing the distance to a dim star, they could work out its intrinsic ('true') brightness.
- Stars are different colours because they are different temperatures. Red is cool, blue is hot.

A Danish astronomer called Ejnar Hertzsprung set about finding if there was any connection between brightness and temperature. He gathered together published data and drew up a chart, which he published in 1911.

An American, Henry Russell, came up with the same idea independently. He was unaware of Hertzsprung's chart, which had been published in a technical journal of photography. Today, the chart is known as the Hertzsprung–Russell diagram, or H–R diagram. A modern version is shown below.

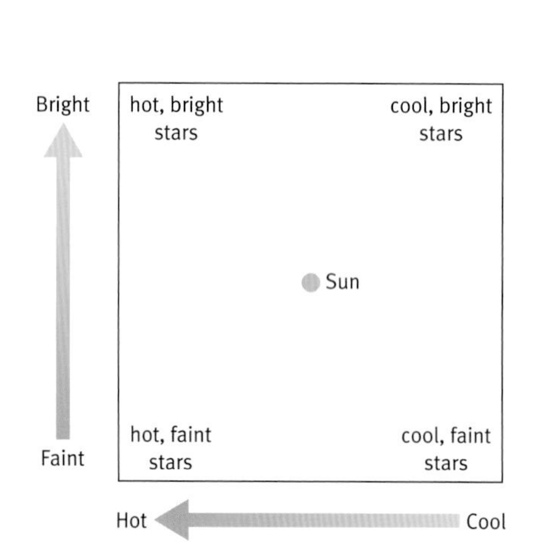

Explaining the axes on the Hertzsprung–Russell diagram.

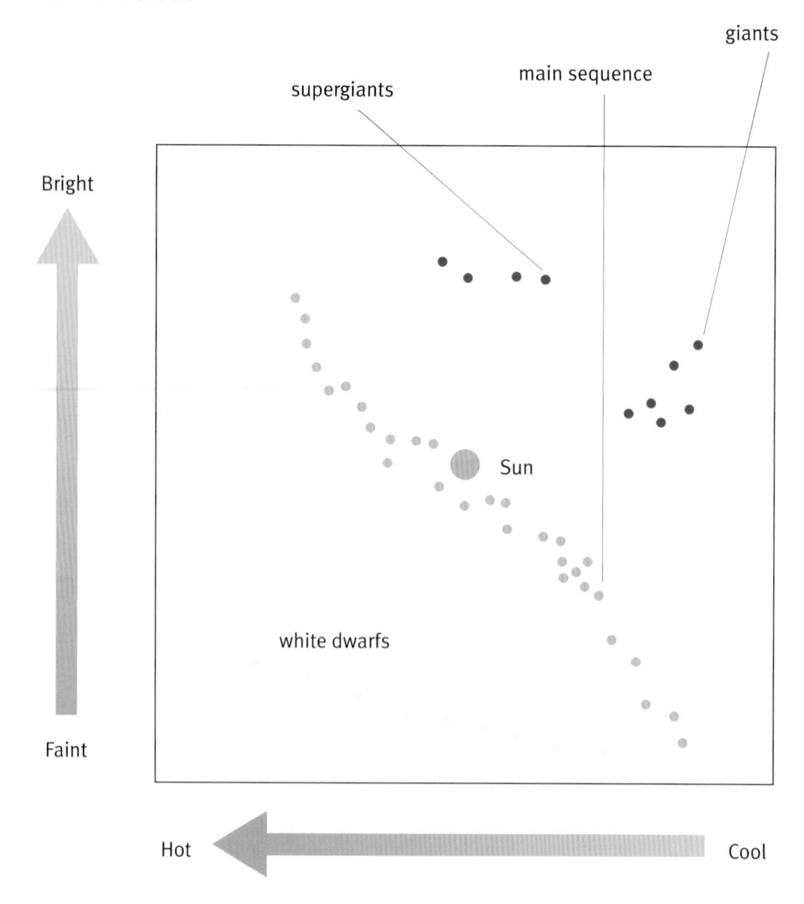

Data for stars – the H–R diagram

Understanding the H–R diagram

This chart has brightness on the vertical axis and temperature on the horizontal axis. It is usually drawn as shown, with temperature *decreasing* along the x-axis. The Sun is roughly at the middle of the chart.

By 1924, over 200 000 stars had been catalogued. When plotted on the H–R diagram, these stars fell into three groups:

- About 90% of stars (including the Sun) fell along a line running diagonally across the diagram. This is known as the main sequence.
- About 10% of stars were white dwarfs, small and hot.
- About 1% of stars were red giants or supergiants, bright but not very hot.

What the H–R diagram reveals

The stars which appear on the H–R diagram are a representative selection of all stars. A first guess might be that there are simply three different, unrelated types of star. However, astronomers now believe that an individual star changes during its lifetime. They interpret the H–R diagram like this:

- Since most stars fall on the main sequence, this suggests that an average star spends most of its lifetime as a main-sequence star.
- A star may spend a small part of its lifetime as a red giant, and/or as a white dwarf.

Astronomers cannot watch individual stars change through a lifetime, because the process is much too slow to see. Instead, they link their observations of star populations to models of how stars work.

In the rest of this section, you will learn how astronomers think stars change during their lifetimes, and how this is reflected in the H–R diagram.

Questions

1 List the colours of stars, from coolest to hottest.

2 If the Sun became dimmer and cooler, how would its position on the H–R diagram change?

3 A star in the bottom left-hand corner of the H–R diagram is dim but hot. Why does this suggest that it is small?

Main-sequence stars

Spacecraft such as *SOHO* have allowed scientists to look in great detail at the surface of the Sun and to measure the rate at which it is pouring energy out into space. Its colour indicates that the surface temperature of the Sun is about 5500 K – which is far too 'cold' for nuclear fusion reactions to take place.

Inside the Sun

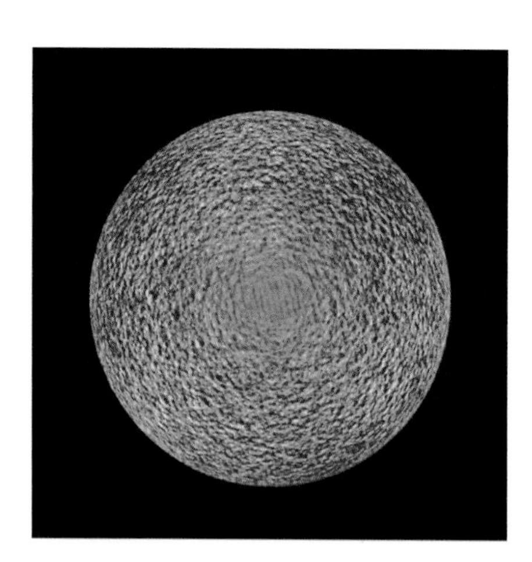

The surface of the Sun, photographed by the *SOHO* satellite, showing the granularity caused by the presence of convective cells.

You cannot tell exactly what is inside the Sun. However, there are some clues that can help physicists to make intelligent guesses:

- Nuclear fusion, the source of the Sun's energy, requires temperatures of millions of degrees.
- Energy leaves the Sun from its surface layer, the **photosphere**, whose temperature is about 5500 K.
- The photosphere has a granular appearance (see the photo), which is continually changing. Something is going on under the surface.
- A star like the Sun can burn steadily for billions of years, so it must radiate energy at the same rate that it generates it from fusion reactions.

Physicists can use these ideas to develop models of the inside of a star. The diagram below shows how they picture the internal structure of the Sun, based on such models.

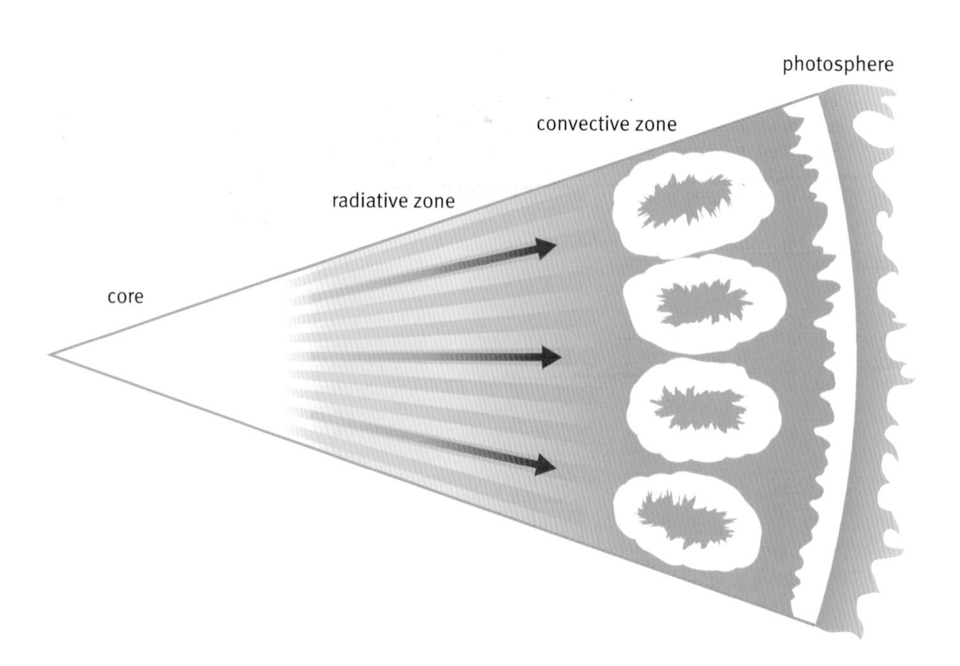

The internal structure of a star like the Sun.

Layer upon layer

- The **core** is the hottest part, with a temperature of the order of 14 million K. This is where nuclear fusion reactions occur. Hydrogen nuclei are fused together to form helium nuclei, releasing energy.
- Photons travel outwards through the **radiative zone**.
- Close to the surface temperature fall to just 1 million K. Matter can flow quite readily, and convection currents are set up, carrying heat energy to the photosphere. This is the **convective zone**. It is the tops of the convective 'cells' that cause the granular appearance of the Sun's surface.
- Electromagnetic radiation is emitted by the photosphere and radiates outwards through the solar atmosphere.

The lifetime of a star depends on its mass and the rate that it radiates energy.

Key words
core
radiative zone
convective zone
photosphere

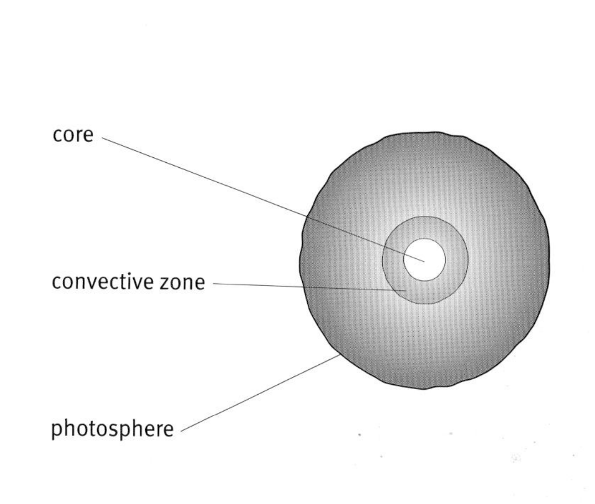

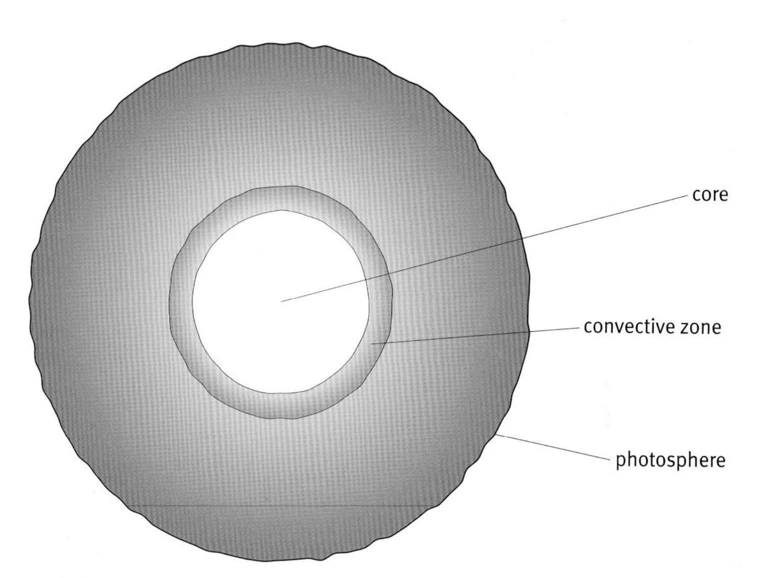

Cross-sections of stars with 1 solar mass and 5 solar masses.

Questions

4 A star like the Sun burns for billions of years, with a steady brightness and temperature. What does this tell you about its position on the H–R diagram?

5 Why is the long-term stability of the Sun important for us?

6 Of what two elements is the Sun mainly composed?

7 Why does hydrogen gas have to be hot for fusion to take place?

8 The diagrams compare the inner structures of a star like the Sun and one which has 5 times the mass. What differences can you identify between them?

Find out about:

- How stars and planetary systems form

4B Protostars

How does a star form? The raw material of stars is hydrogen, and there has been plenty of that since the early days of the Universe. Here is a simplified version of how astronomers think that a star forms:

- A cloud of gas and dust in space starts to contract, pulled together by gravity. Each particle attracts every other particle, so that the cloud collapses towards its centre. It forms a rotating, swirling disc.
- Recall that the temperature of a gas increases when it is compressed. The material at the centre gets hotter and hotter, so that it starts to glow.
- Eventually, the temperature of this material is hot enough for fusion reactions to occur, and a star is born.
- Material further out in the disc clumps together to form planets.

So planets form at the same time as the star that they orbit. In these early stages, as the star forms, it is known as a **protostar**. This stage in the Sun's life is thought to have lasted 100 000 to 1 million years.

A protostar at the centre of a new planetary system.

Getting warmer

Here are two ways to think about when a protostar gets hot enough for fusion to start.

- The *gas* idea. The star starts from a cloud of gas. As you saw on page 238, when a gas is compressed, its temperature rises. In this case, the force doing the compressing is gravity.
- The *particle* idea. Every particle in the cloud attracts every other particle. As they 'fall' inwards, they move faster (gravitational potential energy is being converted to kinetic energy). The particles collide with each other, sharing their energy. The fastest particles are at the centre of the cloud (they have fallen furthest), and fast-moving particles mean a high temperature.

Note that these are *not* competing explanations. They are just different ways of describing what is going on.

Key words

protostar

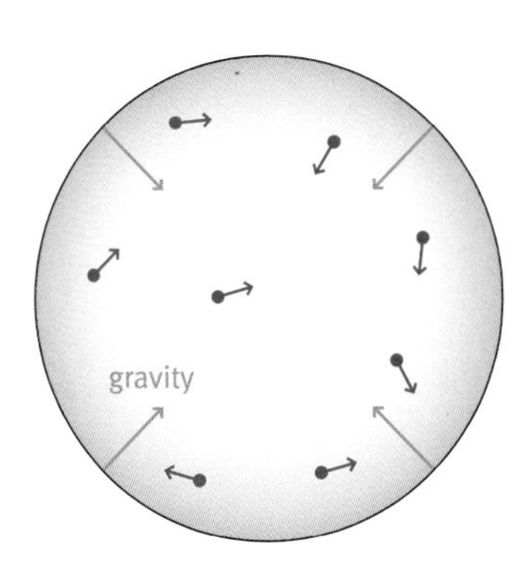

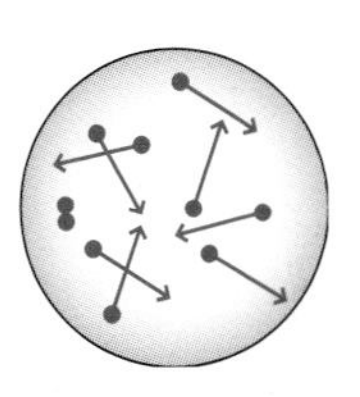

Explaining why a protostar gets hot.

Seek and find

Computer models of star formation can help to explain why the (roughly) spherical material from which a star forms collapses to form a flattened disc. Such models suggest that we will always find that the planets orbiting a star lie in a plane, just as in the Solar System.

Some models of star formation predict that, as a protostar forms, it spins faster and faster. Eventually, it blows out giant jets of hot gas, at right angles to the planetary disc. Large telescopes now have sufficient resolution to allow us to see this going on, as shown in the photo below. Planets travelling around distant stars are generally too small to see directly, but the gas jets travel far out into space and can occasionally be spotted.

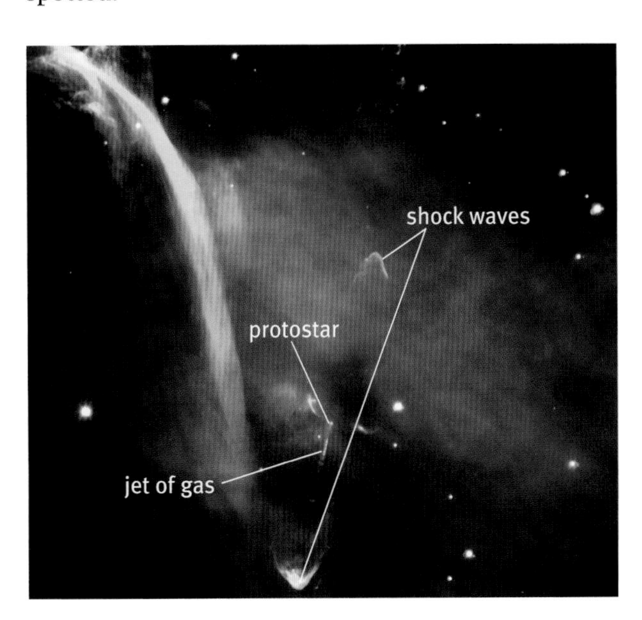

A protostar is forming at the centre of this image. One bright jet of gas can be seen coming downwards from it. Two symmetrical shock waves spread out in opposite directions. Photo taken by the Very Large Telescope in Chile.

Questions

1. From what materials does a protostar form?
2. If astronomers see a protostar glowing, does this indicate that nuclear fusion is taking place?
3. Imagine a sky-rocket exploding in the night sky. A small, hot explosion results in material being thrown outwards.
   - **a** Describe the energy changes that are going on.
   - **b** Now imagine the same scene, but in reverse. How is this similar to the formation of a protostar? How does it differ?

Find out about:

- what happens when hydrogen fusion ends
- how a supernova leaves a neutron star or a black hole

4C Death of a star

Many generations into the future, people can expect the Sun to keep releasing energy at a steady rate. Fusion reactions will continue in its core, as hydrogen is converted to helium. As this happens, the Sun's mass decreases very, very slowly. Einstein's equation $E = mc^2$ says that, if an object radiates energy E, its mass will decrease by an amount m. The constant c is the speed of light.

But this cannot go on for ever, because eventually all of the hydrogen in the Sun's core will be used up. What happens then?

As fusion slows down in the core of any star, its core cools down and there is less pressure, so the core collapses. The star's outer layers, which contain hydrogen, fall inwards, becoming hot. This causes new fusion reactions, making the outer shell expand. At the same time, the surface temperature falls, so that the colour changes from yellow to red. This produces a red giant.

In the case of the Sun, calculations suggest that it may expand sufficiently to engulf the three nearest planets – Mercury, Venus, and Earth.

An artist's impression of the view from a planet when its star has become a red giant. A moon is also shown, for comparison.

RED GIANT STAR
400 million km
orbit of Mars
CORE OF STAR
helium-burning shell
hydrogen-burning shell
carbon–oxygen core

The structure of a red giant.

Inside a red giant

While the outer layers of a red giant star are expanding, its core is contracting and heating up, to 100 million K. This is hot enough for new fusion reactions to start. Helium nuclei have a bigger positive charge than hydrogen nuclei, so there is greater electrical repulsion between them. If they are to fuse, they need greater momentum to overcome this repulsion. When helium nuclei do fuse, they form heavier elements such as carbon, nitrogen, and oxygen, releasing energy.

After a relatively short period of time (a few million years), the outer layers cool and drift off into space. The collapsed inner core remains as a white dwarf, which gradually becomes less bright and hot as fusion stops.

More massive stars

The Sun is a relatively small star. Bigger stars, greater than about 8 solar masses, also expand, to become red supergiants. In these, core temperatures may exceed 3 billion degrees, and more complex fusion reactions can occur, forming even heavier elements and releasing yet more energy. What happens next is described on the next page.

Moving about on the H–R diagram

Picture the life of a star like the Sun on the Hertzsprung–Russell diagram.

- Protostars are to the right of the main-sequence. As it heats up, a protostar moves to a point on the main-sequence, where it stays for billions of years.
- When it becomes a red giant, it moves above the main sequence.
- Finally, as a white dwarf, it appears below the main sequence.

Questions

1 At what point in its life does a star become a red giant?

2 What determines whether a star becomes a red giant or a red supergiant?

3 How many helium-4 nuclei must fuse to give a nucleus of carbon-12? And to give a nucleus of oxygen-16?

4 Use a periodic table to help you decide:

a What element is formed when two helium nuclei fuse?

b And when five fuse?

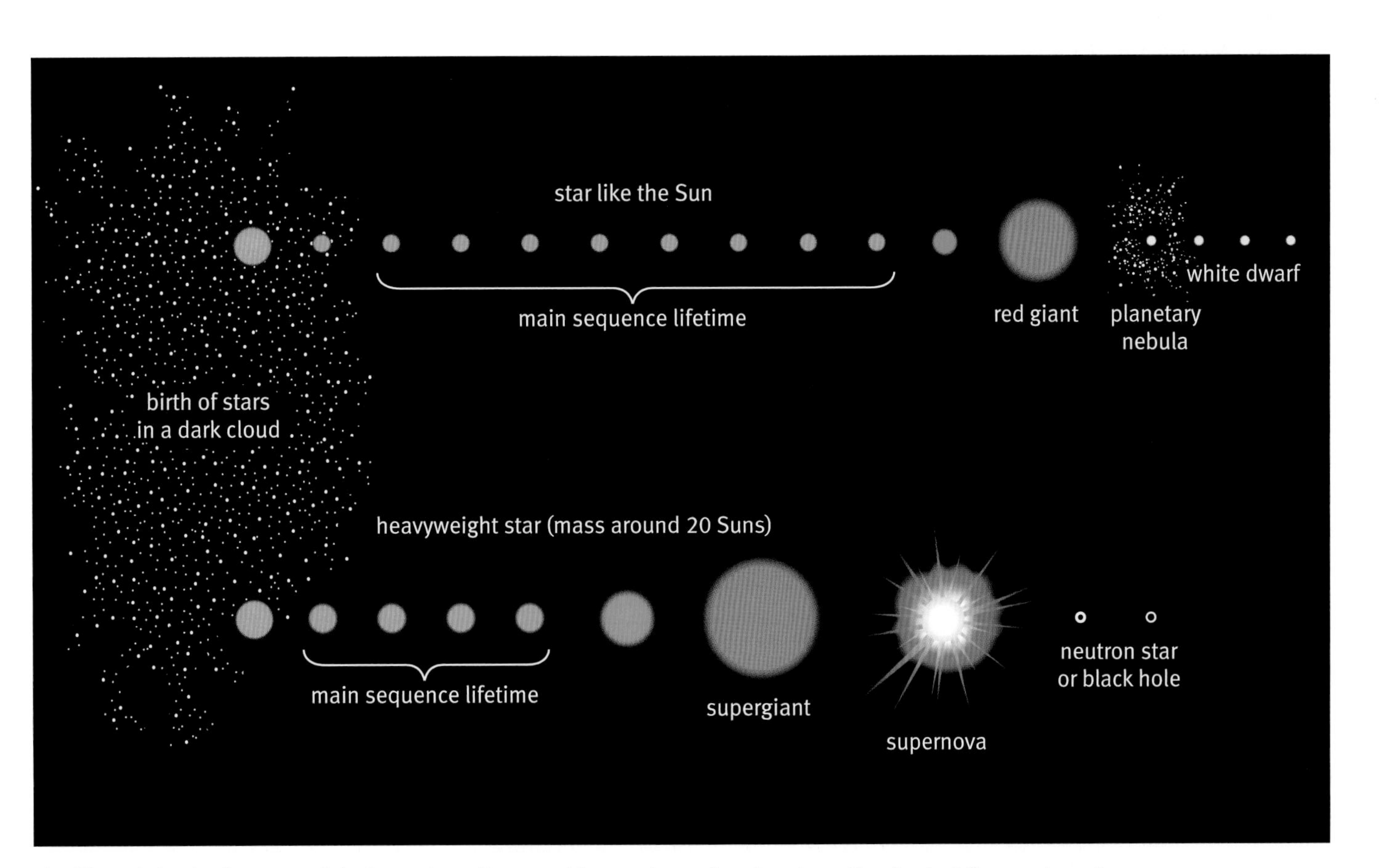

The life and death of a star mainly depends on its mass. The most massive stars have the shortest lives and most spectacular deaths.

A supernova explosion

What happens after the red supergiant phase of a large star is one of the most dramatic events in nature. By fusing lighter elements, the supergiant's core has become largely composed of iron. Iron nuclei absorb energy when they fuse, and so the process slows down. Now the drama starts.

The outer layers of the star are no longer held up by the pressure of the core, and they collapse inwards. The core has become very dense, and the outer material collides with the core and bounces off, flying outwards. The result is a huge explosion called a **supernova**. This is what Tycho Brahe saw in 1572 (page 243)

In the course of the explosion, temperatures rise to 10 billion K, enough to cause the fusion of medium-weight elements and thus form the heaviest elements of all – up to uranium in the periodic table. For a few days, a supernova can outshine a whole galaxy.

The remnants of a supernova in the constellation of Cassiopeia. The cloud is about 10 light-years across. This is a composite image, made using three telescopes to capture infrared, visible, and X-ray data.

Supernova remnants

Supernovas are rare. They are seen about once every century in a typical galaxy. But there are some 100 billion galaxies in the Universe. When astronomers search with their telescopes, they can identify the remnants of supernovas that occurred hundreds or thousands of years ago.

The photograph on the left shows the remnants of a supernova that happened in about 1660. You can see the expanding sphere of dust and gas, formed from the star's outer layers. This material contains all of the elements of the periodic table. As it becomes distributed through space, it may become part of another contracting cloud of dust and gas. A protostar may form with new planets orbiting it, and the cycle starts over again.

Dense and denser

The core of an exploding supernova remains. If its mass is less than about 2.5 solar masses, this central remnant becomes a **neutron star**. This is made almost entirely of neutrons, compressed together like a giant atomic nucleus, perhaps 30 km across.

A more massive remnant collapses even further under the pull of its own gravity, to become a **black hole**. Within a black hole, the pull of gravity is so strong that not even light can escape from it.

Neutron stars are thought to explain the pulsars, discovered by Jocelyn Bell and Anthony Hewish (page 181). As the core of a star collapses to form a neutron star, it spins faster and faster. Its magnetic field becomes concentrated, and this results in a beam of radio waves coming out of its magnetic poles. As the neutron star spins round, this beam sweeps across space and is detected as a regular series of pulses at an observatory on some small, distant planet.

Key words
supernova
neutron star
black hole

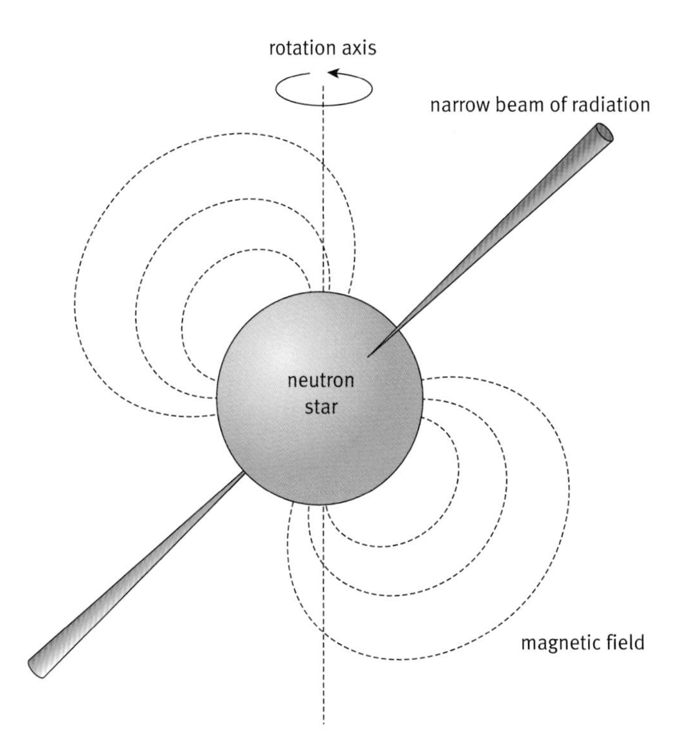

A spinning neutron star sends out a beam of radio waves – the origin of a pulsar.

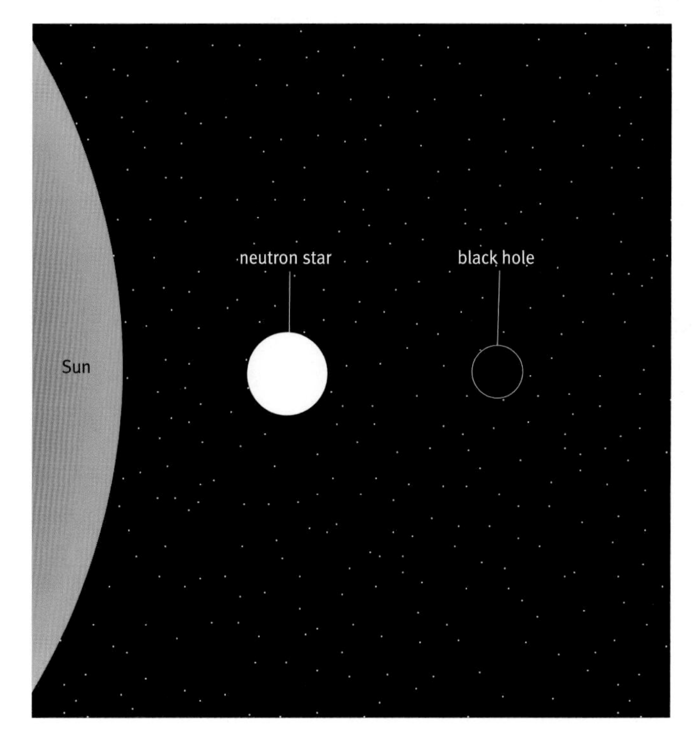

The neutron star and the black hole shown here have the same mass as the Sun.

Questions

5 Put these objects in order, from least dense to most dense:

neutron star, protostar, red supergiant, black hole, main-sequence star

6 Draw a diagram to show the complete life cycle of stars, starting from a protostar. Indicate how stars of different masses end differently.

Astronomers working together at ESO

There are parts of the sky that are not visible from the northern hemisphere. So 11 European countries have set up three observatories in the Atacama Desert high up in the Andes mountains, Chile. Together they, and the organization for astronomical research that runs them, are known as the European Southern Observatory (ESO). Two of ESO's sites are at heights of about 2500 m above sea level, but one, the ALMA array of radio telescopes, is on a 5000 m-high mountain. These remote locations are chosen because the atmosphere is both clear and dry there. This reduces the effects of absorption and refraction of radiation, and of light pollution.

Aerial view of La Silla Observatory, Chile. 'La Silla' means 'the saddle', after the shape of the mountain-top on which it stands

Remote control

Each year, over 1000 astronomers make observations using the ESO telescopes. Half of them do this from their home base in Europe.

ESO does this by making great use of the power of computers and the Internet. The telescopes in Chile are remote. It takes two days of travel to reach them, and two days to get home again. So many of the astronomers who need observations to be made send in their requests, and these are programmed into the ESO control system. Local operators ensure that the observations are made, and the results are sent back to Europe.

One benefit of this is to avoid the 'weather lottery'. Some observations are so sensitive that they require the clearest of skies, which might only be available on 10% of nights in the year. Imagine travelling to Chile for three nights' observing, only to find that the skies were never clear enough! ESO's computer control ensures that your observations are postponed to a later date, while another astronomer's observations are brought forward to take advantage of the available time.

In 2005, ESO's data management system won a top award from Computerworld magazine. The system makes use of the latest database software, and is described as 'end-to-end', because it gathers, stores, and distributes data in a continuous operation. Data is collected, transferred to ESO's headquarters in Germany, and checked for quality. Individual astronomers can access their data over the Internet. They have the exclusive right to use their data for one year after it has been collected, after which it becomes available to any researcher anywhere in the world.

By 2010, the data archive is expected to contain a petabyte of data – that is, 10^{15} bytes of information. It will be accessible via the Grid, the advanced successor to the Internet.

La Silla Hotel, which accommodates visiting astronomers

The archive of DVDs at the ESO headquarters near Munich, Germany

Control room for the new 3.5m New Technology Telescope

Douglas Pierce-Price

ESO people

Douglas Pierce-Price is a British astronomer based in Germany.

How many people work for ESO?

'Technically, I work for the European Organisation for Astronomical Research in the Southern Hemisphere. This organization employs 320 staff members from around the world (many of us work in Germany), as well as 160 local staff recruited in Chile. This means that the organization provides work for local people.

'In addition, there are about 100 students and research fellows who are attached to the organization. So, for UK students, it is possible to join a research team here once they have completed an undergraduate degree.'

What is it like to live and work in a desert on top of a mountain?

'The observatory staff live and work in the Residencia, a futuristic building built partly underground and with a 35-metre-wide glass dome in the roof. It is part of the VLT's "base camp" facility, situated a short distance below the summit of Cerro Paranal.

'Astronomical conditions at Paranal are excellent, but they come at a price. It's a forbidding desert environment; virtually nothing can grow outside. The humidity can be as low as 10%, there are intense ultraviolet rays from the Sun, and the high altitude can leave people short of breath. The nearest town is two hours away, so there is a small paramedic clinic at the base camp.

'Living in this extremely isolated place feels like visiting another planet. Within the Residencia, a small garden and a swimming pool are designed to increase the humidity inside. The building provides visitors and staff with some relief from the harsh conditions outside: there are about 100 rooms for astronomers and other staff, as well as offices, a library, cinema, gymnasium, and cafeteria.'

Up all night

Monika Petr-Gotzens is a German astronomer working at the European Southern Observatory. She is studying how stars form in dense clusters. She is particularly interested in the formation of binary stars. These are pairs of stars that orbit one another. More than half of all stars are in binary pairs.

Monika uses both radio telescopes and optical (light) telescopes in her work.

What is it like to work with a telescope at the top of a mountain?

'The work "at the telescope", i.e. observing during the night, is not as romantic as you might think. The control over the telescope and instrument is 100% computer based and carried out from the control room. These control rooms are hundreds of metres away from the telescope, and you don't even see the telescope during your observation, unless you actively walk into the dome (from where you can't control anything). So it isn't the freezing cold which keeps you awake during the observing night, but the smell of coffee and cookies.'

Monica Petr-Gotzens

What part do computers play in your work?

'Modern astronomy without computer control is unthinkable. It sharpens our view of the Universe by directing the telescope very accurately. Computers also process the data gathered by the telescope to give much higher-quality images and measurements.'

Astronomers today usually work in collaborative teams. Why is this?

'International collaborations are very important. Nowadays, it is not a single clever observation that solves one of the grand questions of the Universe. They are answered through major efforts. For example, surveys of stars across large areas of the sky may be carried out using bigger and bigger telescopes. It is often impossible for individuals to deal with the huge amount of data accumulated from such surveys.'

The ESO telescopes are sited high on mountains. Why is this?

'The factors that influence the choice of site for an observatory depend on the kind of observatory: optical, infrared, millimetre, or radio waves. For optical observatories, a low-turbulent atmosphere (i.e. very good seeing), a dark sky without light pollution, and a high number of clear sky nights are important factors. It's easier for radio telescopes – they can see through clouds.

'The size of the telescope is also important. Independent of the working wavelength, the rule applies that the larger the telescope, the less windy a site must be. Wind means that the density of the air is changing, and this causes radiation to be refracted. (That's why stars appear to twinkle in the night sky.)

'On top of that, sites must also have a reasonable logistical supply. It must be reasonably easy for astronomers to travel to the observatory, and they need accommodation, food, and drink. Natural springs, for example, to be used for the water supply, are an advantage, although not an absolute requirement.

P7 Observing the Universe

Summary

Topic 1 Observatories and telescopes

- Astronomical objects are so distant that their light is effectively parallel.
- Converging (convex) lenses and concave mirrors can be used to focus parallel rays of light.
- power of a lens (in D) = $\frac{1}{\text{focal length (in m)}}$
- A converging lens forms a real image of a distant source of light.
- A simple telescope uses two converging lenses. The eyepiece is more powerful than the objective.
- magnification = $\frac{\text{focal length of objective}}{\text{focal length of eyepiece}}$ H
- Most astronomical telescopes have concave mirrors as their objectives.
- A large telescope is needed to collect the weak radiation from a faint or distant source.
- The aperture of a telescope must be much larger than the wavelength of radiation it detects. H
- Many telescopes are sited on mountains or in space to reduce the effects of the atmosphere, which refracts and absorbs electromagnetic radiation.
- International collaboration can share the cost of an astronomical project, and allows expertise to be shared.
- When an observatory is planned, non-astronomical factors such as cost, environmental impact, and working conditions must be taken into account.

Topic 2 Naked-eye astronomy

- The apparent movements of the Sun, Moon, and stars across the sky can be explained in terms of the rotation of the Earth, the orbit of the Moon around the Earth, and the orbits of the Earth around the Sun.
- Seen from the Earth, the planets move in irregular patterns relative to the fixed stars.
- The apparent motions of the planets can be explained in terms of their orbits around the Sun. H
- The phases of the Moon, and eclipses of the Sun and Moon, can be explained in terms of the relative positions of the Sun, Moon, and Earth.
- Solar eclipses are rare because the orbit of the Moon is tilted relative to the Earth's orbit plane. H
- Different stars are seen in the night sky through a year, as the Earth travels around its orbit.
- A sidereal day is 4 minutes less than a solar day. H
- Nearby stars show parallax: they appear to move relative to more distant stars over the course of a year.
- A parsec is a measure of distance, similar in magnitude to a light-year. Neighbouring stars are typically separated by a few parsecs, galaxies by megaparsecs.
- The luminosity of a star depends on its temperature and size. Observed brightness also depends on distance from Earth.
- Cepheid variables are stars whose brightness varies regularly. The most luminous have the longest periods.
- The changing luminosity of Cepheids allows astronomers to measure their distances, and so measure distances to the galaxies they are in.
- In 1920 two American astronomers took part in a public debate about the scale of the Universe. Within a few years, new evidence conclusively showed that there are galaxies beyond the Milky Way
- Light from distant galaxies is red-shifted. This shows that they are moving away:
  speed of recession = Hubble constant × distance

Topic 3 Inside stars

- The spectrum of a star has a continuous range of frequencies, plus lines that indicate the chemical elements present.
- Line spectra (emission and absorption) arise when electrons within atoms change their energy levels.
- A molecular model can be used to explain why the pressure and volume of a gas vary with temperature.

  temperature in K = temperature in °C + 273
- The alpha scattering experiment showed that an atom has a small, massive, positive nucleus.
- The protons and neutrons in a nucleus are held together by the strong nuclear force, which acts against the electrical repulsion between protons.

Topic 4 The lives of stars

- A protostar forms when gravity compresses a cloud of gas, so that it becomes hot. Planets form around the new star.
- Hydrogen nuclei can fuse together to produce helium, if they are brought close together.
- Hydrogen fusion occurs in the core of a star like the Sun. At later stages in a star's life, further fusion reactions can occur.
- Energy is transferred from a star's core to its surface by a convective zone, and radiated from its photosphere.
- When there is insufficient hydrogen in the core, a star expands to become a red giant or supergiant. When helium fusion ends, a red giant becomes a hot white dwarf, which then fades.
- Fusion stops in a massive star when the core has mostly become iron. The star explodes as a supernova. What remains is a neutron star or a black hole.

Glossary

absolute zero Extrapolating from the behaviour of gases at different temperatures, the theoretically lowest possible temperature, −273 °C. In practice, the lowest temperature achievable is about a degree above this.

absorb (radiation) The radiation which hits an object and is not reflected, or transmitted through it, is absorbed (for example, black paper absorbs light). Its energy makes the object gets a little hotter.

absorption spectrum (of a star) Consists of dark lines superimposed on a continuous spectrum. It is created when light from the star passes through a cooler gas that absorbs photons of particular energies.

activity The rate at which nuclei in a sample of radioactive material decay and give out alpha, beta, or gamma radiation.

actual risk Risk calculated from reliable data.

aerial A wire, or arrangement of wires, that emits radio waves when there is an alternating current in it, and in which an alternating current is induced by passing radio waves. So it acts as a source or a receiver of radio waves.

ALARA The ALARA principle is used when better equipment or procedures can reduce the risks of an activity. These improvements may cost more money. These extra costs must be balanced against the amount by which risk is reduced, for example, it might reduce people's exposure to hazardous chemicals or ionizing radiation.

alpha radiation The least penetrating type of ionizing radiation, produced by the nucleus of an atom in radioactive decay. A high-speed helium nucleus.

alternating current (a.c.) An electric current that reverses direction many times a second.

ammeter A meter that measures the size of an electric current in a circuit.

ampere (or amp, for short) The unit of electric current.

amplifier A device for increasing the amplitude of an electrical signal. Used in radios and other audio equipment.

amplitude For a mechanical wave, the maximum distance that each point on the medium moves from its normal position as the wave passes. For an electromagnetic wave, the maximum value of the varying electric field (or magnetic field).

amplitude modulation (AM) One way in which a radio wave can be made to carry an audio signal. The amplitude of the carrier wave is made to vary in the same way as the sound wave.

analogue signals Signals used in communications in which the amplitude can vary continuously.

angular magnification (of a refracting telescope) The ratio of the angle subtended by an object when seen through the telescope to the angle subtended by the same object when seen with the naked eye. It can be calculated as

$$\frac{\text{focal length of objective lens}}{\text{focal length of eyepiece lens}}$$

aperture (of a telescope) The light-gathering area of the objective lens or mirror.

asteroid A dwarf rocky planet, generally orbiting the Sun between the orbits of Mars and Jupiter.

astrolabe An instrument used for locating and predicting the positions of the Sun, Moon, planets, and stars, as well as navigating and telling the time.

atmosphere The Earth's atmosphere is the layer of gases that surrounds the planet. It contains roughly 78% nitrogen and 21% oxygen, with trace amounts of other gases. The atmosphere protects life on Earth by absorbing ultraviolet solar radiation and reducing temperature extremes between day and night.

attract Pull towards.

average speed The distance moved by an object divided by the time taken for this to happen.

background radiation The low-level radiation, mostly from natural sources, that everyone is exposed to all the time, everywhere.

beta radiation One of several types of ionizing radiation, produced by the nucleus of an atom in radioactive decay. More penetrating than alpha radiation but less penetrating than gamma radiation. A high-speed electron.

big bang An explosion of a single mass of material. This is currently the accepted scientific explanation for the start of the Universe.

black hole A mass so great that its gravity prevents anything escaping from it, including light. Some black holes are the collapsed remnants of massive stars.

carbon cycle The human and natural processes that move carbon and carbon compounds continuously between the Earth, its oceans and atmosphere, and living things.

carrier A steady stream of radio waves produced by an RF oscillator in a radio to carry information.

Cepheid variable A star whose brightness varies regularly, over a period of days.

chain reaction A process in which the products of one nuclear reaction cause further nuclear reactions to happen, so that more and more reactions occur and more and more product is formed. Depending on how this process is controlled, it can be used in nuclear weapons or power station nuclear reactors.

charged Carrying an electric charge. Some objects (such as electrons and protons) are permanently charged. A plastic object can be charged by rubbing it. This transfers electrons to or from it.

climate Average weather in a region over many years.

coding (in communications) Converting information from one form to another, for example changing an analogue signal into a digital one.

comet A rocky lump, held together by frozen gases and water, that orbits the Sun.

compression A material is in compression when forces are trying to push it together and make it smaller.

conservation of energy The principle that the total amount of energy at the end of any process is always equal to the total amount of energy at the beginning – though it may now be stored in different ways and in different places.

constellation A group of stars that form a pattern in the night sky. Patterns recognized are cultural and historical, and are not based on the actual positions of the stars in space.

contamination (radioactive) Having a radioactive material inside the body, or having it on the skin or clothes.

control rod In a nuclear reactor, rods made of a special material that absorbs neutrons are raised and lowered to control the rate of fission reactions.

convective zone (of a star) The layer of a star above its radiative zone, where energy is transferred by convective currents in the plasma.

converging lens A lens that changes the direction of light striking it, bringing the light together at a point.

coolant In a nuclear reactor, the liquid or gas that circulates through the core and transfers heat to the boiler.

core The Earth's core is made mostly from iron, solid at the centre and liquid above.

counter-force A force in the opposite direction to something's motion.

crust A rocky layer at the surface of the Earth, 10–40 km deep.

decoding In communications, converting information back into its original form, for example changing a digital signal back into an analogue one.

decommissioning Taking a power station out of service at the end of its lifetime, dismantling it, and disposing of the waste safely.

detector Any device or instrument that shows the presence of radiation by absorbing it.

diffraction What happens when waves hit the edge of a barrier or pass through a gap in a barrier. They bend a little and spread into the region behind the barrier.

digital code A string of 0s and 1s that can be used to represent an analogue signal, and from which that signal can be reconstructed.

digital signals Signals used in communications in which the amplitude can take only one of two values, corresponding to the digits 0 and 1.

dioptre Unit of lens power, equivalent to a focal length of 1 metre.

direct current (d.c.) An electric current that stays in the same direction.

dispersion The splitting of white light into different colours (frequencies), for example by a prism.

distance–time graph A useful way of summarizing the motion of an object by showing how far it has moved from its starting point at every instant during its journey.

diverging lens A lens that changes the direction of light striking it, spreading it into a wider cone of light.

driving force The force pushing something forward, for example a bicycle.

duration How long something happens for. For example, the length of time someone is exposed to radiation.

earthquake Event in which rocks break to allow tectonic plate movement, causing the ground to shake.

efficiency The percentage of the energy supplied to a device that is transferred to the desired place, or in the desired way. For example, to find the efficiency of a kettle you would divide the gain in energy of the water by the work done on the kettle element by the electricity supply, and multiply by 100.

electric charge A fundamental property of matter. Electrons and protons are charged particles. Objects become charged when electrons are transferred to or from them, for example by rubbing.

electric circuit A closed loop of conductors connected between the positive and negative terminals of a battery or power supply.

electric current A flow of charges around an electric circuit.

electric field A region where an electric charge experiences a force. There is an electric field around any electric charge.

electromagnetic induction The name of the process in which a potential difference (and hence often an electric current) is generated in a wire, when it is in a changing magnetic field.

electromagnetic spectrum The 'family' of electromagnetic waves of different frequencies and wavelengths.

electromagnetic wave A wave consisting of vibrating electric and magnetic fields, which can travel in a vacuum. Visible light is one example.

electron A tiny, negatively charged particle which is part of an atom. Electrons are found outside the nucleus. Electrons have negligible mass and one unit of charge.

electrostatic attraction The force of attraction between objects with opposite electric charges. A positive ion, for example, attracts a negative ion.

emission spectrum (of an element) The electromagnetic frequencies emitted by an excited atom as electron energy levels fall.

erosion The movement of solids at the Earth's surface (for example, soil, mud, rock) caused by wind, water, ice, and gravity, or living organisms.

exoplanet The planet of any star other than the Sun.

extended object An astronomical object made up of many points, for example the Moon or a galaxy. By contrast, a star is a single point.

eyepiece lens (of an optical telescope) The lens nearer the eye, which will have a higher power. Often called a telescope 'eyepiece'.

focal length The distance from the optical centre of a lens or mirror to its focus.

focus (of a lens or mirror) The point at which rays arriving parallel to its principal axis cross each other. Also called the 'focal point'.

focusing Adjusting the distance between lenses, or between the eyepiece lens and a photographic plate (or CCD), to obtain a sharp image of the object.

force A push or a pull experienced by an object when it interacts with another. A force is needed to change the motion of an object.

fossil The stony remains of an animal or plant that lived millions of years ago, or an imprint it has made (for example, a footprint) in a surface.

fossil fuel Natural gas, oil, or coal.

frequency modulation (FM) One way in which a radio wave can be made to carry an audio signal. The frequency of the carrier wave is made to vary in the same way as the sound wave.

friction The force exerted on an object due to the interaction between it and another object that it is sliding over, or tending to slide over. It is caused by the roughness of both surfaces at a microscopic level.

fuel rod A container for nuclear fuel, which enables fuel to be inserted into, and removed from, a nuclear reactor while it is operating.

galaxy A collection of thousands of millions of stars held together by gravity.

gamma radiation (gamma rays) The most penetrating type of ionizing radiation, produced by the nucleus of an atom in radioactive decay. The most energetic part of the electromagnetic spectrum.

globular cluster A cluster of hundreds of thousands of old stars.

greenhouse effect The atmosphere absorbs infrared radiation from the Earth's surface and radiates some of it back to the surface, making it warmer than it would otherwise be.

greenhouse gas Gases that contribute to the greenhouse effect. Includes carbon dioxide, methane, and water vapour.

half-life The time taken for the amount of a radioactive element in a sample to fall to half its original value.

high level waste A category of nuclear waste that is highly radioactive and hot. Produced in nuclear reactors and nuclear weapons processing.

Hubble constant The ratio of the speed of recession of galaxies to their distance, with a value of about 72 km/s per Mpc.

in parallel A way of connecting electric components that makes a branch (or branches) in the circuit so that charges can flow round more than one loop.

in series A way of connecting electric components so that they are all in a single loop. The charges pass through them all in turn.

infrared Electromagnetic waves with a frequency lower than that of visible light, beyond the red end of the visible spectrum.

instantaneous speed The speed of an object at a particular instant. In practice, its average speed over a very short time interval.

intensity (of light in a star's spectrum) The amount of a star's energy gathered by a telescope every second, per unit area of its aperture.

interaction What happens when two objects collide, or influence each other at a distance. When two objects interact, each experiences a force.

interaction pair Two forces that arise from the same interaction. They are equal in size and opposite in direction, and each acts on a different object.

interference What happens when two waves meet. If the waves have the same frequency, an interference pattern is formed. In some places, crests add to crests, forming bigger crests; in other places, crests and troughs cancel each other out.

intermediate level waste A category of nuclear waste that is generally short-lived but requires some shielding to protect living organisms, for example contaminated materials that result from decommissioning a nuclear reactor.

intrinsic brightness (of a star) A measure of the light that would reach a telescope if a star were at a standard distance from the Earth.

inverted image An image that is upside down compared to the object.

ionizing radiation Radiation with photons of sufficient energy to remove electrons from atoms in its path. Ionizing radiation, such as ultraviolet, X-rays, and gamma rays, can damage living cells.

irradiation Being exposed to radiation from an external source.

isotope Atoms of the same element which have different mass numbers because they have difference numbers of neutrons in the nucleus.

Kelvin scale A scale of temperature in which 0 K is absolute zero, and the triple point of water (where solid, liquid, and gas phases co-exist) is 273 K.

kinetic energy The energy which something has owing to its motion.

kinetic model of matter The idea that a gas consists of particles (atoms or molecules) that move around freely, colliding with each other and with the walls of any container, with most of the volume of gas being empty space.

light-dependent resistor (LDR) An electric circuit component whose resistance varies depending on the brightness of light falling on it.

light pollution Light created by humans, for example street lighting, that prevents city dwellers from seeing more than a few bright stars. It also cause problems for astronomers.

light-year The distance travelled by light in a year.

longitudinal wave A wave in which the particles of the medium vibrate in the same direction as the wave is travelling. Sound is an example.

luminosity (of a star) The amount of energy radiated into space every second. This can be measured in watts, but astronomers usually compare a star's luminosity to the Sun's luminosity.

low-level waste A category of nuclear waste that contains small amounts of short-lived radioactivity, for example paper, rags, tools, clothing, and filters from hospitals and industry.

lunar eclipse When the Earth comes between the Moon and the Sun, and totally or partially covers the Moon in the Earth's shadow as seen from the Earth's surface.

magnification (of an optical instrument) The process of making something appear closer than it really is.

mantle A thick layer of rock beneath the Earth's crust, which extends about halfway down to the Earth's centre.

mass extinction Event in the history of the Earth when many species became extinct at the same time.

medium A material through which a wave travels. The plural is 'media'.

megaparsec (Mpc) A million parsecs.

microwaves Radio waves of the highest frequency (shortest wavelength), used for mobile phones and satellite TV.

Milky Way The galaxy in which the Sun and its planets including Earth are located. It is seen from the Earth as an irregular, faintly luminous band across the night sky.

modulate To vary the amplitude or frequency of the carrier waves produced in a radio so that they carry the information in a sound wave.

momentum A property of any moving object. Equal to mass multiplied by velocity. The plural is momenta.

mountain chain A group of mountains that extend along a line, often hundreds or even thousands of kilometres. Generally caused by the movement of tectonic plates.

nanometres A unit used for microscopic measurements. $1\text{nm} = 0.001\mu\text{m} = 0.000001\text{mm}$.

negative A label used to name one type of charge, or one terminal of a battery. It is the opposite of positive.

neutron star The collapsed remnant of a massive star, after a supernova explosion. Made almost entirely of neutrons, they are extremely dense.

noise Unwanted electrical signals that get added on to radio waves during transmission, causing additional modulation. Sometimes called 'interference'.

non-ionizing radiation Radiation with photons that do not have enough energy to ionize molecules.

normal An imaginary line drawn at right angles to the point at which a ray strikes the boundary between one medium and another. Used to define the angle of the ray that strikes or emerges from the boundary.

nuclear fission The process in which a nucleus of uranium-235 breaks apart, releasing energy, when it absorbs a neutron.

nuclear fuel In a nuclear reactor, each uranium atom in a fuel rod undergoes fission and releases energy when hit by a neutron.

nuclear fusion The process in which two small nuclei combine to form a larger one, releasing energy. An example is hydrogen combining to form helium. This happens in stars, including the Sun.

objective lens (of an optical telescope) The lens nearer the object, which will have a lower power. Often called a telescope 'objective'.

observed brightness (of a star) A measure of the light reaching a telescope from a star.

oceanic ridge A line of underwater mountains in an ocean, where new seafloor constantly forms.

ohm The unit of electrical resistance. Symbol Ω

Ohm's law The result that the current, *I*, through a resistor, *R*, is proportional to the voltage, *V*, across the resistor, provided its temperature remains the same. Ohm's law does not apply to all conductors.

optical fibres Thin glass fibres, down which a light beam can travel. The beam is reflected at the sides by total internal reflection, so very little escapes. Used in modern communications, for example to link computers in a building into a network.

ozone layer A thin layer in the atmosphere, about 30 km up, where oxygen is in the form of ozone molecules. The ozone layer absorbs ultraviolet radiation from sunlight.

parallax The apparent shift of an object against a more distant background, as the position of the observer changes. The further away an object is, the less it appears to shift. This can be used to measure how far away an object is, for example to measure the distance to stars.

parallax angle When observed at an interval of six months, a star will appear to move against the background of much more distant stars. Half of its apparent angular motion is called its parallax angle.

parsec (pc) A unit of astronomical distance, defined as the distance of a star which has a parallax angle of one arcsecond. Equivalent to 3.1×10^{12} km.

peak frequency The frequency with the greatest intensity.

peer review The process whereby scientists who are experts in their field critically evaluate a scientific paper or idea before and after publication.

penumbra An area of partial darkness in a shadow, for example places in the Moon's path where the Earth only partially blocks off sunlight. Some sunlight still reaches these places because the Sun has such a large diameter.

phases (of the Moon) Changing appearance, due to the relative positions of the Earth, Sun, and Moon.

photons Tiny 'packets' of electromagnetic radiation. All electromagnetic waves are emitted and absorbed as photons. The energy of a photon is proportional to the frequency of the radiation.

photosphere The visible surface of a star, which emits electromagnetic radiation.

planet A very large, spherical object that orbits the Sun, or other star.

positive A label used to name one type of charge, or one terminal of a battery. It is the opposite of negative.

potential difference (p.d.) The difference in potential energy (for each unit of charge flowing) between any two points in an electric circuit.

power In an electric circuit, the rate at which work is done by the battery or power supply on the components in a circuit. Power is equal to current × voltage.

precautionary principle Take steps to minimize the risks associated with specific human actions when no one knows how serious the risks are.

pressure (of a gas) The force a gas exerts per unit area on the walls of its container.

primary energy source A source of energy not derived from any other energy source, for example fossil fuels or uranium.

principal axis An imaginary line perpendicular to the centre of a lens or mirror surface.

proton A positively charged particle found in the nucleus of atoms. The relative mass of a proton is 1 and it has one unit of charge.

protostar The early stages in the formation of a new star, before the onset of nuclear fusion in the core.

quarks Fundamental particles that make up neutrons, protons, and other sub-atomic particles.

radiation A flow of information and energy from a source. Light and infrared are examples. Radiation spreads out from its source, and may be absorbed or reflected by objects in its path. It may also go (be transmitted) through them.

radiation dose A measure, in millisieverts, of the possible harm done to your body, which takes into account both the amount and type of radiation you have been exposed to.

radiative zone (of a star) The layer of a star surrounding its core, where energy is transferred by photons to the convective zone.

radio waves Electromagnetic waves of a much lower frequency than visible light. They can be made to carry signals and are widely used for communications.

radioactive Used to describe a material, atom, or element, that produces alpha, beta, or gamma radiation.

radioactive dating Estimating the age of an object such as a rock by measuring its radioactivity. Activity falls with time, in a way that is well understood.

radioactive decay The spontaneous change in an unstable element, giving out alpha, beta, or gamma radiation. Alpha and beta emission result in a new element.

radiotherapy Using radiation to treat a patient.

ray diagram A way of representing how a lens or telescope affects the light that it gathers, by drawing the rays (which can be thought of as very narrow beams of light) as straight lines.

reaction (of a surface) The force exerted by a hard surface on an object that presses on it.

reflection What happen when a wave hits a barrier and bounces back off it. If you draw a line at right angles to the barrier, the reflected wave has the same angle to this line as the incoming wave. For example, light is reflected by a mirror.

reflector A telescope that has a mirror as its objective. Also called a reflecting telescope.

refraction Waves change their wavelength if they travel from one medium to another in which their speed is different. For example, when travelling into shallower water, waves have a smaller wavelength as they slow down.

refractor A telescope that has a lens as its objective, rather than a mirror.

renewable energy source Resources that can be used to generate electricity without being used up, such as the wind, tides, and sunlight.

repel Push apart.

resistance The resistance of a component in an electric circuit indicates how easy or difficult it is to move charges through it.

resolving power The ability of a telescope to measure the angular separation of different points in the object that is being viewed. Resolving power is limited by diffraction of the electromagnetic waves being collected.

resultant force The sum, taking their directions into account, of all the forces acting on an object.

retrograde motion An apparent reversal in a planet's usual direction of motion, as seen from the Earth against the background of fixed stars. This happens periodically with all planets beyond the Earth's orbit.

risk The probability of an outcome which is seen as undesirable, associated with some behaviour or process.

rock cycle Continuing changes in rock material, caused by processes such as erosion, sedimentation, compression, and heating.

sampling In the context of physics, measuring the amplitude of an analogue signal many times a second in order to convert it into a digital signal.

seafloor spreading The process of forming new ocean floor at oceanic ridges.

secondary energy source Energy in a form that can be distributed easily but is manufactured by using a raw energy resource such as a fossil fuel or wind. Examples of secondary energy sources are electricity, hot water used in heating systems, and steam.

selective absorption Some materials absorb some forms of electromagnetic radiation but not others. For example, glass absorbs infrared but is transparent to visible light.

sidereal day The time taken for the Earth to rotate 360°: 23 hours and 56 minutes.

signal Information carried through a communication system, for example by an electromagnetic wave with variations in its amplitude or frequency, or being rapidly switched on an off.

solar day The time taken for the Earth to rotate so that it fully faces the Sun again: exactly 24 hours.

solar eclipse When the Moon comes between the Earth and the Sun, and totally or partially blocks the view of the Sun as seen from the Earth's surface.

Solar System The Sun and objects which orbit around it – planets and their moons, comets, and asteroids.

source An object that produces radiation.

spectrometer An instrument that divides a beam of light into a spectrum and enables the relative brightness of each part of the spectrum to be measured.

spectrum One example is the continuous band of colours, from violet to red, produced by shining white light through a prism. Passing light from a flame test through a prism produces a line spectrum.

speed of light 300 000 kilometres per second – the speed of all electromagnetic waves in a vacuum.

speed of recession The speed at which a galaxy is moving away from us.

star life cycle All stars have a beginning and an end. Physical processes in a star change throughout its life, affecting its appearance .

static electricity Electric charge that is not moving round a circuit but has built up on an object such as a comb or a rubbed balloon.

strong (nuclear) force A fundamental force of nature that acts inside atomic nuclei.

Sun The star nearest Earth. Fusion of hydrogen in the Sun releases energy which makes life on Earth possible.

supernova A dying star that explodes violently, producing an extremely bright astronomical object for weeks or months.

tectonic plate Giant slabs of rock (about 12, comprising crust and upper mantle) which make up the Earth's outer layer.

telescope (from Greek, meaning 'far-seeing') An instrument that gathers electromagnetic radiation, to form an image or to map data, from astronomical objects such as stars and galaxies. It makes visible things that cannot be seen with the naked eye.

tension A material is in tension when forces are trying to stretch it or pull it apart.

thermistor An electric circuit component whose resistance changes markedly with its temperature. It can therefore be used to measure temperatures.

total internal reflection (TIR) What can happen when a wave hits a boundary with a medium in which it moves faster (for example, light going from glass into air). If the wave hits the boundary at an angle greater than the critical angle, it is reflected.

transformer An electrical device, consisting of two coils of wire wound on an iron core. An alternating current in one coil causes an ever-changing magnetic field which induces an alternating current in the other. Used to 'step' voltage up or down to the level required.

transmitted (transmit) When radiation hits an object, it may go through it. It is said to be transmitted through it. We also say that a radio aerial transmits a signal. In this case, transmits means 'emits' or 'sends out'.

transverse wave A wave in which the particles of the medium vibrate at right angles to the direction in which the wave is travelling. Water waves are an example.

ultraviolet (UV) Electromagnetic waves with frequencies higher than those of visible light, beyond the violet end of the visible spectrum.

umbra An area of total darkness in a shadow. For example, places in the Moon's path where the Earth completely blocks off sunlight.

Universe All things (including the Earth and everything else in space).

unstable The nucleus in radioactive isotopes is not stable. It is liable to change, emitting one of several types of radiation. If it emits alpha or beta radiation, a new element is formed.

velocity The speed of an object in a given direction. Unlike speed, which only has a size, velocity also has a direction.

velocity-time graph A useful way of summarizing the motion of an object by showing its velocity at every instant during its journey.

vibrates Moves rapidly and repeatedly back and forth.

volcano A vent in the Earth's surface that erupts magma, gases, and solids.

voltage The voltage marked on a battery or power supply is a measure of the 'push' it exerts on charges in an electric circuit. The 'voltage' between two points in a circuit means the 'potential difference' between these points.

wave speed The speed at which waves move through a medium.

wavelength The distance between one wave crest (or wave trough) and the next.

work Work is done whenever a force makes something move. The amount of work is force multiplied by distance moved in the direction of the force. This is equal to the amount of energy transferred.

X-rays Electromagnetic waves with high frequency, well above that of visible light.

A

B

C

D

E

F

Q

R

S

T

U

V

W

X

Y

Z

Acknowledgements

Publishers acknowledgements

Oxford University Press wishes to thank the following for their kind permission to reproduce copyright material

P8/9 Data courtesy Marc Imhoff of NASA GSFC and Christopher Elvidge of NOAA NGDC. Image by Craig Mayhew and Robert Simmon, NASA GSFC./NASA; **p12l** Jack Sullivan/Alamy, **p12c** Enzo & Paolo Ragazzini/Corbis UK Ltd., **p12r** Sinclair Stammers/Science Photo Library; **p14t** Theowulf Mähl/Photolibrary Group, **p14b** Bettmann/Corbis UK Ltd.; **p16** Dr Ken MacDonald/Science Photo Library; **p21** Stephen & Donna O'Meara/Science Photo Library; **p22l** Eckhard Slawik/Science Photo Library, **p22c** Charles O'Rear/Corbis UK Ltd., **p22r** JPL/NASA; **p23t** Pierre Thomas/Laboratoire de Sciences de la Terre - ENS de Lyon, **p23b** David Brodie; **p24** D. Van Ravensway/Science Photo Library; **p25** Mike Widdowson; **p26l** N.A.Sharp, NOAO/AURA/NSF/National Optical Astronomy Observatories, **p26r** N.A.Sharp/NSO/Kitt Peak FTS/AURA/NSF/National Optical Astronomy Observatories; **p27t** NASA/Zooid Pictures, **p27b** David Malin/Anglo-Australian Observatory; p28 Jerry Lodriguss/Science Photo Library; p29t Zooid Pictures, **p29b** NACO/VLT/ESO/European Southern Observatory HQ; **p30** Data courtesy Marc Imhoff of NASA GSFC and Christopher Elvidge of NOAA NGDC. Image by Craig Mayhew and Robert Simmon, NASA GSFC./NASA; **p31t** NASA/Zooid Pictures, **p31b** Two Micron All Sky Survey (2MASS); **p33** Science Photo Library; **p34/35** Data courtesy Marc Imhoff of NASA GSFC and Christopher Elvidge of NOAA NGDC. Image by Craig Mayhew and Robert Simmon, NASA GSFC./NASA; **p38l** Trevor Worden/Photolibrary Group, p38c Gildo Nicolo Spadoni/Images.Com/Photolibrary Group, **p38r** Stephanie Sinclair/Corbis UK Ltd.; **p40t** Ralph A. Clevenger/Corbis UK Ltd., **p40b** Pictor International/ImageState/Alamy; p41t NASA/Science Photo Library, **p41b** Solent News and Photos/Rex Features; **p42** Gideon Mendel/Corbis UK Ltd.; **p43** David Wrench/Leslie Garland Picture Library/Alamy; **p44tc** David Turnley/Corbis UK Ltd., **p44bc** Philipp Mohr/Alamy, **p44t** Simon Belcher/Alamy, **p44b** CNRI/Science Photo Library; **p45** Mike Hill/Alamy; **p47** John Nordell/Index Stock Imagery/Photolibrary Group; **p48** Janine Wiedel/Janine Wiedel Photolibrary/Alamy; **p50** Martyn F. Chillmaid; **p52t** Image Source/Alamy, **p52b** Advertising Archives **p57l** Oxford University Press, **p57c** KJ Pictures/The Flight Collection/Alamy, **p57tr** Yves Forestier/Sygma/Corbis UK Ltd., **p57br** Martin Bond/Photofusion Picture Library/Alamy; **p58** D.A. Peel/Science Photo Library; **p59** Steve Morgan/Alamy; **p61** David Marsden/Rex Features; **p66** Derek Croucher/Corbis UK Ltd.; **p69** Julia Hedgecoe; **p72** Photolibrary Group; **p73** Prof. Richard Lawson/Central Manchester and Manchester Children's University Hospitals NHS Trust ; **p74** Mike Derer/AP Photo; **p75** Geoff Tompkinson/Science Photo Library; **p76** Davies & Starr/The Image Bank/Getty Images; **p77l** Matthias Kulka/Corbis UK Ltd., **p77r** Peter Thorne, Johnson Matthey/Science Photo Library; **p79** US Department Of Energy/Science Photo Library; **p80l** Jerry Mason/Science Photo Library, **p80r** Keith Beardmore/The Point/British Nuclear Fuels Limited; **p81** ARGONNE NATIONAL LABORATORY/Science Photo Library; **p83l** Steve Allen/Science Photo Library, **p83r** Keith Beardmore/The Point/British Nuclear Fuels Limited; **p86l** Richard Folwell/Science Photo Library, **p86c** Peter Bowater/Alamy, **p86r** Peter Bowater/Science Photo Library; **p87** British Nuclear Fuels Limited ; **p94** Magrath Photography/Science Photo Library ; **p96l** NASA, **p96r** Sutton Motorsport Images; **p97** Peter Turnley/Corbis UK Ltd. ; **p98l** David Parker/Science Photo Library, **p98r** Andrew Syred/Science Photo Library; **p102cl&l&cr&r** Zooid Pictures; **p103tl&tr** Essex Police, **p103bl** Mark Seymour/Oxford University Press, **p103br** Adam Hart-Davis/Science Photo Library; **p106** Zooid Pictures; **p108** Empics; **p109** Cambridge Science Media; **p110** EuroNCAP Partnership; **p113** Corbis UK Ltd.; **p114** Tom Stewart/Corbis UK Ltd.; **p116** Alton Towers; **p120/121** Scottish Power; **p122t** Kent Wood/Science Photo Library, **p122b** Charles D. Winters/Science Photo Library; **p125** Ron Chapple/Thinkstock/Alamy; **p129** Anthony Redpath/Corbis UK Ltd.; **p130** Zooid Pictures; **p133t** The Image Bank/Getty Images, **p133b** sciencephotos/Alamy; **p134** Ralph Krubner/Stock Connection, Inc./Alamy; **p135** Peter Dazeley/Alamy; **p139** Vic Singh Studio/Alamy; **p140t** Martyn F. Chillmaid/Science Photo Library, **p140b** Andrew Lambert Photography/Science Photo Library; **p142** Barry Batchelor/PA Photos; **p144t** Anthony Vizard/Eye Ubiquitous/Corbis UK Ltd., **p144b** Scottish Power; **p145l** British Energy, **p145c** Nicholas Bailey/Rex Features, **p145r** Scottish Power; **p146/147** Scottish Power; **p150** Robert Shaw/Corbis UK Ltd.; **p154t&b** Peter Gould; **p155t&b** Peter Gould, **p155c** Scottish Power; **p159** Amanda Friedman/The Image Bank/Getty Images; **p160** D. Boone/Corbis UK Ltd.; **p162l** Pascal Goetgheluck/Science Photo Library, **p162r** JPL/NASA; **p163** Lester V. Bergman/Corbis UK Ltd.; **p164t** Geoff Tompkinson/Science Photo Library, **p164b** Shout; **p165** Professor Robin Millar ; **p166** Victor De Schwanberg/Science Photo Library; **p167** Zooid Pictures; **p169** BBC Points West/BBC Bristol; **p171** Dr Jeremy Burgess/Science Photo Library; **p172t** Alfred Pasieka/Science Photo Library, **p172b** Pure Digital; **p173** A. Tovy/Robert Harding Picture Library Ltd/Alamy.

OXFORD
UNIVERSITY PRESS

Great Clarendon Street, Oxford OX2 6DP

Oxford University Press is a department of the University of Oxford.
It furthers the University's objective of excellence in research, scholarship,
and education by publishing worldwide in

Oxford New York

Auckland Cape Town Dar es Salaam Hong Kong Karachi
Kuala Lumpur Madrid Melbourne Mexico City Nairobi
New Delhi Shanghai Taipei Toronto

With offices in

Argentina Austria Brazil Chile Czech Republic France Greece
Guatemala Hungary Italy Japan Poland Portugal Singapore
South Korea Switzerland Thailand Turkey Ukraine Vietnam

First published 2006

British Library Cataloguing in Publication Data

Data available

ISBN-13: 978-0-19-915051-9

10 9 8 7 6 5 4

Typeset by IFA Design Ltd, Plymouth, UK; Q2A Design, Delhi, India; and Oxford University Press

Printed by Printplus, China

Project Team acknowledgements

These resources have been developed to support teachers and students undertaking a new OCR suite of GCSE science specifications, *Twenty First Century Science*.

Many people from schools, colleges, universities, industry, and the professions have contributed to the production of these resources. The feedback from over 75 Pilot Centres was invaluable. It led to significant changes to the course specifications, and to the supporting resources for teaching and learning.

The University of York Science Education Group (UYSEG) and Nuffield Curriculum Centre worked in partnership with an OCR team led by Mary Whitehouse, Elizabeth Herbert, and Emily Clare to create the specifications, which have their origins in the *Beyond 2000* report (Millar & Osborne, 1998) and subsequent Key Stage 4 development work undertaken by UYSEG and the Nuffield Curriculum Centre for QCA. Bryan Milner and Michael Reiss also contributed to this work, which is reported in: *21st Century Science GCSE Pilot Development: Final Report* (UYSEG, March 2002).

Sponsors

The development of *Twenty First Century Science* was made possible by generous support from:

- The Nuffield Foundation
- The Salters' Institute
- The Wellcome Trust

wellcome trust